Chemical Vapor Deposition of Refractory Metals and Ceramics II

MATERIALS RESEARCH SOCIETY SYMPOSIUM PROCEEDINGS   VOLUME 250

Chemical Vapor Deposition of Refractory Metals and Ceramics II

Symposium held December 4-6, 1991, Boston, Massachusetts, U.S.A.

EDITORS:

Theodore M. Besmann
Oak Ridge National Laboratory, Oak Ridge, Tennessee, U.S.A.

Bernard M. Gallois
Stevens Institute of Technology, Hoboken, New Jersey, U.S.A.

James W. Warren
Composite Innovation Corporation, Woodland Hills, California, U.S.A.

MATERIALS RESEARCH SOCIETY
Pittsburgh, Pennsylvania

This work was supported by the Air Force Office of Scientific Research, Air Force Systems Command, USAF, under Grant Number AFOSR 91-0407.

Research sponsored by the U.S. Department of Energy, Assistant Secretary for Conservation and Renewable Energy, Advanced Industrial Concepts Division, Advanced Industrial Concepts Materials Program, under subcontract number 347067, and under contract DE-AC05-84OR21400 with Martin Marietta Energy Systems, Inc.

Single article reprints from this publication are available through
University Microfilms Inc., 300 North Zeeb Road, Ann Arbor, Michigan 48106

CODEN: MRSPDH

Copyright 1992 by Materials Research Society.
All rights reserved.

This book has been registered with Copyright Clearance Center, Inc. For further information, please contact the Copyright Clearance Center, Salem, Massachusetts.

Published by:

Materials Research Society
9800 McKnight Road
Pittsburgh, Pennsylvania 15237
Telephone (412) 367-3003
Fax (412) 367-4373

Library of Congress Cataloging in Publication Data

Chemical vapor deposition of refractory metals and ceramics II : symposium held December 4-6, 1991, Boston, Massachusetts, U.S.A. / editors, Theodore M. Besmann, Bernard M. Gallois, James Warren.

    p. cm. — (Materials Research Society symposium proceedings, ISSN 0272-9172 ; v. 250)
    Includes bibliographical references and index.
    ISBN 1-55899-144-1
    1. Refractory coating—Congresses.   2. Vapor-plating—Congresses.
   I. Besmann, Theodore M.   II. Gallois, Bernard M.   III. Warren, James, 1932-
   IV. Series: Materials Research Society symposium proceedings : v. 250.
TS695.9.C492   1992   92-6323
671.7'35—dc20   CIP

Manufactured in the United States of America

Contents

PREFACE ix

ACKNOWLEDGMENTS xi

MATERIALS RESEARCH SOCIETY SYMPOSIUM PROCEEDINGS xii

PART I: FUNDAMENTALS/MODELING

*THERMOCHEMISTRY IN CVD - ON THE CHOICE OF HALIDE GAS
SPECIES 3
  C. Bernard and R. Madar

*CHEMICAL VAPOR DEPOSITION MODELING FOR HIGH TEMPERATURE
MATERIALS 17
  Suleyman A. Gokoglu

THEORETICAL STUDY OF GAS-PHASE THERMODYNAMICS RELEVANT
TO SILICON CARBIDE CHEMICAL VAPOR DEPOSITION 29
  Mark D. Allendorf and Carl F. Melius

KINETIC MODELLING OF THE DEPOSITION OF SiC FROM
METHYLTRICHLOROSILANE 35
  Stratis V. Sotirchos and George D. Papasouliotis

KINETICS OF CHEMICAL VAPOR DEPOSITION OF SiC BETWEEN 750 AND
850°C AT 1 BAR TOTAL PRESSURE 41
  Dieter Neuschütz and Farzin Salehomoum

KINETIC CHARACTERISTICS OF Si_3N_4 CVD 47
  W.Y. Lee, J.R. Strife, and R.D. Veltri

ONSET CONDITIONS FOR GAS PHASE REACTION AND NUCLEATION IN
THE CVD OF TRANSITION METAL OXIDES 53
  J. Collins, D.E. Rosner, and J. Castillo

RAPID GROWTH OF CERAMIC FILMS BY PARTICLE-VAPOR
CODEPOSITION 59
  Robert H. Hurt and Mark D. Allendorf

STEP COVERAGE MODELING OF THIN FILMS DEPOSITED BY CVD
USING FINITE ELEMENT METHOD 65
  Ching-Yi Tsai and Seshu B. Desu

PYROLYTIC CARBON DEPOSITION ON GRAPHITIC SURFACES 71
  Ismail M.K. Ismail

PART II: IN SITU DIAGNOSTICS

SURFACE CHEMISTRY OF FLUORINE-CONTAINING MOLECULES
RELATED TO CVD PROCESS ON SILICON NITRIDE: SiF_4, XeF_2, AND HF ... 79
  Duane A. Outka

DETERMINATION OF THE ROUGHNESS OF CVD SURFACES BY LASER
SCATTERING 85
  Max Klein and Bernard Gallois

*Invited Paper

IN-SITU LIGHT-SCATTERING MEASUREMENTS DURING THE CVD OF
POLYCRYSTALLINE SILICON CARBIDE 93
    Brian W. Sheldon, Philip A. Reichle, and Theodore M. Besmann

INTERACTIVE USE OF ELECTRON MICROSCOPY AND LIGHT
SCATTERING AS DIAGNOSTICS FOR PYROGENIC AGGREGATES 101
    Richard A. Dobbins

TUNABLE DIODE LASER ABSORPTION SPECTROSCOPY OF THE
PYROLYSIS OF METHYLSILAZANE 107
    H.C. Sun, Y.W. Bae, E.A. Whittaker, and B. Gallois

AXIAL CONCENTRATION PROFILE OF H2 PRODUCED IN THE CVD
OF Si_3N_4 113
    Stephen O. Hay and Ward C. Roman

LASER INDUCED FLUORESCENCE FOR TEMPERATURE MEASUREMENT
IN REACTING FLOWS 119
    R.G. Joklik

IN-SITU OPTICAL EMISSION SPECTRA OF Ti, TiN AND $TiSi_2$ PLASMA
DURING THIN FILM GROWTH BY PULSED LASER EVAPORATION 125
    S. Pramanick and J. Narayan

MOLECULAR PRECURSORS TO BORON NITRIDE THIN FILMS: THE
REACTIONS OF DIBORANE WITH AMMONIA AND WITH HYDRAZINE
ON Ru(0001) 131
    Charles M. Truong, José A. Rodriguez, Ming-Cheng Wu, and
    D.W. Goodman

NEAR-ROOM TEMPERATURE DEPOSITION OF W AND WO_3 THIN FILMS BY
HYDROGEN ATOM ASSISTED CHEMICAL VAPOR DEPOSITION 137
    Wei William Lee and Robert R. Reeves

PART III: MICROSTRUCTURE-PROCESS-PROPERTY RELATIONSHIPS

*EFFECT OF PROCESS CONDITIONS AND CHEMICAL COMPOSITION
ON THE MICROSTRUCTURE AND PROPERTIES OF CHEMICALLY VAPOR
DEPOSITED SiC, Si, ZnSe, ZnS, AND ZnS_xSe_{1-x} 145
    Michael A. Pickering, Raymond L. Taylor, Jitendra S. Goela, and
    Hemant D. Desai

CVD OF SILICON NITRIDE PLATE FROM $HSiCl_3$-NH_3-H_2-MIXTURES 161
    J.W. Lennartz and M.B. Dowell

CHEMICAL VAPOR DEPOSITION OF MULTIPHASE BORON-CARBON-
SILICON CERAMICS 167
    E. Michael Golda and B. Gallois

MODIFICATION OF OPTICAL SURFACES EMPLOYING CVD BORON
CARBIDE COATINGS 173
    Richard A. Lowden, Laura Riester, and M. Alfred Akerman

SiC THIN FILMS BY CHEMICAL CONVERSION OF SINGLE CRYSTAL Si 179
    Chien C. Chiu, Chi Kong Kwok, and Seshu B. Desu

EFFECT OF HIGH TEMPERATURE ANNEALING ON THE
MICROSTRUCTURE OF SCS-6 SiC FIBERS 187
    X.J. Ning, P. Pirouz, and R.T. Bhatt

*Invited Paper

LOW-TEMPERATURE PACVD SILICON CARBIDE COATINGS 193
W. Halverson, G.D. Vakerlis, D. Garg, and P.N. Dyer

THIN FILM PROPERTIES OF LPCVD TiN BARRIER FOR SILICON DEVICE
TECHNOLOGY 199
Rama I. Hegde, Robert W. Fiordalice, Edward O. Travis, and
Philip J. Tobin

PART IV: CHEMICAL VAPOR INFILTRATION

*ADVANCES IN MODELING OF THE CHEMICAL VAPOR INFILTRATION
PROCESS 207
Thomas L. Starr

X-RAY TOMOGRAPHIC MICROSCOPY OF NICALON PREFORMS AND
CHEMICAL VAPOR INFILTRATED NICALON/SILICON CARBIDE
COMPOSITES 215
M.D. Butts, S.R. Stock, J.H. Kinney, T.L. Starr, M.C. Nichols,
C.A. Lundgren, T.M. Breunig, and A. Guvenilir

EFFECTS OF FIBER ORIENTATION AND OVERLAPPING OF KNUDSEN,
TRANSITION, AND ORDINARY REGIME DIFFUSION IN FIBROUS
SUBSTRATES 221
Manolis M. Tomadakis and Stratis V. Sotirchos

CONTRIBUTION OF GAS-PHASE REACTIONS TO THE DEPOSITION OF
SiC BY A FORCED-FLOW CHEMICAL VAPOR INFILTRATION PROCESS 227
Ching-Yi Tsai and Seshu B. Desu

FIBER-REINFORCED TUBULAR COMPOSITES BY CHEMICAL VAPOR
INFILTRATION 233
D.P. Stinton, R.A. Lowden, and T.M. Besmann

EFFECT OF BN INTERFACIAL COATING ON THE STRENGTH OF A
SILICON CARBIDE/SILICON NITRIDE COMPOSITE 239
Kirk P. Norton and Holger H. Streckert

MICROWAVE ASSISTED CHEMICAL VAPOR INFILTRATION 245
D.J. Devlin, R.P. Currier, R.S. Barbero, B.F. Espinoza, and N. Elliott

CVD OF SILICON CARBIDE ON STRUCTURAL FIBERS:
MICROSTRUCTURE AND COMPOSITION 251
Lisa C. Veitch, Francis M. Terepka, and Suleyman A. Gokoglu

THE CVD OF CERAMIC PROTECTIVE COATINGS ON SiC
MONOFILAMENTS 257
K.L. Choy and B. Derby

CVD COATING OF CERAMIC MONOFILAMENTS 263
Jason R. Guth

FACILITY FOR CONTINUOUS CVD COATING OF CERAMIC FIBERS 269
Arthur W. Moore

THE OXIDATION STABILITY OF BORON NITRIDE THIN FILMS ON
MgO AND TiO_2 SUBSTRATES 275
Xiaomei Qiu, Abhaya K. Datye, Robert T. Paine, and Lawrence F. Allard

*Invited Paper

PART V: ORGANOMETALLIC CVD

*CVD OF SiC AND AlN USING CYCLIC ORGANOMETALLIC PRECURSORS   283
L.V. Interrante, D.J. Larkin, and C. Amato

CHEMICAL VAPOR DEPOSITION OF COPPER OXIDE THIN FILMS   291
Yuneng Chang and Glenn L. Schrader

MOCVD GROWTH OF COPPER AND COPPER OXIDE FILMS FROM BIS-β-DIKETONATE COMPLEXES OF COPPER. THE ROLE OF CARRIER GAS ON DEPOSIT COMPOSITION   297
William S. Rees, Jr. and Celia R. Caballero

CHEMICAL VAPOR DEPOSITION OF TUNGSTEN AND MOLYBDENUM FILMS FROM $M(\eta^3\text{-}C_3H_5)_4 (M=Mo,W)$ 303
Rein U. Kirss, Jian Chen, and Robert B. Hallock

THE CHEMICAL VAPOR DEPOSITION OF PURE NICKEL AND NICKEL BORIDE THIN FILMS FROM BORANE CLUSTER COMPOUNDS   311
Shreyas Kher and James T. Spencer

DEPOSITION OF TUNGSTEN NITRIDE THIN FILMS FROM $(^tBuN)_2W(NH^tBu)_2$ 317
Hsin-Tien Chiu and Shiow-Huey Chuang

DEPOSITION AND CHARACTERIZATION OF METALORGANIC CHEMICAL VAPOR DEPOSITION ZrO_2 THIN FILMS USING $Zr(thd)_4$ 323
Jie Si, Chien H. Peng, and Seshu B. Desu

ADVANCED DIELECTRICS DEPOSITED BY LPCVD   331
Andrew P. Lane, Arthur Chen, Neal P. Sandler, Dean W. Freeman, and Barry S. Page

PART VI: DIAMOND FILMS

ROUTES TO DIAMOND HETEROEPITAXY   339
Andrzej Badzian and Teresa Badzian

DIAMOND DEPOSITION BY A NONEQUILIBRIUM PLASMA JET   351
D.G. Keil, H.F. Calcote, and W. Felder

GROWTH AND CHARACTERIZATION OF PECVD DIAMOND FILMS   357
J.A. Mucha and L. Seibles

CORRELATION OF RAMAN SPECTRA AND BONDING IN DLC FILMS DEPOSITED BY LASER ABLATION AND LASER-PLASMA ABLATION TECHNIQUES   367
A. Rengan, J. Narayan, J.L. Park, and M. Li

AUTHOR INDEX   373

SUBJECT INDEX   375

*Invited Paper

Preface

The papers contained in this volume were originally presented at the Symposium on the Chemical Vapor Deposition of Metals and Ceramics held at the Fall Meeting of the Materials Research Society in Boston, Massachusetts on December 4-6, 1991. This symposium was sponsored by the Directorate of Electronic and Materials Sciences - Air Force Office of Scientific Research and the U.S. Department of Energy.

The volume leads off with papers on the fundamentals of CVD. Highlighting these are the invited papers by C. Bernard and R. Madar, and by Suleyman Gokoglu. They, respectively, discuss the thermochemistry of CVD and the choice of halide gas species, and CVD modeling for high-temperature materials. Bernard and Madar noted the importance of thermochemical analysis in the choice of reactants. Gokoglu commented on the relative difficulty of modeling and the hierarchy of what is understood today. Other papers on fundamentals discussed kinetics and mass transport in CVD, with an emphasis on silicon carbide deposition.

A new and exciting area involves in situ diagnostics in CVD. The papers emphasize both the importance of in situ diagnostics and how this area is just now emerging. The chapter on microstructure-process-property relationships reveals the ability to control the CVD coatings to a level not seen before. Coatings of more than one phase are now routinely deposited to obtain material with a variety of properties. Chemical vapor infiltration to produce composite materials is the subject of a chapter and includes considerable discussion of the modeling of the infiltration process, which is becoming quite advanced.

The chapter on organometallic CVD focuses on a number of issues related to precursors and the mechanisms of deposition. In light of the interest in high-temperature superconductors, a good bit of the reported work revolved around copper precursors. The last section is on the CVD of diamond. The papers report on nucleation studies, the nature of the plasma in plasma CVD of diamond, and on novel techniques for growth.

T.M. Besmann
B.M. Gallois
J.W. Warren

February 4, 1992

Acknowledgments

The symposium co-chairs would like to thank the session chairs who did an excellent job of helping to obtain papers and run the sessions. These very helpful individuals were:

> T.L. Starr
> B.W. Sheldon
> D. Burgess, Jr.
> R.A. Lowden
> M. Headinger
> K. Gonsalves
> R.E. Clausing
> A. Badzian

The invited papers by leaders in the field of chemical vapor deposition required considerable effort. The co-chairs would like to thank those authors particularly for their contribution to this volume:

> C. Bernard
> R. Madar
> S.A. Gokoglu
> T.L. Starr
> M.A. Pickering
> R.L. Taylor
> J.S. Goela
> H.D. Desai
> L.V. Interrante

We are also greatly indebted to J.R. Lowe who organized, revised, mailed, and did everything necessary to prepare this book for publication.

Finally, we sincerely appreciate the financial support provided by the Directorate of Electronic and Materials Sciences - Air Force Office of Scientific Research and the U.S. Department of Energy, Assistant Secretary for Conservation and Renewable Energy, Advanced Industrial Concepts Division, Advanced Industrial Concepts Materials Program.

MATERIALS RESEARCH SOCIETY SYMPOSIUM PROCEEDINGS

Volume 219—Amorphous Silicon Technology—1991, A. Madan, Y. Hamakawa, M. Thompson, P.C. Taylor, P.G. LeComber, 1991, ISBN: 1-55899-113-1

Volume 220—Silicon Molecular Beam Epitaxy, 1991, J.C. Bean, E.H.C. Parker, S. Iyer, Y. Shiraki, E. Kasper, K. Wang, 1991, ISBN: 1-55899-114-X

Volume 221—Heteroepitaxy of Dissimilar Materials, R.F.C. Farrow, J.P. Harbison, P.S. Peercy, A. Zangwill, 1991, ISBN: 1-55899-115-8

Volume 222—Atomic Layer Growth and Processing, Y. Aoyagi, P.D. Dapkus, T.F. Kuech, 1991, ISBN: 1-55899-116-6

Volume 223—Low Energy Ion Beam and Plasma Modification of Materials, J.M.E. Harper, K. Miyake, J.R. McNeil, S.M. Gorbatkin, 1991, ISBN: 1-55899-117-4

Volume 224—Rapid Thermal and Integrated Processing, M.L. Green, J.C. Gelpey, J. Wortman, R. Singh, 1991, ISBN: 1-55899-118-2

Volume 225—Materials Reliability Issues in Microelectronics, J.R. Lloyd, P.S Ho, C.T. Sah, F. Yost, 1991, ISBN: 1-55899-119-0

Volume 226—Mechanical Behavior of Materials and Structures in Microelectronics, E. Suhir, R.C. Cammarata, D.D.L. Chung, 1991, ISBN: 1-55899-120-4

Volume 227—High Temperature Polymers for Microelectronics, D.Y. Yoon, D.T. Grubb, I. Mita, 1991, ISBN: 1-55899-121-2

Volume 228—Materials for Optical Information Processing, C. Warde, J. Stamatoff, W. Wang, 1991, ISBN: 1-55899-122-0

Volume 229—Structure/Property Relationships for Metal/Metal Interfaces, A.D Romig, D.E. Fowler, P.D. Bristowe, 1991, ISBN: 1-55899-123-9

Volume 230—Phase Transformation Kinetics in Thin Films, M. Chen, M. Thompson, R. Schwarz, M. Libera, 1991, ISBN: 1-55899-124-7

Volume 231—Magnetic Thin Films, Multilayers and Surfaces, H. Hopster, S.S.P. Parkin, G. Prinz, J.-P. Renard, T. Shinjo, W. Zinn, 1991, ISBN: 1-55899-125-5

Volume 232—Magnetic Materials: Microstructure and Properties, T. Suzuki, Y. Sugita, B.M. Clemens, D.E. Laughlin, K. Ouchi, 1991, ISBN: 1-55899-126-3

Volume 233—Synthesis/Characterization and Novel Applications of Molecular Sieve Materials, R.L. Bedard, T. Bein, M.E. Davis, J. Garces, V.A. Maroni, G.D. Stucky, 1991, ISBN: 1-55899-127-1

Volume 234—Modern Perspectives on Thermoelectrics and Related Materials, D.D. Allred, G. Slack, C. Vining, 1991, ISBN: 1-55899-128-X

Volume 235—Phase Formation and Modification by Beam-Solid Interactions, G.S. Was, L.E. Rehn, D. Follstaedt, 1992, ISBN: 1-55899-129-8

Volume 236—Photons and Low Energy Particles in Surface Processing, C Ashby, J.H. Brannon, S. Pang, 1992, ISBN: 1-55899-130-1

Volume 237—Interface Dynamics and Growth, K.S. Liang, M.P. Anderson, R.F. Bruinsma, G. Scoles, 1992, ISBN: 1-55899-131-X

me 238—Structure and Properties of Interfaces in Materials, W.A.T. Clark, C.L. Briant, U. Dahmen, 1992, ISBN: 1-55899-132-8

MATERIALS RESEARCH SOCIETY SYMPOSIUM PROCEEDINGS

Volume 239—Thin Films: Stresses and Mechanical Properties III, W.D. Nix, J.C. Bravman, E. Arzt, L.B. Freund, 1992, ISBN: 1-55899-133-6

Volume 240—Advanced III-V Compound Semiconductor Growth, Processing and Devices, S.J. Pearton, D.K. Sadana, J.M. Zavada, 1992, ISBN: 1-55899-134-4

Volume 241—Low Temperature (LT) GaAs and Related Materials, G.L. Witt, R. Calawa, U. Mishra, E. Weber, 1992, ISBN: 1-55899-135-2

Volume 242—Wide Ban-Gap Semiconductors, T.D. Moustakas, J.I. Pankove, Y. Hamakawa, 1992, ISBN: 1-55899-136-0

Volume 243—Ferroelectric Thin Films II, A.I. Kingon, E.R. Myers, B. Tuttle, 1992, ISBN: 1-55899-137-9

Volume 244—Optical Waveguide Materials, M.M. Broer, H. Kawazoe, G.H. Sigel, R.Th. Kersten, 1992, ISBN: 1-55899-138-7

Volume 245—Advanced Cementitious Systems: Mechanisms and Properties, F.P. Glasser, P.L. Pratt, T.O. Mason, J.F. Young, G.J. McCarthy, 1992, ISBN: 1-55899-139-5

Volume 246—Shape-Memory Materials and Phenomena—Fundamental Aspects and Applications, C.T. Liu, M. Wuttig, K. Otsuka, H. Kunsmann, 1992, ISBN: 1-55899-140-9

Volume 247—Electrical, Optical, and Magnetic Properties of Organic Solid State Materials, L.Y. Chiang, A.F. Garito, D.J. Sandman, 1992, ISBN: 1-55899-141-7

Volume 248—Complex Fluids, D. Weitz, E.Sirota, T. Witten, J. Israelachvili, 1992, ISBN: 1-55899-142-5

Volume 249—Synthesis and Processing of Ceramics: Scientific Issues, W.E. Rhine, T.M. Shaw, R.J. Gottschall, Y. Chen, 1992, ISBN: 1-55899-143-3

Volume 250—Chemical Vapor Deposition of Refractory Metals and Ceramics II, T.M. Besman, B.M. Gallois, J. Warren, 1992, ISBN: 1-55899-144-1

Volume 251—Pressure Effects on Materials Processing and Design, K. Ishizaki, E. Hodge, 1992, ISBN: 1-55899-145-X

Volume 252—Tissue-Inducing Biomaterials, M. Flanagan, L. Cima, E. Ron, 1992, ISBN: 1-55899-146-8

Volume 253—Applications of Multiple Scattering Theory to Materials Science, W.H. Butler, P.H. Dederichs, A. Gonis, R. Weaver, 1992, ISBN: 1-55899-147-6

Volume 254—Specimen Preparation for Transmission Electron Microscopy of Materials III, R. Anderson, J. Bravman, B. Tracy, 1992, ISBN: 1-55899-148-4

Volume 255—Hierarchically Structured Materials, I.A. Aksay, E. Baer, M. Sarikaya, D.A. Tirrell, 1992, ISBN: 1-55899-149-2

Volume 256—Light Emission from Silicon, S.S. Iyer, L.T. Canham, R.T. Collins, 1992, ISBN: 1-55899-150-6

Prior Materials Research Society Symposium Proceedings available by contacting Materials Research Society.

PART I

Fundamentals/Modeling

THERMOCHEMISTRY IN C.V.D. - ON THE CHOICE OF HALIDE GAS SPECIES

C. Bernard and R. Madar*
Laboratoire de Thermodynamique et Physico-Chimie Métallurgiques, ENSEEG, BP.75 38402 St Martin d'Hères - France. *Laboratoire des Matériaux et du Génie Physique, ENSPG, BP.46, 38402 Saint Martin d'Hères - France.

ABSTRACT

The production of thin or thick films of metals or ceramics by chemical vapour deposition has often been achieved by the use of halide gas precursors. In certain cases, this choice was made purely for reasons of simplicity: gas cylinder available, gas species already used in another field, etc. Experience has subsequently shown, however, that this choice can give rise to significant changes in the nature and proportions of deposited phases. These are highly dependent upon:
- the value of the oxidiser:reducer ratio in the gas phase,
- the degree of metal oxidation in the halide considered,
- possible competition between two reducing agents designed to reduce the halide.

These factors, among others, strongly influence the thermochemistry of the deposition reaction. Their roles must therefore be clearly understood, interpreted and predicted by the thermochemical analysis. Based on examples relating to silicide, nitride and boride deposits, an attempt will be made to determine the sensitive parameters and to deduce selection criteria.

INTRODUCTION

In the deposition processes currently used for producing thin or thick films, a wide variety of chemical species and reactions is involved: thermal decomposition of metal carbonyls, metal-organic compounds, hydrides, halides, hydrogen reduction of various gas carriers, disproportioning reactions, etc. Of all these various possibilities, the deposition processes based on halides have high potential for industrial development. Some relatively recent examples give undeniable evidence of this: thin films of Si [1] or W [2] in electronics, protective SiC [3] or TiN [4] deposits for coatings. Halide compounds are not however always used in the best manner, especially in the semi-conductor field. With a view to improving the approach to these systems, this article aims to bring to light the most sensitive parameters and to propose selection criteria.

THERMOCHEMICAL ASSESSMENT INSTRUMENT

The standard halide molecule can be written MX_x. In this work, M may be a transition metal in group III_a to II_b in the periodic table or even an element belonging to group III_b to VI_b, where X is the halogen F, Cl, Br or I. The first basic question to be asked regarding the feasibility of a CVD process is simply how stable is the molecule? This is one of the key factors in the CVD reaction since, if this molecule is to be transported, it must have a sufficiently high saturation vapour pressure at relatively low temperatures. This assumes, therefore, a fairly stable molecule. But as it must also be capable of decomposing on the substrate at medium temperatures (the current trend being towards lowering this temperature), the molecule obviously must not be too stable. To study the relative stability of different gas molecules, it is possible, where specific process requirements permit, to try different types of metal M from different groups in the periodic table (cf. Table 1), to vary the nature of M or X within the same group in the table (c f. Table 2), or to change the value of x (cf. Table 3).

From these three tables, it is clear that there is no clear-cut way of finding trends or fixing general rules, except for the well-known reduction in stability for a given M and x when changing from fluoride to chloride to bromide to iodide.

To obtain the broadest possible view of the thermochemistry of the systems considered, it will therefore be necessary to simulate the complex chemical equilibria involved [5], [6] and, using the various representation methods available, such as CVD diagrams, or efficiency diagrams, study the changes observed from one halide to the next as well as the modifications caused, for a given system, by varying such parameters as partial and total pressure or temperature. The software used for these simulations is based on minimisation of the Gibbs total free energy of the system [7].

TABLE I

Gibbs energy of formation at 1000 K of certain gaseous chlorides MCl_2 of the fourth line in the periodic table.

Species ΔG kJ mol⁻¹	$TiCl_2$	VCl_2	$MnCl_2$	$FeCl_2$	$CoCl_2$	$ZnCl_2$
	-260	-242	-293	-187	-137	-272

TABLE II

Gibbs energy of formation at 1000 K of gaseous chlorides MCl₄ of groups IV_A and V_A of titanium halides TiX₃ in kJ mol⁻¹.

Group IV_A	TiCl₄	ZrCl₄	HfCl₄	
	-642	-755	-771	
Group VI_A	CrCl₄	MoCl₄	WCl₄	
	-323	-285	-239	
Group VII_B	TiF₃	TiCl₃	TiBr₃	TiI₃
	-1143	-490	-381	-204

TABLE III

Gibbs energy of formation at 1000 K of the gaseous chlorides TiCl$_x$, WCl$_x$ and MoCl$_x$ in kJ mol⁻¹.

x	1	2	3	4	5	6
TiCl$_x$	49	-260	-490	-642		
WCl$_x$	439			-239	-234	-220
MoCl$_x$				-285	-264	-170

RELIABILITY OF THE INSTRUMENT

The thermodynamic approach is, of course, only the first stage in the optimisation of a CVD experiment [8], and which is often performed for a closed system, supposedly at equilibrium. However, with regard to the choice of reacting species and sensitivity of the chemical system to forces induced by reactional parameters, this preliminary approach already provides a mine of information.

To illustrate the reliability of the thermodynamic analysis, a favourable case was chosen. However, the complexity of the case studied indicates just how powerful this approach can be. Figure 1 shows part of the tungsten silicide deposition diagram corresponding to the phases deposited from a WCl₄-SiH₄-H₂-Ar mixture for a pressure of 1.31 10⁻³Atm and a temperature de 873 K, with an argon dilution of $\bar{P}_{Ar} = 1.18.10^{-3}$Atm [9]. To the now classical domains have been added the shaded areas which correspond to the impact of the combined effect of uncertainties concerning thermodynamic data involved in the calculation. The boundaries of the various domains are known to a satisfactory degree in this case. The points A, B, C, D, E were then selected for experimental testing of the thermodynamic analysis predictions. The theoretical and experimental results presented in Table 4 show excellent agreement.

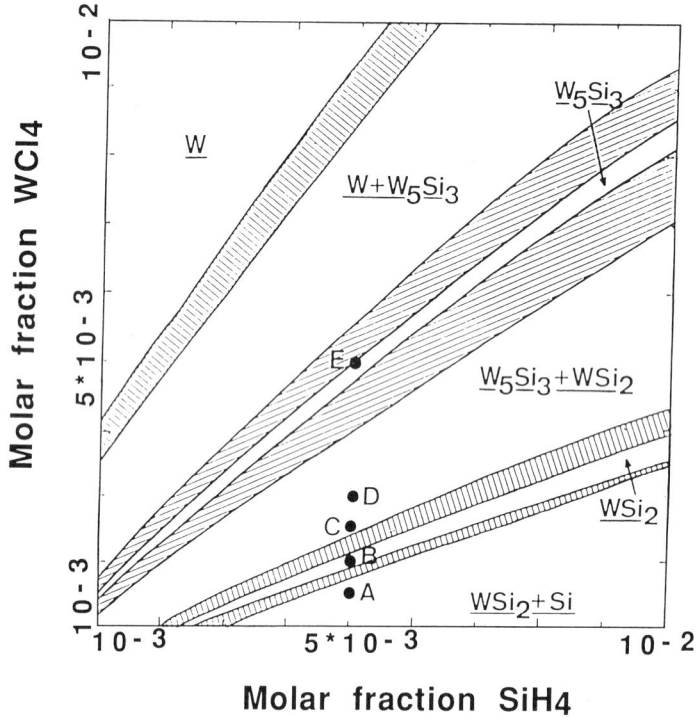

Figure 1: Computed equilibrium CVD phase diagram at 873 K for a WCl$_4$-SiH$_4$-H$_2$-Ar gas mixture. Total pressure 1.31 10^{-3} atm, argon dilution 1.18 10^{-3} atm. [9]

TABLE IV

Solid phases predicted by the thermodynamic simulation compared to experimental results.

Position in CVD phase diagram	Computed phases	Experimentally identified (XRD) phases
A	Si-WSi$_2$	Si-WSi$_2$-W$_5$Si$_3$
B	WSi$_2$	WSi$_2$
C	WSi$_2$-W$_5$Si$_3$	WSi$_2$-W$_5$Si$_3$
D	WSi$_2$-W$_5$Si$_3$	WSi$_2$-W$_5$Si$_3$
E	W$_5$Si$_3$	W$_5$Si$_3$-W

IMPORTANCE OF CHOICE OF M AND X

The study for selecting the most suitable disilicide to replace polycrystalline silicon is a good example in more ways than one [10]. For a $MX_x + SiH_4 + H_2$ gas mixture, this study highlighted the following:

- the repercussions of the choice of M on the ease with which it will be possible to obtain M_2Si. In this respect, the size of the respective deposition domains of WSi_2, $TaSi_2$ and $TiSi_2$ on figures 2, 3, 4 is worthy of note,

- the importance of the choice of X with, in certain specific cases, an additional indication of the role that could be played by the halide which sometimes forms after decomposition of the halide initially introduced in the reactor. For example, the replacement of WCl_6 by WF_6, apart from its catastrophic effect on the size of the WSi_2 deposition domain, has considerably increased the size of the W deposition domain (cf. Figure 2 and 5). This is not due to the increase in stability of the WF_6 molecule with respect to WCl_6 since, for both of these cases, under the stated calculation conditions, all the available W is deposited in one form or another. The modification in fact originates from the lower stability of the gaseous $SiCl_4$ halide formed which consumes much less silicon than the SiF_4 molecule,

- the difficulties encountered in depositing titanium by CVD from $TiCl_4$, despite the fact that the deposition of tungsten, whether selective or not, from halides is reputed to be easy (cf. figures 2, 3 and 4). The case of tantalum is an intermediate one as only a $Ta + Ta_3Si$ co-deposit is obtained under the conditions studied. This throws some light on the manner of changing from one case to another and it may also show how experimental parameters, such as temperature or pressure, modify this changeover. Figures 6 and 7 show how sensitive the extent of the $Ta + Ta_3Si$ co-deposition range is to temperature changes [11].

IMPORTANCE OF CHOICE OF x

Table 3 shows that the stability of the halide molecule used may increase with the degree of oxidation x - this is the case for titanium chlorides - or, conversely, decrease - as is the case for molybdenum chlorides, or remain relatively constant, as for WCl_4, WCl_5 and WCl_6. Depending on the stated objective, it will be possible to benefit from such trends. For example, to deposit titanium nitride from a $TiCl_4$-N_2-H_2 mixture on a temperature-sensitive substrate, it is relatively difficult to reduce the substrate temperature below a certain threshold. This result was obtained by choosing a less stable halide, $TiCl_3$, which is formed by pre-reducing $TiCl_4$ on titanium [12]. The transformation caused by this simple change in value of x can be seen on the deposition diagrams: figures 8 and 9.

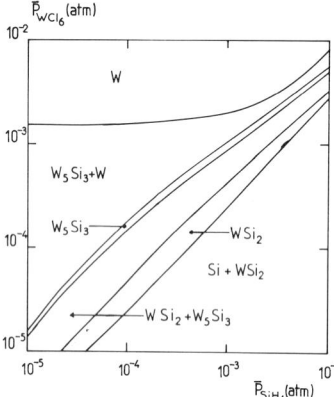

Figure 2: Computed equilibrium CVD phase diagram at 1000 K, 1 atm total pressure, and 0.9 atm Ar partial pressure gas system: W-Si-Cl-H-Ar. [10]

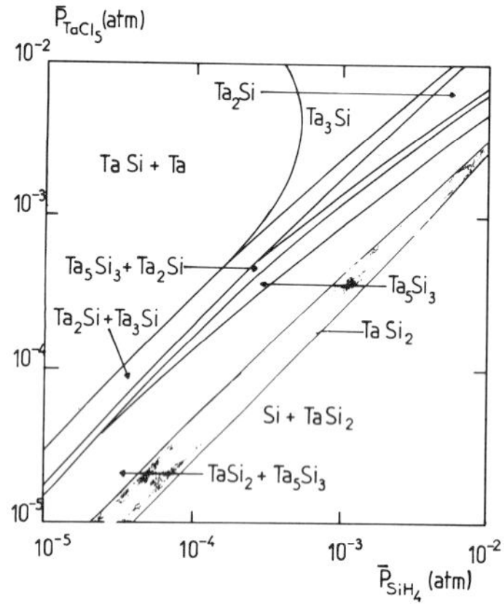

Figure 3: Gas system Ta-Si-Cl-H-Ar, same conditions as in fig. 2. [10]

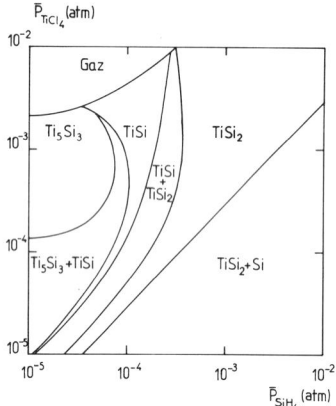
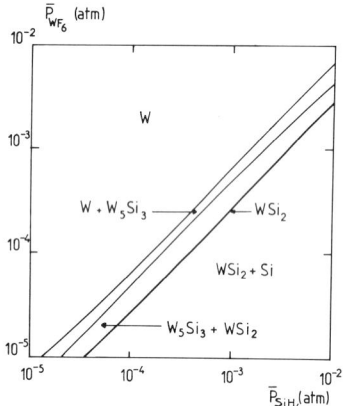

Figure 4: Gas system Ti-Si-Cl-H-Ar, same conditions as in fig. 2. [10]

Figure 5: Gas system W-Si-F-H-Ar, same conditions as in fig. 2. [10]

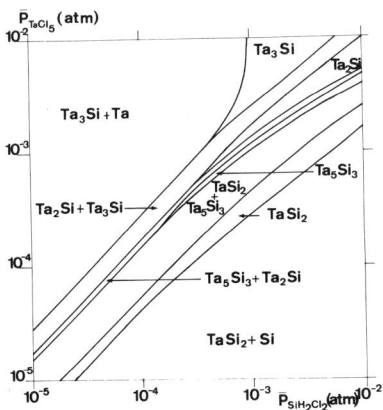

Figure 6: Gas system Ta-Si-Cl-H-Ar at 1000 K, same conditions as in fig. 2 except SiH$_4$ which is replaced by SiH$_2$Cl$_2$.[11]

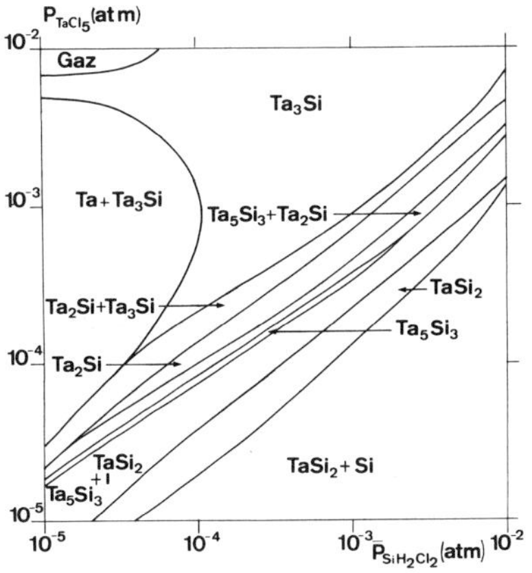

Figure 7: Gas system Ta-Si-Cl-H-Ar, same conditions as in fig. 6 except the temperature which is equal to 850 K.[11]

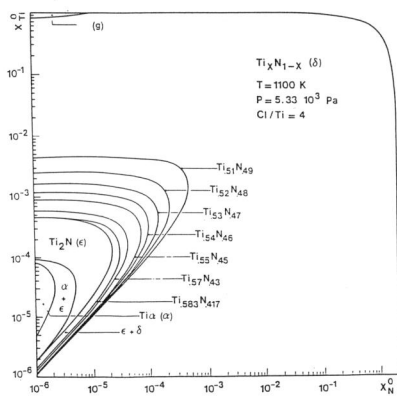
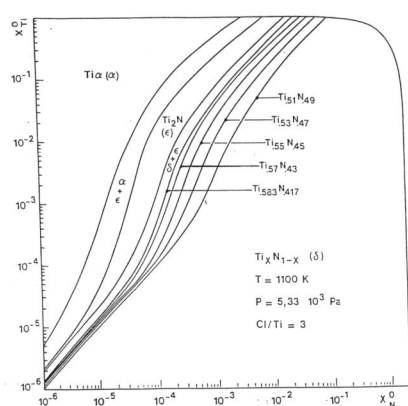

Figure 8: Computed equilibrium CVD phase diagram at 1100 K, 5.25 10^{-2}atm total pressure. The initial molar fractions ploted correspond to a TiCl$_4$-N$_2$-H$_2$ gas mixture where Cl/Ti=4. [12]

Figure 9: Computed equilibrium CVD phase diagram, same conditions as in fig. 8 except the ratio Cl/Ti=3. [12]

CONTROLLING PARTIAL PRESSURES OF HALOGEN CARRIERS

When a halide is available in gas cylinder form, eg. WF_6, BCl_3, the possibility of controlling its partial pressure by means of a simple mass flow-meter is most attractive. This simple logic is, in many cases, the main reason why certain industrial choices are made and which occasionally prove to be disastrous. Moreover, certain halides are in the liquid phase at room temperature, $TiCl_4$ for example, with vapour pressures allowing quantitative entrainment thanks to an optimised reflux column [13]. However, many of these halides are solid at room temperature and their sublimation proves difficult to control. In such cases, in-situ halogenation would seem to be an interesting alternative and has already been successfully practised in the synthesis of group III-V materials.

The major problems raised by the implementation of such a technique are as follows:
- existence of multiple degrees of oxidation which, at first sight, create problems for quantitative monitoring of the reaction,
- the halides formed, often stable, may co-exist with the metal in the halogenation reactor and may redeposit in the pipes between the halogenation cell and the deposition chamber.

The two apparent problems described above, by occurring simultaneously, can in fact provide a very high performance means of controlling the partial pressures involved, once the drawbacks of spurious deposition have been overcome by minimising transport distances and by adequate heating of the transports ducts.

Take, for example, the case of the in-situ chlorination of tungsten which occurs during production of WSi_2 for electronics applications [14], cf. figure 10. Calculations show that a flow of chlorine with the addition of argon over tungsten between 600 and 1200 K, under a total pressure of $1.31 \; 10^{-3}$ Atm, is accompanied by the formation of $WCl_2(s)$, a solid in its stability range (it is supposed to decompose around 862 K), and several gaseous chlorides which, by order of importance, are WCl_4, WCl_2, WCl_5 and WCl_3.

The first reaction to this is to try to avoid the presence of condensed chloride, either by working at a higher temperature or by lowering the total pressure to below the saturating partial pressure. On further thought, however, it may be noted that the formation of $WCl_2(s)$ in contact with $W(s)$ enables a step-by-step control of all the partial pressures of gaseous tungsten halides present:

$WCl_2(s) \leftrightarrows WCl_2(g)$

at constant temperature $\bar{P}WCl_2(g)$ is fixed

$2WCl_2(s)$ or $(g) \leftrightarrows WCl_4(g) + W(s)$

$\bar{P}WCl_4(g)$ is fixed by this equilibrium.

$2WCl_4(g) \leftrightarrows WCl_2(g) + WCl_6(g)$

$\bar{P}WCl_6(g)$ is in turn fixed, and so on ...

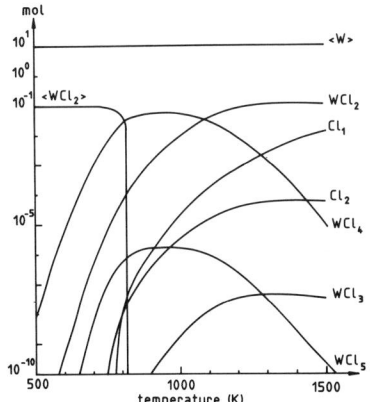

Figure 10: Equilibrium mole numbers of tungsten chlorides versus chlorination temperature for a Cl_2-Ar gas mixture. Total pressure $1.31 \cdot 10^{-3}$ atm, $1.31 \cdot 10^{-4}$ atm Cl_2 partial pressure and an excess of 10 moles of W. [14]

To summarise, in a such a reactor, the experimental worker in fact has total control of the reaction, in spite of, or rather thanks to the complexity of the phases involved.

This type of action has been used in more complex systems such as the activation cementation system where the formation of gaseous metallic halides at the surface of a donor pre-alloy (or cement) is based on the same principle [15].

First of all, to establish the parallel with tungsten, a simple aluminisation system may be considered. It suffices to place a surplus amount of AlF_3(s), activating material, in the presence of Al(s), donor alloy. As before, at a given temperature, the partial pressure of the gaseous halides AlF_3, AlF, AlF_2, Al_2F_6 can be blocked. If several elements are to be co-deposited, an alloy of suitable composition must be used in place of aluminium. This composition must correspond to a two-phase domain for a binary donor alloy or to a three-phase domain for a tertiary donor alloy.

Take for example the co-deposition of aluminium and hafnium on nickel alloy foil [15]. At constant temperature we have:

AlF_3(s) ⇋ AlF_3(g)

$\bar{P}_{AlF_3}$(g) is fixed

$2AlF_3$(g) + <<Al>> → $3AlF_2$(g)

If the aluminium activity in the donor alloy is blocked, $a_{<<Al>>}$ is constant and $\bar{P}_{AlF_2}$(g) is fixed. The same can be said for all the partial pressures of aluminium fluoride species. It then remains to determine the hafnium activity which will enable the final optimum concentration to be obtained, by transport. For a required aluminium activity in the alloy, the computer software gives the partial pressures of the gaseous species formed, and HfF$_4$(g) in particular - the only quantifiable gaseous hafnium halide present - as a function of an arbitrarily fixed hafnium activity, cf. figure 11. With this type of representation, the value of the most beneficial activities can be fixed and the composition of the corresponding donor alloy found.

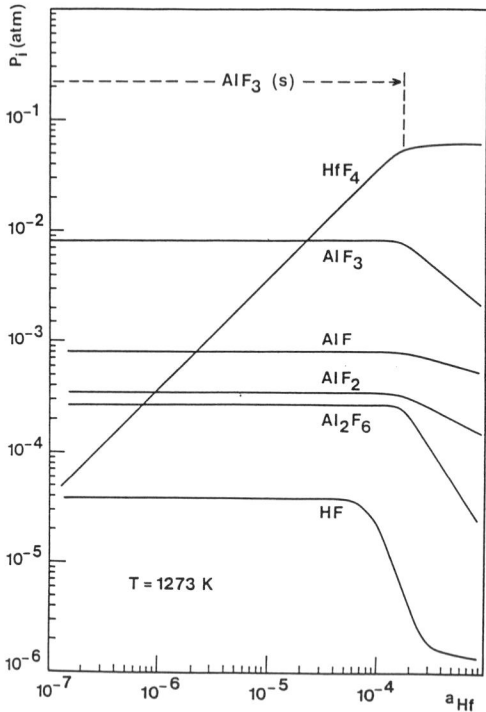

Figure 11: Evolution of the equilibrium partial pressure in the gas phase as a function of the hafnium activity in the cement, for an Al activity equal to 2.1 10^{-3}, total pressure 6.58 10^{-2} atm. [15]

CHOICE OF REACTION MECHANISM

To decompose a gaseous halide carrier in CVD processes, two types of reaction are often used: reduction by the substrate material, or reduction by another species present in the gas phase. For example, tungsten coatings for electronics are obtained by the reduction of WF$_6$(g) by the silicon in the substrate. This reaction stops when the tungsten layer has reached a critical thickness of about 500 Å. It is also possible to deposit this metal by using a WF$_6$(g)-H$_2$(g) mixture. In this case, there is no thickness limit but, in the initial stage, there is no doubt competition between the two reducing agents, hydrogen and silicon [16].

In the light of this example, it would seem to be of interest to determine for which range of experimental parameters these two mechanisms co-exist, and whether there are specific domains for each of them. In this respect, considerations made on neighbouring systems could possibly throw some useful light on an apparently complex case. This is illustrated by the case of amorphous boron deposition on titanium from BCl$_3$(g)-H$_2$(g) mixtures [17].

Experimentally, titanium and its alloys are considered in the literature as having little or no potential as substrates for boron deposition [18]. A calculation performed at 1300 K for the normal gaseous compositions of BCl$_3$(g)-H$_2$(g) mixtures in the presence of titanium shows a high substrate reactivity with the abundant formation of TiCl$_3$(g) and TiCl$_4$(g) and a very low H$_2$ activity with the formation of HCl.

Moreover, it is known that tungsten is the substrate most commonly used as support in the CVD process for boron fibre preparation. The same calculation as for titanium was therefore performed for the case of tungsten, cf. figure 12. This figure indicates the percentages of initial BCl$_3$(g) that are either reduced by the H$_2$(g) in HCl(g), or combined in metal halide form after substrate attack, depending on the BCl$_3$(g) dilution in the initial mixture. Two domains are apparent: one with low dilutions with BCl$_3$(g) ratio value greater than 10^{-1}, where the two reactions are competing, and the other with higher dilution values

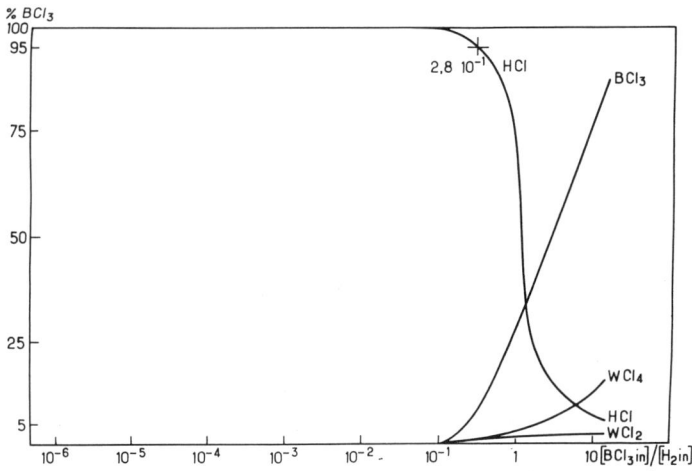

Figure 12: Percentage of BCl$_3$ either reduced by hydrogen to HCl or combined as metallic halides or curreacted at 1300 K and 1 atm. [17]

where only the hydrogen reduction mechanism is effective. Armed with this information, the case of titanium was recalculated with a very high dilution value. The results show that, for dilution ratios greater than 10^{-5}, it is possible to return to the hydrogen reduction regime only. This enabled the experimenters to develop a deposition method in which the initial stage consisted "roughly" in using hydrogen doped with BCl_3 [17].

CONCLUSIONS

Compared with the numerous studies devoted to global optimisation of CVD processes, studies which are currently giving extremely interesting results, the purpose of this study is on a far more modest scale. By focussing on the initial thermodynamic analysis, and more particularly on the role of halides in processes in which they are involved, the authors wanted to demonstrate that, by a careful choice, it is possible to select: the ad hoc molecule, the ranges of experimental parameters promoting its optimum effect and, in certain cases, the type of mechanism whereby it takes part in the deposition reaction. If performed systematically when developing processes involving halides, this type of analysis could avoid a number of errors, help in making the choices required to define the process and occasionally facilitate the understanding of some of the phenomena involved.

REFERENCES

[1] J.O. Borland, in Proc. 10th International Conference on Chemical Vapor Deposition, G.W. Cullen Ed. (The Electrochemical Society, Pennington, NJ, 1987), PP.307-316.

[2] M.L. Hitchman, A.D. Jobson and L.F.T. Kwakman, Appl. Surf. Science, 38, 312-337 (1989).

[3] F. Langlais and C. Prebende, in Procd. 11th International Conference on Chemical Vapor Deposition, K.E. Spear and G.W. Cullen Eds. (The Electrochemical Society, Pennington, N.J., 1990), PP. 686-695.

[4] D.G. Bhat, in Procd. 11th International Conference on Chemical Vapor Deposition, K.E. Spear and G.W. Cullen Eds (The Electrochemical Society, Pennington, N.J., 1990), PP. 648- 655.

[5] C. Bernard and R. Madar, in Chemical Vapor Deposition of Refractory Metals and Ceramics, T.M. Besmann and B.M. Gallois Eds. (Mat. Res. Soc. Proc. 168, Pittsburgh, PA 1990) PP. 3-17.

[6] K.E. Spear and R.R. Dirkx, in Chemical Vapor Deposition of Refractory Metals and Ceramics, T.M. Besmann and B.M. Gallois Eds. (Mat. Res. Soc. Proc. 168, Pittsburgh, PA 1990), PP. 19-30.

[7] J.N. Barbier and C. Bernard, Proc. 15th Calphad Meeting, L. Kaufman Ed., Calphad, 10, 203-238 (1986).
[8] C. Bernard, High Temp. Sci., 27, 131-142 (1990).
[9] N. Thomas, P. Suryanarayana, E. Blanquet, C. Vahlas, R. Madar and C. Bernard, to be published in J. Appl. Phy.
[10] C. Bernard, R. Madar and Y. Pauleau, Sol. State Tech., 32, 79-84 (1989).
[11] E. Blanquet, D.E.A., Université de Grenoble, 1987.
[12] B. Drouin-Ladouce, Thesis, Université d'Orléans, 1990.
[13] F. Teyssandier, M. Ducarroir and C. Bernard, Ann. Chim., 11, 543-555 (1986).
[14] N. Thomas, E. Blanquet, C. Vahlas, C. Bernard and R. Madar, Mat. Res. Soc. Symp. Procd. 204, Pittsburgh, PA 1991, PP. 451-456.
[15] G. Leprince, Thesis, Université d'Orléans, 1989.
[16] E.K. Broadbent and C.L. Ramiller, J. Electrochem. Soc., 131, 1427-1433 (1984).
[17] J. Thebault, R. Naslain and C. Bernard, in High Temperature Chemistry of Inorganic and Ceramic Materials, F.P. Glasser and P.E. Potter Eds. (the Chemical Society, Burlington House, London, 1976), PP. 146-153.
[18] R. Bonetti, D. Conte and H.E. Hintermann, in Proc. 5th International Conference on Chemical Vapor Deposition, J.M. Blocher, H.E. Hintermann and L.H. Hall Eds. (The Electrochemical Society, Pennington, N.J., 1975), PP. 495-504.

CHEMICAL VAPOR DEPOSITION MODELING FOR HIGH TEMPERATURE MATERIALS

SULEYMAN A. GOKOGLU
NASA Lewis Research Center, Cleveland, OH 44135

ABSTRACT

The formalism for the accurate modeling of chemical vapor deposition (CVD) processes has matured based on the well established principles of transport phenomena and chemical kinetics in the gas phase and on surfaces. The utility and limitations of such models are discussed in practical applications for high temperature structural materials. Attention is drawn to the complexities and uncertainties in chemical kinetics. Traditional approaches based on only equilibrium thermochemistry and/or transport phenomena are defended as useful tools, within their validity, for engineering purposes. The role of modeling is discussed within the context of establishing the link between CVD process parameters and material microstructures/ properties. It is argued that CVD modeling is an essential part of designing CVD equipment and controlling/optimizing CVD processes for the production and/or coating of high performance structural materials.

INTRODUCTION

Among the available methods for fabricating materials from the gas phase, chemical vapor deposition (CVD) provides many diverse opportunities for commercial application because it is more economical to develop and easier to scale up. Indeed, its versatility allows CVD to be routinely employed in various areas of materials science and technology. However, as the interest and demand in more sophisticated advanced materials grows to meet today's more stringent performance requirements, it is becoming clearer that the CVD technique should be better exploited.
Structural materials have not generally received the same degree of attention in terms of the precision of the CVD process during their fabrication as the electronic and optical materials. However, this is no longer the case.
A sufficient understanding of the interactive physico-chemical phenomena involved in CVD is required to efficiently produce novel materials with superior properties. Therefore, modeling of such complex systems is now recognized to be an essential and integral part of CVD research.
The controlled implementation of the CVD process is an interdiciplinary effort involving many different fields of science and engineering. A comprehensive analysis should naturally include gas phase and surface chemical reaction kinetics, heat and multicomponent mass transport, fluid physics, and thermodynamics [1]. The simultaneous description of the coupled phenomena in multidimensions with multireaction schemes obviously requires highly efficient computational software and hardware [2]. The analysis should be fully supported by the measurement, testing and characterization techniques that are

available to specialists in heat transfer, fluids, chemistry and materials science.

The modeling of the interactions of such multiparameter systems is indeed a challenge. One has only the directly controllable parameters of the system to manipulate the properties of the coated or fabricated material. These available CVD input variables can be the substrate temperature, reactor pressure, reactant gas composition, or the total flow rate. However, the factors that affect the material property more directly are the growth rate, chemical composition, doping or impurity levels, and microstructure of the deposit material, all of which are also interrelated. All of these latter factors are, in turn, influenced by the convective, thermal and chemical species' concentration fields that prevail inside CVD reactors based on the directly controlled operating parameters. The role of CVD modeling in assisting the process optimization and control is to shed some light onto the transport and chemical processes evolving inside CVD reactors and governing the key factors that determine the resulting material properties. The modest capabilities of CVD modeling today is mostly restricted to accurate growth rate and some deposit chemical composition predictions for only limited materials, and is far from being able to correctly predict doping/impurity levels in the deposit or provide an indication of evolving microstructures. Yet, before undertaking the challenge of improving the material microstructure/property at the chemical and/or materials science level, there are still many useful ways of utilizing modeling in terms of obtaining more rational flow and thermal fields inside CVD reactors. In that respect, by assisting CVD equipment manufacturers for better designs, modeling has taken the established CVD technology back almost a decade!

So far most sophisticated CVD models have been developed for electronic material applications. A review of the current status of CVD modeling is given in Refs. 1 and 3. With increasing interest on advanced structural materials and emphasis for high temperature applications, modeling efforts addressing these issues specifically have recently started to appear in the literature. Besides the papers related to modeling chemical vapor infiltration (CVI) processes, which this paper will not discuss, examples of such efforts are the recent publications related to the CVD synthesis or coating of continuous fibers, which are used for the reinforcement of (inter)metallic and ceramic matrices [4-6].

Because of the emphasis on high temperature applications of structural materials, exposure of the materials to high temperatures during the CVD process is not necessarily disadvantageous. In fact, it may even be desirable to ascertain that the material/coating will survive temperatures as high as at least the intended service temperature. Therefore, some of the advantages of low temperature, such as plasma-enhanced, CVD for electronic material applications do not apply to materials which are expected to be used, for example, in the hot sections of the turbine engine for aeropropulsion applications. Therefore, this paper is limited to modeling only subatmospheric and atmospheric thermal CVD processes and excludes others such as very low pressure, plasma-enhanced or photo-assisted CVD.

APPROACHES TO CVD MODELING

The sophistication levels of the approaches taken to model CVD processes vary depending on the objectives of the effort, the level of complexity, and the availability of thermodynamic, transport and/or chemical kinetic data. Generally, CVD modeling efforts can be categorized as either approaches based on thermochemical equilibrium calculations or dynamic approaches where rate issues are addressed based on the transport and chemical kinetic phenomena.

Thermochemical Equilibrium

Inspired by the conventional phase diagrams, thermochemical equilibrium calculations are used to obtain "CVD phase diagrams". Computer programs, such as those based on free-energy minimization (e.g., SOLGASMIX and NASA CEC [7,8]), are utilized frequently to predict possible windows of operation to deposit the desired materials for given parameter ranges of temperature, pressure and elemental composition.

The limitations and reliability of such an equilibrium-based approach are, naturally, due to finite rate phenomena such as gas phase transport and/or chemical kinetics. Besides their inability to predict the rates of deposition, such approaches also fail to account for the shifts of the deposit material from the equilibrium phases and compositions. A trivial example of the deficiency of the method can be given as follows: although one can predict the deposition of silicon from silane even at room temperature, it does not occur in real life because of kinetic barriers. Conversely, successful deposition of silicon nitride can be accomplished from silicon tetraflouride and ammonia precursors as demonstrated at the United Technologies Research Center: here conditions that are substantially outside the predicted region for stable silicon nitride formation have been used [9]. Numerous other examples of such departures from equilibrium behavior are observed and reported in the literature.

However, thermochemical equilibrium calculations can still be useful tools for initial feasibility studies, providing guidance for the formation of possible gaseous species and condensed phases of different chemical composition. In fact, their capability to explain the observed deposit chemical composition variation along the deposition surface has been demonstrated, ifjudiciously applied by considering the **local** variations of temperature, pressure and gas composition [10]. Such treatments are expected to be even more accurate at higher temperature thermal CVD applications, i.e. for high temperature structural materials, where chemical equilibria are approached. Furthermore, it may be the only available rational approach for complex multicomponent chemical systems, as is the case for the deposition of superconducting materials [11,12] or the codeposition of different compounds [13]. The severe lack of chemical kinetic information for many systems of interest and the uncertainties associated with the available kinetic data render the equilibrium approach still very useful.

Gas Phase Transport Phenomena

With the increasing availability of elaborate transport models based on the well established principles of fluid dynamics, heat, and mass transfer, accurate predictions of flow, thermal and concentration fields in complex shaped CVD reactors are quite possible today [14]. In fact, the capability to predict flow and thermal fields in reactors with steep temperature gradients (~1000K/cm) has probably been the largest contribution of CVD modeling in helping the equipment manufacturers design better reactors. Providing answers to such questions as the proper location of gas inlet and exit ports, the proper positioning of the substrate, the proper operating conditions and geometrical considerations to avoid bouyancy-driven free convection flows, and the proper heating and cooling of the reactor system to obtain more uniform thermal fields around the substrate are only a few examples where modeling has had a considerable impact to improve performances of different CVD reactors.

Figure 1 is a demonstration of how a computational fluid dynamics code may be used to predict the complex flow structures that can develop in a laboratory-scale cold-wall reactor with an inductively heated rectangular substrate. It is shown (by the presence (top case) and absence (bottom case) of the terrestrial gravitational field) that gravitationally-induced free convection flows are responsible for both the recirculation cell created in front of the substrate and the three-dimensional helical flow around the substrate. It is interesting to note that the incoming flow is directed towards the bottom of the hot substrate by the presence of the recirculation cell and will, therefore, supply uneven amounts of source gases to top and bottom surfaces, which will lead to asymmetric deposit thickness profiles.

Despite the maturity of the computational transport codes, one has to still be very cautious of the assumptions involved for their application to each specific case. The validity of using the Boussinesque approximation (versus using the explicit temperature dependence of gas density in the gravitational body force term of the momentum conservation equation), the treatment of the dependence of the fluid transport and thermodynamic properties on temperature and composition, the inclusion of the effects of Soret diffusion, and the proper accounting for radiative heat transfer are typical questions to be addresssed for evaluating the suitability of available models. Also, for nondilute source gas mixtures, one has to consider the heats of reaction for both gas phase and surface reactions, as well as the nonzero gas velocity normal to the deposition surface, the so-called Stefan flow. Such additional complications naturally make the models more cumbersome and increase the coupling among the momentum, energy and species mass conservation equations. As a consequence, one pays a severe penalty in computational complexity and time. It is, therefore, imperative to select the level of sophistication of the model judiciously.

Sophisticated models can give quite inaccurate solutions because of the improper implementation of boundary conditions. Therefore, it is critical to realize, for example, that the use of constant temperature or adiabatic wall boundary conditions does not correctly describe the role of radiation or the coupled interaction of the environment with the CVD system [15]. Asymmetric and other unexpected flow and temperature fields are observed and can be predicted by using more realistic boundary conditions [16]. For a correct description of the temperature distribution on the solid surfaces inside a CVD reactor, conjugate heat transfer analyses (where the coupled gas and solid phase temperature profiles are simultaneously solved) are becoming increasingly necessary.

In the case of CVD processing of fibers, continuous operation is a practical consideration to produce economically feasible quantities of continuous lengths. Therefore, models should be capable of treating the time-dependent growth process coupled with the associated dimensional changes of the fiber.

In cases where the prevailing molecular mean free path becomes comparable to the characteristic dimensions of interest (e.g. fiber diameter), the applicability of the transport equations based on the continuum approximations starts becoming questionable. This is analogous to the Knudsen diffusion situation for CVI processes where the pore size becomes comparable to or smaller than the molecular mean free path. Considering the increasing interest in CVD processing of smaller diameter fibers (<~25μm), especially for ceramic matrix composites, it is doubtful if conventional models can accurately describe the transport processes in high temperature and low pressure CVD systems involving such small diameter fibers. For such cases, molecular trajectory calculation techniques can be utilized by employing, for example, the Monte Carlo direct simulation method.

Gas Phase and Surface Chemical Kinetics

The availability and reliability of gas phase and surface kinetic information are currently the most limiting factors in CVD modeling. The mechanisms, pathways and kinetics of many of the reaction systems of interest to the CVD community are scarcely known. Yet, models should be able to account for reactive species inventories and depletion rates in the gas phase. Surface phenomena such as adsorption, diffusion, nucleation, incorporation, or desorption must appear as boundary conditions in such simulations. Because of the difficulties associated with obtaining experimental kinetic data for the extensive sets of possible reactions under relevant conditions, theoretical techniques utilizing quantum chemistry and electronic structures are being increasingly employed for the required thermochemistry. Table I lists the number of gas phase and surface reactions that have been used in recent CVD models for some of the material systems of interest. Excess hydrogen is the carrier gas for all cases.

Table I. Number of Reactions Used in Recent CVD Models

Material	Source Gases/H_2	Gas Phase	Surface	Ref./Year
Si	SiH_4	27	13	17/1989
GaAs	$TMG+AsH_3$	232	115	18/1989
SiC	$SiH_4+C_3H_8$	166	36	19/1991
Diamond	CH_4+O_2	158	52	20/1991
W	WF_6	8	65	21/1991

As can be seen from Table I, the number of simultaneous gas phase and surface reactions, and therefore, the species, necessary to properly model even single element deposit materials can be as high as tens to hundreds. The kinetic rate expressions of these reactions are obtained from theoretical calculations. Usually, the numerical difficulties associated with significant differences among the characteristic times of reactions and transport processes (the so-called "stiff" systems) have been known to create additional challenges. Such expected difficulties in CVD can not always be circumvented by the schemes and algorithms usually developed for the computational fluid dynamics of high-speed, chemically reacting flows. Furthermore, unlike the simpler geometries used by the references listed in Table I, most realistic industrial applications will involve three-dimensional systems demanding prohibitive computer power, which even today's "personal" Cray's can not meet with so many species and reactions! Therefore, the demand for "smarter" methodologies, not only relying on numerical techniques but also exploiting the available physico-chemical information, will indeed be growing. To that effect, an approach, for example, utilizing a statistical design of computational experiments can be used in order to reduce the excessively large sets of reactions down to a more feasible and managable size which can mimic the actual behavior of the chemical system [22].

Another issue of concern for CVD modeling is the degree of uncertainty regarding the chemical kinetic parameters of reaction systems. Besides the questions related to the low pressure limits of the rate constants, one also has to be careful about the temperature ranges of applicability of the reported rate constants for high temperature CVD systems because many of these values are obtained for lower temperature electronic material applications. Even for the most studied case of silicon deposition from silane, there seems to be no consensus on the reaction kinetics of either gas phase or surface kinetics. Table II lists the different rate constants reported only in the most recent literature for the gas phase dissociation of silane into silylene and hydrogen in a hydrogen bath gas.

Table II. Reported rate constants, k, for silane dissociation, $SiH_4 \rightarrow SiH_2 + H_2$.
$k = AT^\beta \exp(-E/RT)$

$A(sec^{-1})$	β	E(kcal/mole)	Ref./Year
1.09 E25	-3.37	61.2	17/1989
3.3 E15	-0.5	55.9	23/1990
6.671E29	-4.795	63.45	19/1991
6.17 E15	0	59.99	24/1991
5.2 E13	0	52.54	25/1991

Similar to gas phase chemistry, the controversies and inconsistencies related to surface kinetics for silicon deposition from silane have been recently discussed by Ref. 26, i.e. the experimentally observed reaction efficiencies vary by more than two orders of magnitude for nominally identical conditions and the deposition rates obtained at low temperatures and very low pressures are comparable to those obtained at high temperatures and atmospheric pressure. The disagreements have continued in 1990 and 1991 as evidenced by the different expressions published by Refs. 25 and 27 for the reactive sticking coefficients of SiH_4 and Si_2H_6 on silicon surfaces.

Under the circumstances, expecting CVD models to be able to accurately predict the minute levels of dopant or impurity concentrations evolving in the deposit material during their in-situ deposition is currently unrealistic. Yet, it should be mentioned that the limitation is not due to the lack of a plausible formalism to incorporate such treatments into CVD models but is due to

the unavailability and/or uncertainty of chemical kinetic information.
It seems that CVD modeling can be most helpful in areas where chemical kinetic limitations due to gas phase and/or surface reactions are minimal, i.e. in the gas phase mass transport controlled regime. After all, it would certainly be desirable to operate under such conditions where the deposition rates and source gas utilization would be maximized. The limitations of chemical kinetic barriers can be circumvented by increasing CVD temperatures. In principle, this should be less of a concern for high temperature materials (cf. electronic materials) up to their intended service temperatures or the limit of endurance for the substrate material. Provided the kinetic barriers are overcome, one can then use thermochemical equilibrium calculations at surface conditions to define the chemical composition of the deposit and the boundary conditions for the species mass transport equations [28].
However, at higher temperatures one has to consider another potential problem of gas phase powder formation. The prevailing supersaturation levels can be sufficiently high for the reactive precursors created at high temperatures to lead to gas phase nucleation and particle formation. Therefore, one needs to define the safe operation boundaries to preclude the usually very sudden onset of gas phase nucleation [29].

AN EXAMPLE CASE: FIBER CVD

CVD modeling of thin fibrous substrates encompasses specific problems that are not encountered for other typical substrates with flat surfaces [4-6]. The chemical and structural complexity of the Textron SCS-6 SiC fiber grown by a continuous CVD process, as is schematically depicted in Figure 2, is a good example of how changes in CVD conditions may lead to cross-sectional variations in chemical compositions and microstructures [30]. In an endeavor to study the relationships between process variables and deposit compositions, microstructures and properties, we modeled a miniature Textron CVD reactor built in our laboratory. It is an upflow batch reactor made of a quartz tube of 1.8cm I.D. and adjustable length. The monofilament is resistively heated. Silicon deposition from silane is chosen because of its better studied gas phase and surface chemistry and relevance to other silicon-based CVD materials of high temperature interest. The results are presented for the first 20cm from the inlet. A more detailed discussion is given Ref. 6.
The axisymmetric geometry of the reactor with a thin hot fiber at the center creates very steep temperature gradients (~10000K/cm versus 1000K/cm in conventional geometries) near the fiber surface. This results in such atypical Soret diffusion effects that Soret becomes one of the primary gas transport mechanisms, i.e. predicted growth rates change by a few hundred percent with and without its inclusion in the model (Figure 3). Hence, accurate information for such transport coefficients is essential for the reliability of fiber CVD models. Furthermore, the sensitivity of predicted rates on fiber temperature indicates that the accurate measurement and control of fiber temperature is of utmost importance to controlling rates.
The higher temperature region within ~250μm from the fiber surface is also the chemically active region where most of the SiH_2 production, which governs the deposition rate of silicon,

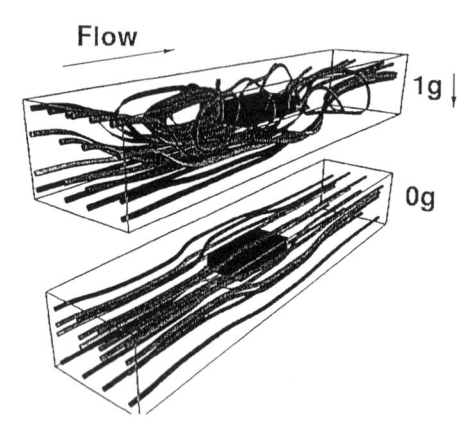

Figure 1 - A schematic of flow patterns in a cold-wall CVD reactor with an inductively heated rectangular substrate. The 3-dimensional complex flows in the presence of gravity is due to bouyancy-driven free convection effects.

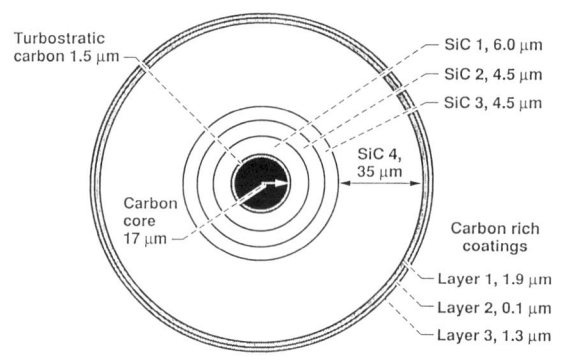

Figure 2 - Schematic of Textron's SCS-6 silicon carbide fiber microstructure [30].

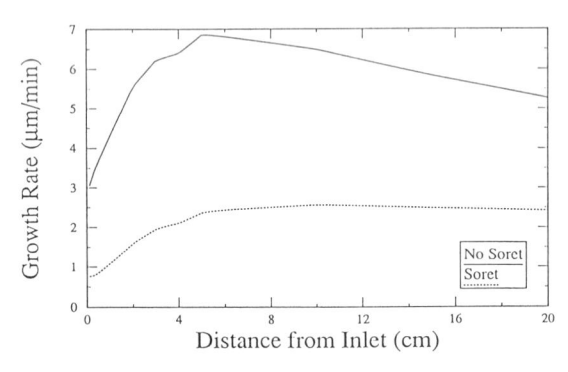

Figure 3 - Predicted silicon growth rates with and without Soret diffusion versus axial distance along fiber [6].

occurs. Therefore, the above discussed uncertainties of the chemical kinetic parameters, which may not give significant differences in predicted Si deposition rates for conventional flat plate-like geometries, result in unacceptably magnified rate predictions depending on whose data is used from Table II [6].

The significance of the steep temperature gradients near the fiber surface and, consequently, the amplified importance of Soret diffusion and chemical kinetics increase even more as the fiber diameter gets smaller. Therefore, the correct prediction of temperature fields inside the cold wall fiber reactor becomes of paramount importance. In predicting the wall temperatures for typical operating conditions of our fiber CVD reactor, it is determined that radiation has a strong contribution especially for low thermal conductivity gases (Figure 4). However, if radiation is fully accounted for, accurate predictions of wall temperatures are indeed possible [31].

Radial growth rates on the fiber will vary with respect to the fiber diameter. However, besides this expected change, which should scale with the available deposition surface area, there are other less obvious interactions of growth rate with fiber diameter as a consequence of the above mentioned temperature profile, Soret and chemistry effects (Figure 5).

The specific issues pertaining to cold wall fiber CVD for high temperature applications creates more challenges for CVD modeling, for both batch and continuous modes of reactor operation. Furthermore, with increasing interest in smaller diameter fibers, most of these issues, such as the atypical importance of Soret diffusion, the unusually stringent accuracy requirement for the chemical kinetic parameters, or the complex interaction of deposition rates with fiber diameter, get even more difficult to tackle.

FUTURE RECOMMENDATIONS AND CLOSING REMARKS

Despite the present limitations of CVD modeling in providing immediate solutions to the complex problems associated with various reactors, different chemical systems and applications, the utility of CVD modeling can not be disputed in assisting the resolution of many existing issues. It can provide invaluable insights into the complex physicochemical phenomena taking place inside CVD reactors, provided that one is consciencious about its current capabilities and cognizant of its anticipated future challenges. It is certainly helpful for confirming and enhancing our understanding of the process. CVD models can be routinely used for designing more efficient reactor geometries and optimizing gas inlet, outlet and substrate configurations. They can go beyond predicting qualitative trends and are increasingly being used to obtain accurate quantitative predictions of deposition rates and deposit chemical compositions.

The usefulness and beauty of presenting information in a "universal" format, for example by using dimensionless variables, have long been recognized by scientists and engineers. The use of CVD phase diagrams is partly such an approach which condenses the various operating conditions of different systems into a general form by utilizing the fundamentals of equilibrium thermodynamics. The limitations of such an approach due to transport and chemical kinetic phenomena are discussed above. Attempts have been made in the literature to incorporate the effects of transport-induced shifts into CVD phase diagrams [32]. However,

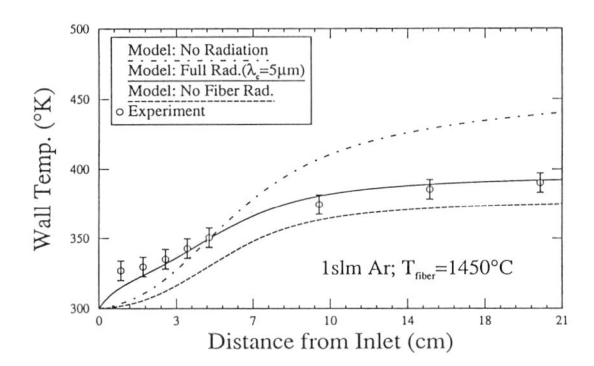

Figure 4 - Experimentally measured and predicted reactor wall temperatures for fiber CVD [31]. A sophisticated radiation model is necessary for improved accuracy.

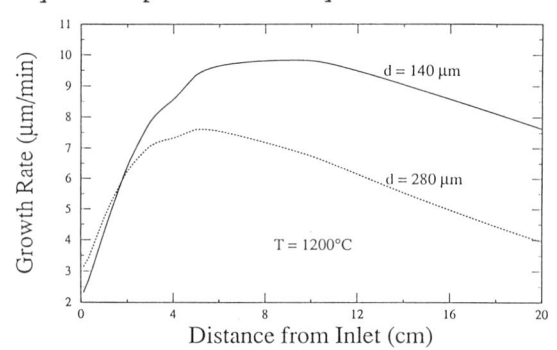

Figure 5 - Predicted silicon growth rates versus axial distance along fiber for two fiber diameters [6].

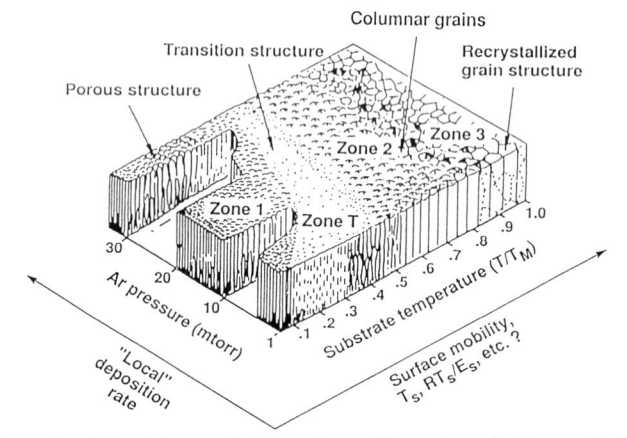

Figure 6 - Application of Thornton Structural Zone Model [33] (developed for PVD) to CVD by recommended new coordinates.

their inability to give information about rates and be applicable to different reactor geometries still limit their utility for practical applications. The growing volume of CVD literature is certainly increasing the burden of gleaning useful information from publications to apply to a different situation. Currently, technical papers related to CVD are being published at a rate of about two articles a day. The need to find more general ways of reporting information in the CVD literature is greater than ever today. Such an approach is also essential for scaling up laboratory research to industrial size production. In an attempt to unify the efforts to establish the link between process parameters and material microstructures, it is suggested that a generic plot, as is shown in Figure 6, be explored, at least for the same material systems using the same chemical precursors as source gases. It is inspired by the structural zone model proposed by Thornton [33] for physical vapor deposition systems. The suggested coordinates could be, for example, the local growth rates normal to the deposition surface and the surface temperature, which could be nondimensionalized with respect to the activation energy of the relevant surface phenomenon. A similar graph is given in Ref. 9. Even if the data of different investigators fail to collapse into a common form when cast into such a format, it is anticipated that the differences will be reduced and the analysis will be simplified. Once, and if, such a "universal" diagram is constructed for a chemical system by complimentary experimentation, then the loop would be closed by reducing the goal of CVD modeling to accurate rate and deposit composition predictions which is more feasible.

The accuracy and reliability of CVD models will undoubtedly depend on their verification by carefully controlled experiments and the quality of the transport, thermodynamic and chemical kinetic data supplied to them. Thus, developing high fidelity models and improving the confidence levels for their use are directly coupled with experimentation for testing the models and refining such input information. This can be more efficiently accomplished via more focused and synergistic theoretical and experimental efforts. Modeling geared to address specific aspects of a complex phenomenon, where the analyses are justifiably reduced to a more manageable size with a negligible sacrifice in information, will indeed be more effective. Similarly, smaller scale experimental efforts in better defined environments, set up to provide answers to individual thermochemical questions, can be more meaningful. Future CVD research should, therefore, include both experiments and modeling as integral parts of a coordinated program.

REFERENCES

1. K.F. Jensen, in Microelectronics Processing: Chemical Engineering Aspects, edited by D.W. Hess and K.F. Jensen (Advances in Chemistry Series, No.221, American Chemical Society, Washington, D.C., 1989) pp. 199-263.
2. Supercomputer Research in Chemistry and Chemical Engineering, edited by K.F. Jensen and D.G. Truhlar (American Chemical Soc., Washington, D.C., 1987.)
3. S.A. Gokoglu, in Chemical Vapor Deposition XI, edited by K.E. Spear and G.W. Cullen (The Electrochemical Soc., 1990) pp.1-9.

4. S.A. Gokoglu, M. Kuczmarski, L.C. Veitch, P. Tsui, and A. Chait, in *Chemical Vapor Deposition XI*, edited by K.E. Spear and G.W. Cullen (The Electrochemical Soc., 1990) pp. 31-37.
5. J.H. Sholtz, J. Gatica, H.J. Viljoen, and V. Hlavacek, J. Crystal Growth, 108, 190 (1991).
6. S.A. Gokoglu, M. Kuczmarski, and L.C. Veitch, J. Materials Research, submitted 1991.
7. G. Eriksson, Chem. Scr., 8, 100 (1975).
8. S. Gordon and B.J. McBride, NASA SP-273, 1976.
9. J.R. Strife, et al., presented at the First Thermal Structures Conference, Univ. of Virginia, Charlottesville, VA, Nov.15, 1990.
10. C.F. Wan and K.E. Spear, CALPHAD, 7, 149 (1983).
11. M. Ottosson, A. Harsta, and J.O. Carlsson, in *Chemical Vapor Deposition XI*, edited by K.E. Spear and G.W. Cullen (The Electrochemical Soc., 1990) pp. 180-187.
12. C. Vahlas and T.M. Besmann, in *Chemical Vapor Deposition XI*, edited by K.E. Spear and G.W. Cullen (The Electrochemical Soc., 1990) pp. 188-194.
13. W.J. Lackey, A.W. Smith, D.M. Dillard, and D.J. Twait, in *Chemical Vapor Deposition X*, edited by G.W. Cullen (The Electrochemical Soc., 1987) pp. 1008-1027.
14. K.F. Jensen, E.O. Einset, and D.I. Fotiadis, Annu. Rev. Fluid Mech., 23, 197 (1991).
15. D.I. Fotiadis, M. Boekholt, K.F. Jensen and W. Richter, J. Crystal Growth, 100, 577 (1990).
16. P.J. Roksnoer, C. Van Opdorp, J.W.F.M. Maes, M. De Keijser and C. Weber, J. Electrochem Soc., 136, 2427 (1989).
17. M.E.Coltrin, R.J. Kee, G.H. Evans, J. Electrochem. Soc., 136, 819 (1989).
18. M. Tirtowidjojo and R. Pollard, J. Crystal Growth, 98, 420 (1989).
19. M.D. Allendorf and R.J. Kee, J. Elecrochem. Soc., 138, 841 (1991).
20. M. Frenklach and H. Wang, American Physical Soc., Physical Review B, 43, 1520 (1991).
21. R. Arora and R. Pollard, J. Electrochem. Soc., 138, 1523 (1991).
22. A.R. Marsden,Jr., M. Frenklach, and D.D. Reible, J. Air Pollution Control Assoc., 37, 370 (1987).
23. C.J. Guinta, R.J. McCurdy, J.D. Chapple-Sokol, and R.G. Gordon, J. Appl. Phys., 67, 1062 (1990).
24. H.K. Moffat, K.F. Jensen, and R.W. Carr, J. Phys. Chem., 95, 145 (1991).
25. C.R. Kleijn, J. Electrochem. Soc., 138, 2190 (1991).
26. J.H. Comfort and R. Reif, J. Electrochem. Soc., 136, 2386 (1989).
27. W.G. Breiland and M.E. Coltrin, J. Electrochem. Soc., 137, 2313 (1990).
28. S.A. Gokoglu, J. Electrochem. Soc., 135, 1562 (1988).
29. J. Collins, D.E. Rosner, and J. Castillo, ibid, (1991).
30. X.J. Ning and P. Pirouz, J. Mater. Res., 6, 2234 (1991).
31. M. Kassemi, S.A. Gokoglu, C. Panzerella, and L. Veitch, National Heat Transfer Conf., San Diego, CA, 1992, submitted.
32. D.E. Rosner and J. Collins, in *Chemical Vapor Deposition XI*, edited by K.E. Spear and G.W. Cullen (The Electrochemical Soc.,1990) pp. 49-60.
33. J.A. Thornton, Ann. Rev. Mater. Sci., 7, 239 (1977).

THEORETICAL STUDY OF GAS-PHASE THERMODYNAMICS RELEVANT TO SILICON CARBIDE CHEMICAL VAPOR DEPOSITION[*]

MARK D. ALLENDORF AND CARL F. MELIUS
Combustion Research Facility, Sandia National Laboratories, Livermore, CA 94551-0969

ABSTRACT

Equilibrium calculations are reported for conditions typical of silicon carbide (SiC) deposition from mixtures of silane and hydrocarbons. Included are 34 molecules containing both silicon and carbon, allowing an assessment to be made of the importance of organosilicon species (and organosilicon radicals in particular) to the deposition process. The results are used to suggest strategies for improved operation of SiC CVD processes.

I. INTRODUCTION

The excellent resistance of silicon carbide (SiC) to corrosive, oxidizing or high-temperature environments, in addition to its outstanding hardness and semiconductor properties, make it of considerable industrial interest. Comprehensive models including detailed chemical reactions that occur in the gas-phase and on the surface, coupled with reactor fluid mechanics are required to optimize new deposition processes. Recent work provides several indications that gas-phase chemistry is an important element in the chemical vapor deposition (CVD) of SiC. First, a comparison of measured deposit compositions with thermodynamic predictions for mixtures containing silicon, carbon, and hydrogen showed that deposits typically contain excess silicon in regions where pure SiC is predicted at equilibrium[1]. This is attributed to the suspected higher surface reactivity of gas-phase silicon-containing species relative to that of stable hydrocarbons (such as CH_4 and C_2H_4). Several experimental and theoretical studies have confirmed this difference in reactivity [2,3]. Second, the reactivity of gas-phase hydrocarbons themselves with silicon substrates varies widely, depending on the degree of unsaturation in the molecule. Saturated species, such as C_3H_8, which is commonly used to deposit epitaxial SiC, have reactive sticking coefficients on Si(111) (whose crystal structure is the same as that of β-SiC) on the order of 10^{-5} to 10^{-4} [3], while unsaturated molecules such as C_2H_2 react with a higher probability [4] on the order of 0.001 - 0.01. Radical species are even more reactive, with sticking coefficients near unity [2,5]. This low reactivity of the initial precursors relative to their decomposition products implies that some decomposition by gas-phase pyrolysis may be required to achieve efficient deposition with these reactants. Indeed, at the high gas-phase temperatures typical of SiC-CVD (> 1500 K), hydrocarbons such as C_3H_8 are known to decompose, yielding large amounts of CH_4, C_2H_4, and C_2H_2 [6]. This conversion could be the rate-limiting step in SiC deposition under some conditions. Clearly, with such large variations (a factor of 10^4) in surface reactivity, the exact composition of the gas phase can make a large difference in the deposition rate. Finally, surface temperatures lower than 1500-1600 K are desirable in order to deposit SiC on thermally sensitive substrates such as metals [7]. To develop predictive models of such processes, a much better understanding of the role of gas-phase chemical reactions is required, since lower deposition temperatures tend to produce gas-phase compositions that are far from equilibrium.

Equilibrium calculations are useful for determining which species are likely to be stable in the event that sufficient thermal energy is available for their initial formation. They can also show how reactor conditions can be altered to favor formation of particular species. Earlier studies of gas-phase equilibria under SiC CVD conditions used limited

---

[*] This work was supported by the U. S. Department of Energy Advanced Industrial Concepts Materials Program.

numbers of gas-phase species and included few, if any, molecules containing both silicon and carbon [1,8-10]. In this work, we report gas-phase equilibrium calculations performed for typical epitaxial SiC CVD conditions. Calculations were also extended to low-pressure, inert-gas environments now under study for low-temperature SiC CVD. The calculations make use of new thermodynamic data recently determined by us for 34 compounds in the Si-C-H system [11]. This expanded species set thus allows the importance of organosilicon species in SiC deposition to be evaluated.

II. THEORETICAL METHODS

Equilibrium calculations were performed using the STANJAN code developed by Reynolds et al [12]. Recently reported heats of formation for Si_2H_n molecules of Ho and Melius were used [13], which are somewhat different from those published in an earlier paper [14]. Data for hydrocarbon species were obtained from ab initio calculations and are in excellent agreement with accepted experimental values [14,15]. Data for SiC, SiC_2, and Si_2C were obtained from the JANAF Tables [16]. Thermodynamic data for the remaining organosilicon species were also determined by ab initio calculations [11].

III. RESULTS

Figure 1 shows the results of equilibrium calculations of the gas-phase species mole fractions expected for typical epitaxial growth conditions and allowing SiC, Si, or graphite to form as a solid phase. The results indicate that most of the initial gas-phase reactants are converted to the thermodynamically stable SiC (> 90% conversion to SiC) with only a few stable species such as CH_4 and SiH_4 present in significant quantities in the gas-phase. Previous equilibrium calculations have reached similar conclusions [10]. As discussed above, it is very unlikely that this uninteresting result is representative of actual gas-phase conditions

A more realistic simulation of the gas-phase can be obtained if one assumes that surface reactions are rate-limiting, allowing the gas-phase to come to partial equilibrium by not allowing solid phases to form. The effect of this change is shown in Figure 2, again for typical epitaxial growth conditions. Now the gas-phase contains many more species, with both saturated and unsaturated radical species present in significant quantities. The hydrocarbon in highest concentration at all temperatures is CH_4, with C_2H_4 and C_2H_2 becoming more important at temperatures above 1400 K. The two most important silicon species are SiH_2 and Si_3, which at temperatures of approximately 1150 and 1300 K, respectively, appear in concentrations high enough to affect the deposition.

In addition to species containing only Si or C as the heavy atom, two other molecules containing both Si and C appear in significant quantities above 1300 K. The Si_2C molecule was shown previously to form under these conditions [10]. However, $SiCH_2$, which is one of the molecules added from the BAC-MP4 calculations, also appears, though in smaller quantities. The minimum amount of such radicals that must be present in the gas-phase to have an effect on the deposition process can be estimated by assuming that they will have surface reactivities approaching 1.0, while an abundant hydrocarbon such as CH_4 has a reactive sticking coefficient of only 5 x 10^{-5} [3]. Thus, mole fractions of Si_2C and $SiCH_2$ of only about 2 x 10^{-8} would be required for these radicals to deposit the same amount of carbon as CH_4. The concentrations of both these molecules is well above this level for temperatures greater than 1300 K, implying that an important role in the deposition process would be played by these species if conditions leading to gas-phase pseudo-equilibrium were obtained.

As the concentration of hydrogen decreases, Si_2C and $SiCH_2$ remain the organosilicon radicals with the highest mole fraction, but the temperature at which these radicals reach high enough concentrations to affect deposition rates decreases. For example, for H_2/Si = 1000, Si_2C/CH_4 reaches 0.01 at 1400 K and continues to increase with temperature. With the same conditions as in Figure 2 but with a H_2:Si ratio of 10 (data not shown), the ratio Si_2C/CH_4 reaches 0.01 at approximately 1200 K (again continuing to increase with temperature.) This indicates that the reactive species such as

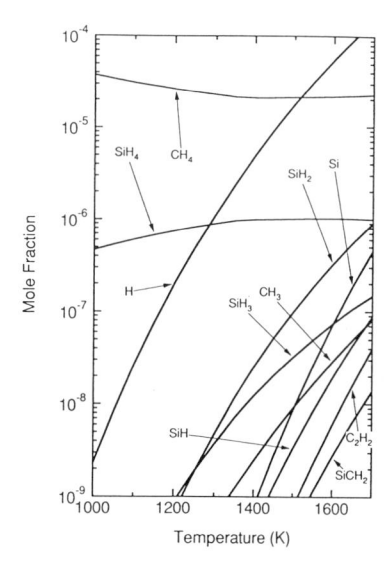

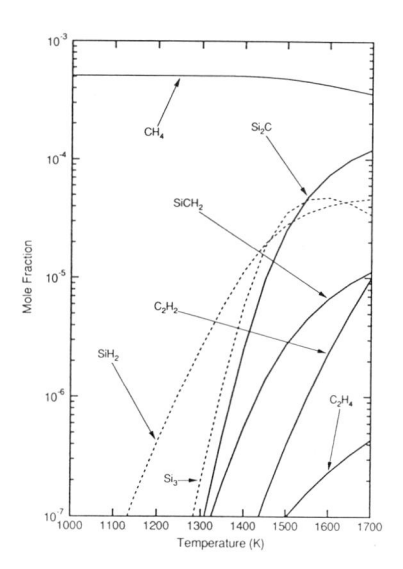

Figure 1: Mole fractions of gas-phase species as a function of temperature, allowing SiC, Si, and graphite to deposit as solid phases. Initial conditions: pressure = 760 torr; moles SiH_4 = 0.49; moles CH_4 = 0.51, moles H_2 = 1000.

Figure 2: Mole fractions of gas-phase species as a function of temperature, but with no solid phases allowed to form. Initial conditions are the same as for Figure 1.

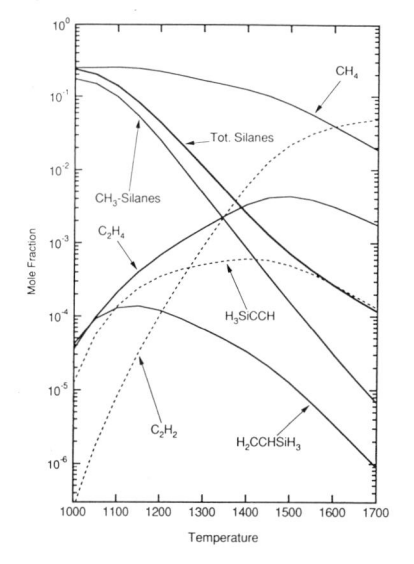

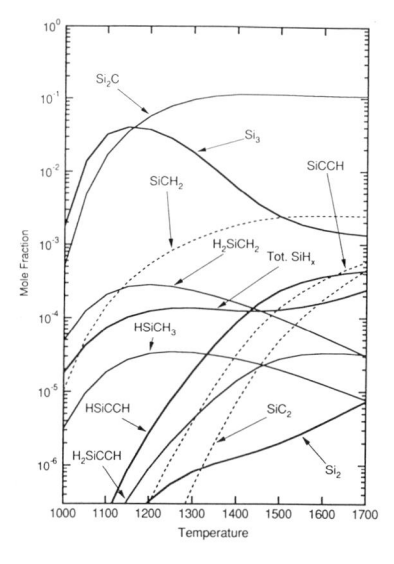

Figure 3: Mole fractions of gas-phase species as a function of temperature, but with no solid phases allowed to form. Initial conditions: pressure = 760 torr; moles SiH_4 = 0.49; moles CH_4 = 0.51, moles H_2 = 0. (a) stable species; (b) radicals.

Si$_2$C are more likely to contribute significantly to carbon deposition when hydrogen carrier gas concentrations are low. Thus, these calculations suggest that SiC deposition rates may increase at low temperatures if the concentration of hydrogen carrier gas is decreased.

If hydrogen carrier gas is eliminated, so that the only hydrogen in the system comes from the reactants SiH$_4$ and CH$_4$, then the number of organosilicon species increases dramatically. This is shown in Figure 3, where the mole fractions of both stable (Figure 3A) and radical (Figure 3B) species are shown as a function of temperature. In this case, the concentration of organosilicon radicals is comparable to that of stable species such as CH$_4$, methylsilanes, C$_2$H$_4$, and C$_2$H$_2$. In addition to Si$_2$C and SiCH$_2$, several other radicals are present, including HSiCCH, H$_2$SiCCH, SiCCH, and SiC$_2$. These species could be formed by reaction between SiH$_2$ (the initial product of SiH$_4$ decomposition) and C$_2$H$_4$ or C$_2$H$_2$ to form H$_2$C=CHSiH$_3$ and H$_3$SiC≡CH, respectively. The radical species would form by loss of hydrogen. Since both H$_2$C=CHSiH$_3$ and H$_3$SiC≡CH are present in appreciable concentrations at equilibrium (Fig. 3A) and the reactions producing them are near the diffusion-controlled limit [17], these two molecules are favored both kinetically and thermodynamically. It is thus likely that some or all of the radicals resulting from H$_2$C=CHSiH$_3$ and H$_3$SiC≡CH decomposition are present in the gas-phase in concentrations high enough to affect the deposition in the absence of H$_2$ carrier gas.

The substitution of an inert carrier gas such as argon or helium for hydrogen produces markedly different gas-phase concentration profiles. Figure 4 presents results of equilibrium calculations for conditions again typical of epitaxial SiC CVD, but with hydrogen carrier gas replaced by argon (760 torr, Si/C~1.0.) Now, C$_2$H$_2$ is the stable hydrocarbon at most temperatures, with the concentration of CH$_4$ exceeding that of C$_2$H$_2$ only at the lowest temperatures. More interestingly, Si$_2$C is the molecule in the highest concentration. Other organosilicon species that were shown to be stable under low hydrogen conditions are also present, such as SiCH$_2$, SiC$_2$, and HSiCCH. The very high concentration of Si$_2$C is unlikely to be realized in practice at the low end of the temperature range, since this species is probably very reactive. However, this result does point out the strong ability of hydrogen to suppress the amounts of radical species at low temperatures (below 1300 K) by converting them to more stable forms such as CH$_4$. Figure 4 also shows that organosilicon species are more stable in general under these conditions than radicals containing only silicon, such as Si$_3$ and Si. This is another indication that organosilicon radicals may serve to transport carbon from the gas-phase to the growing deposit.

Another interesting result of these calculations is that high concentrations of hydrogen suppress the formation of radicals of all kinds, except at the highest temperatures (compare Figure 4 with Figure 2.) The presence of high concentrations of silicon-containing radicals has been suggested as the reason for the occurrence of homogeneous nucleation ("reactor snow") when SiH$_4$ is used to deposit epitaxial silicon [18]. Thus, a beneficial effect of hydrogen carrier gas on SiC CVD may be reduction in the probability for homogeneous nucleation, which would improve the quality and reproducibility of deposits.

The potential for SiC deposition at much lower temperatures (< 1273 K) has been recently explored by Komiyama and co-workers [7]. In their experiments, mixtures of disilane (Si$_2$H$_6$) and C$_2$H$_4$ or C$_2$H$_2$ at low pressures (≤ 1 torr) and temperatures (≤ 1273 K) were used to deposit SiC in a hot-wall reactor. No carrier gas was used. Their results [7] showed that SiC can be deposited, but that at temperatures below 1273 K, the deposits were silicon-rich. Increasing the concentration of C$_2$H$_4$ by a factor of 10 had little effect on the carbon content of the deposits, indicating either that carbon deposition was limited by a slow reaction occurring on the surface, or that an efficient gas-phase reaction to form an organosilicon species was responsible for most of the carbon deposition. They proposed that two likely reactions would be SiH$_2$ + C$_2$H$_4$ → H$_2$C=CHSiH$_3$ and SiH$_2$ + C$_2$H$_2$ → HC≡CSiH$_3$ and analyzed the temperature dependence of carbon deposition using a simple gas-phase mechanism. Although the results they obtained were consistent with their observations, this now appears to have been fortuitous since the rates used for SiH$_2$ insertion into C$_2$H$_4$ and C$_2$H$_2$ have since been considerably revised. They also did not account for fall-off in the decomposition rate of Si$_2$H$_6$. Nevertheless, their results point to the possibility that organosilicon species could be important to SiC deposition at low pressures and temperatures.

Equilibrium calculations are of limited value at the lowest temperatures (873 K) used in the experiments of Komiyama et al., since the gas-phase chemical reactions will be slow due to high activation barriers for Si$_2$H$_6$ and C$_2$H$_4$ decomposition. However, at the higher temperatures used by them, sufficient thermal energy is available for Si$_2$H$_6$ decomposition. Given the thermal energy required to surmount this barrier, other reactions that are either exothermic or have lower activation barriers can proceed toward equilibrium. In particular, SiH$_2$ insertion into C$_2$H$_4$ and C$_2$H$_2$ is quite exothermic (-56.6 and -66.0 kcal mol^{-1}, respectively) and the barriers to reaction are small. Sufficient energy should thus be present in the system for these molecules to continue to decompose (most likely by H$_2$ elimination) to form other stable gas-phase species. We have therefore extended our equilibrium calculations to the experimental conditions used by Komiyama et al [7] in order to identify those organosilicon species that are the most stable under their conditions.

Figure 5 shows equilibrium mole fractions assuming a ten-fold excess of carbon, 0.76 torr, and no H$_2$ carrier gas, as used in one set of experiments by Tanaka and Komiyama [7]. In this case, C$_2$H$_2$ is the most stable hydrocarbon at all but the lowest temperatures. In addition, the radical Si$_2$C appears in high concentrations. This molecule would not be a product of the SiH$_2$ insertion into C$_2$H$_4$ and C$_2$H$_2$ discussed above. Two species that would form from such reactions, HC≡CSiH$_3$ and HC≡CSiH, are present, however. Possible decomposition products of these molecules are also present (SiC$_2$ and SiCCH). This indicates that SiH$_2$ insertion into unsaturated hydrocarbons is a probable route to organosilicon radicals that could deposit carbon during the SiC CVD process, as suggested previously.

Figure 4: Mole fractions of gas-phase species as a function of temperature, but with no solid phases allowed to form. Initial conditions: pressure = 760 torr; moles SiH$_4$ = 0.49; moles CH$_4$ = 0.51; Ar:Si = 1000.

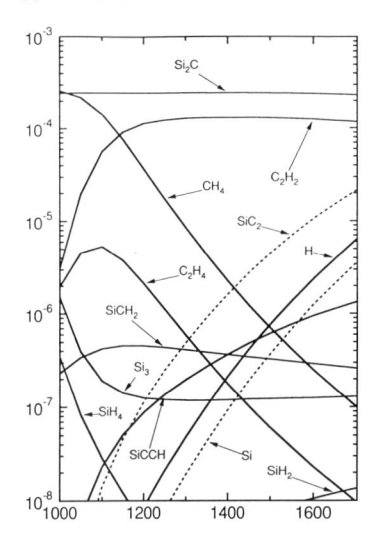

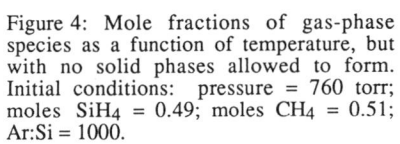

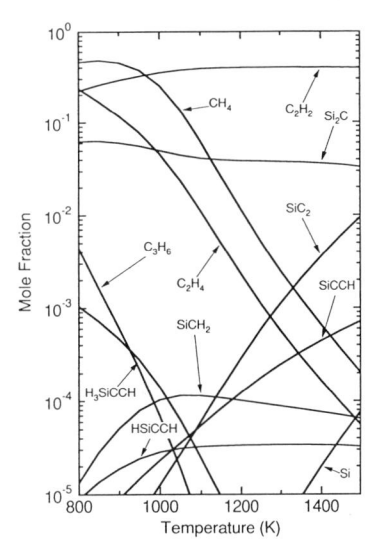

Figure 5: Mole fractions of gas-phase species as a function of temperature, but with no solid phases allowed to form. Initial conditions: pressure = 0.76 torr; moles Si$_2$H$_6$ = 1.0; moles C$_2$H$_4$ = 11.0.

The equilibrium calculations also provide an explanation why stoichiometric SiC can be deposited at such low (1073 K) temperatures. Figure 5 shows that radicals containing only Si (with or without hydrogen) are present in very low concentrations (only Si atoms are present in mole fractions above 10^{-5}.) The excess carbon in the system is responsible for the increased thermodynamic stability of the organosilicon species. Thus, the number of gas-phase species that can deposit excess silicon is reduced.

IV. SUMMARY

Equilibrium calculations are reported for conditions currently used to deposit silicon carbide (SiC) from mixtures of silane and hydrocarbons using an expanded species set that includes organosilicon species. The results suggest that, under conditions typically used to deposit epitaxial SiC (760 torr, high H_2 concentrations, and temperatures above 1600 K) organosilicon species play little if any role. At lower temperatures or in the absence of hydrogen carrier gas, the concentration of organosilicon radicals at equilibrium increases substantially. Several of these species, including SiC_2 and $HSiC\equiv CH$, are expected to be products of the reaction between SiH_2 and unsaturated hydrocarbons such as C_2H_2 and C_2H_4. Although the rates of gas-phase reactions will determine the actual concentrations of these species, the large exothermicity of the SiH_2 insertion reactions indicates that SiC_2 and $HSiC\equiv CH$ may be produced in quantities sufficient to contribute significantly to carbon deposition. Thermodynamic data are now in place to permit development of kinetic models to proceed.

References

[1]. Fischman, G. S.; Petuskey, W. T. *J. Am. Ceram. Soc.* **1985**, *68*, 185.
[2]. Robertson, R. M.; Rossi, M. J. *J. Chem. Phys.* **1989**, *91*, 5037.
[3]. Stinespring, C. D.; Wormhoudt, J. C. *J. Appl. Phys.* **1989**, *65*, 1733.
[4]. Mogab, C. J.; Leamy, H. J. *J. Appl. Phys.* **1974**, *45*, 1075.
[5]. Noorbatcha, I.; Raff, L. M.; Thompson, D. L. *J. Chem. Phys.* **1984**, *81*, 3715.
[6]. Allendorf, M. D.; Kee, R. J. *J. Electrochem. Soc* **1990**, *138*, 841.
[7]. Tanaka, S.; Komiyama, H. *J. Am. Ceram. Soc.* **1990**, *73*, 3046.
[8]. Harris, J. M.; Gatos, H. C.; Witt, A. F. *J. Electrochem. Soc.* **1971**, *118*, 338.
[9]. Kingon, A. I.; Lutz, L. J.; Liaw, P.; Davis, R. F. *J. Am. Ceram. Soc.* **1983**, *66*, 558.
[10]. Stinespring, C. D.; Wormhoudt, J. C. *J. Crystal Growth* **1988**, *87*, 481.
[11]. Allendorf, M. D.; Melius, C. F. *J. Phys. Chem.* **1991**, accepted for publication.
[12]. Reynolds, W. C. *The Element Potential Method for Chemical Equilibrium Analysis:* Stanford Univ. 1986.
[13]. Ho, P.; Melius, C. F. *J. Phys. Chem.* **1990**, *94*, 5120-5127.
[14]. Ho, P.; Coltrin, M. E.; Binkley, J. S.; Melius, C. F. *J. Phys. Chem.* **1986**, *90*, 3399.
[15]. Melius, C. F. in Bulusu, S. N., Ed. *Chemistry and Physics of Energetic Materials*; Kluwer Academic Publishers: Dorderecht, 1990, p. 21.
[16]. *The JANAF Tables.*, *J. Phys. Chem. Ref. Data* **1985**, *14, Suppl 1*.
[17]. Chu, J. O.; Beach, D. B.; Jasinski, J. M. *J. Phys. Chem.* **1987**, *91*, 5340.
[18]. Allen, K. D.; Sawin, H. H. *J. Electrochem. Soc.* **1986**, *133*, 421.

KINETIC MODELLING OF THE DEPOSITION OF SiC FROM METHYLTRICHLOROSILANE

Stratis V. Sotirchos and George D. Papasouliotis
Department of Chemical Engineering
University of Rochester
Rochester, NY 14627

ABSTRACT

A kinetic model is presented for the deposition of silicon carbide through decomposition of methyltrichlorosilane (MTS). The developed model includes gas phase (homogeneous) reactions that lead to formation of deposition precursors and surface (heterogeneous) reactions that lead or can lead to deposition of silicon carbide, silicon, and carbon. The kinetic model is incorporated in a transport and reaction model for a tubular hot-wall reactor, and the overall reactor model is used to obtain some preliminary results on the effects of pressure and distance in the reactor on the rate of deposition and the composition of the deposit. The results show that the model can reproduce most of the experimental observations of the literature.

INTRODUCTION

Because of the various chemical, mechanical, and electronic properties of silicon carbide, its study has been a subject of considerable interest. A large number of experimental studies has been conducted on the deposition of SiC films through decomposition of various precursors [1]. Deposition from methyltrichlorosilane (CH_3SiCl_3) has been one of the most frequently studied and used routes for SiC deposition. However, the results of the various experimental investigations vary widely from one research group to another, indicating that the details of the deposition process (and hence, the deposition rate) are a strong function of the experimental conditions and of the experimental arrangement (reactor configuration) used to carry out the deposition process [1-3].

Our own experimental investigation of SiC deposition from MTS showed that the deposition rate depends on the distance in the reactor, suggesting, among other things, that the decomposition products may affect the deposition process and its rate [22]. Subsequent studies revealed a strongly inhibitory effect of HCl on the deposition process, in agreement with results obtained by other investigators [3]. Needless to say, any kinetic model for the deposition of silicon carbide through the decomposition of MTS should account for these experimental trends, apart from the observed dependence of the deposition rate on temperature, pressure, and flow rate.

Presently, most of the published kinetic modelling studies for SiC deposition are for codeposition processes, the precursors being mixtures of silanes and hydrocarbons. In most cases, more effort was put into modelling the transport phenomena in the CVD reactor than the chemistry of the reacting system under consideration [5,6]. One of the most complete kinetic modelling studies was presented by Allendorf and Kee [4], who studied the deposition of silicon carbide from silane and propane. They combined detailed kinetic mechanisms for the decomposition of the two precursor compounds, and incorporated the homogeneous chemistry into the reaction and transport model of a rotating disk. The surface chemistry was modelled by using a large number of heterogeneous reactions, all leading, directly or indirectly, to growth of stoichiometric silicon carbide. Hence, the model cannot account for the deviations from stoichiometry that have been both theoretically predicted and experimentally observed [12,14].

A kinetic model for Si, C, and SiC deposition from MTS is proposed in this study. The mechanism includes all the chemical species found in significant quantities in experimental studies of silicon carbide deposition from MTS, and both the homogeneous and heterogeneous reactions that it encompasses are supposed to be reversible. The kinetic model is incorporated in a model for transport and reaction in a tubular hot-wall reactor, and the overall model is used to obtain some preliminary results on the spatial variation of the concentrations of the gaseous species and of the deposition rate and on the effects of reaction pressure on the stoichiometry of the deposit. More results, as well as a more detailed description of the mathematical model, will be given in future publications.

MECHANISM FOR THE DEPOSITION OF SiC FROM MTS
Gas Phase Chemistry

The compounds included in the kinetic mechanism are those found in appreciable quantities by Ivanova and Pletyushkin [13] in their MTS decomposition study. Their results are in relative agreement with thermodynamic calculations and the experimental findings of Besmann and Johnson [2]. The radical species included in the mechanism are either gas phase decomposition products, intermediates for the generation of stable compounds, or deposition precursors.

The reactions used to model the homogeneous chemistry of the system are given in Table 1. The forward rates were determined using the results of various gas phase kinetic studies, while the reverse rates were calculated from the equilibrium constants, which in turn were found using data from the JANAF Thermochemical Tables [26] and Barin [27]. The equilibrium constants of reactions R10 and R11 in the temperature range of our computations were determined by extrapolating thermodynamic data for C_2H_5 given in [11].

Table 1. Homogeneous Reactions

R1. $CH_3SiCl_3 \rightarrow SiCl_3 + CH_3$
R2. $SiCl_3 + H_2 \rightarrow SiHCl_3 + H$
R3. $SiHCl_3 \rightarrow SiCl_2 + HCl$
R4. $H_2 \rightarrow 2H$
R5. $CH_3 + H_2 \rightarrow CH_4 + H$
R6. $SiCl_3 + HCl \rightarrow SiCl_4 + H$
R7. $CH_3 + H \rightarrow CH_2 + H_2$
R8. $CH_2 + CH_3 \rightarrow C_2H_4 + H$
R9. $2CH_3 \rightarrow C_2H_6$
R10. $C_2H_6 + H \rightarrow C_2H_5 + H_2$
R11. $C_2H_5 + M \rightarrow C_2H_4 + H + M$
R12. $C_2H_4 \rightarrow C_2H_2 + H_2$

Table 2. Heterogeneous Reactions

SR1. $SiCl_2 + [S] \rightarrow [SiCl_2]_s$
SR2. $C_2H_2 + 2[S] \rightarrow 2[CH]_s$
SR3. $C_2H_4 + 2[S] \rightarrow 2[CH_2]_s$
SR4. $[CH]_s + H \rightarrow [CH_2]_s$
SR5. $CH_3 + [S] \rightarrow [CH_3]_s$
SR6. $[CH_3]_s + H \rightarrow [CH_2]_s + H_2$
SR7. $SiCl_3 + [S] \rightarrow [SiCl_3]_s$
SR8. $[SiCl_3]_s + H \rightarrow [SiCl_2]_s + HCl$
SR9. $[CH_2]_s \rightarrow C + H_2$
SR10. $[SiCl_2]_s + H_2 \rightarrow Si + 2HCl$
SR11. $[CH_2]_s + [SiCl_2]_s \rightarrow SiC + 2HCl$

The rate constant of the methyltrichlorosilane decomposition reaction was taken from Burgess and Lewis [7], and that of R2 was assumed to be of the same magnitude. The reaction sequence leading to tri- and tetrachlorosilane was proposed by Ashen et al. [8], and the $SiCl_2$ generating reaction by Sirtl et al. [24]. Dichlorosilane was not included in the mechanism because in studies for the deposition of silicon from chlorosilanes [24,25], its concentration was found to be one order of magnitude lower than those of the other chlorosilanes.

An approximate value for the rate constant of R3 was taken from the kinetic study of Clark and Tedder [9], who gave values for the rate constant of the trichloromethane decomposition reaction. The rate constant for R6 was obtained from the chlorosilane reduction study of Ashen et al. [8], and the constant used for the recombination of the hydrogen radicals was the one used in the methane decomposition study of Orechkina et al. [15].

The rate constants for the methane and acetylene generation reactions were obtained from the study of Allendorf and Kee [4]. Methane is the main carbon bearing species at atmospheric pressure, while the formation of acetylene is favored as the temperature increases and the pressure drops. Ethylene is an intermediate species for the generation of C_2H_2, and two alternative routes were considered for its production. The first is reported in [4] (attributed to Miller and Melius) and consists of reactions R7 and R8. The rate constants used for these two reactions are the same as in [4]. The second route consisting of R9, R10, and R11 was postulated by Cao and Back [11], who used the proposed mechanism to interpret both their experimental results and those of previous studies. Westbrook and Pitz [10] proposed the same sequence and the reaction rate constants were obtained form their work. Our simulation results showed that because of the very low concentration of the methylene radical in the gas phase, the first route behaves as a kinetic 'bottle-neck' within the range of conditions employed in our simulations. Actually, the rate of ethylene production through the first route is so low that the mathematical model with both routes employed gives results identical, for all practical purposes, to those obtained with the second route only.

Heterogeneous Reaction Chemistry

The heterogeneous reactions that were included in the mechanism are listed in Table 2. All the gas phase species, except MTS, CH_4, and $SiCl_4$, were assumed to participate in deposition leading reactions. Because of their molecular structure, methane and tetrachlorosilane have a very low probability of sticking on the surface. The sticking probability values obtained from the literature [16,17,20] for these two species are by more than two orders of magnitude lower than those of the other gases. Methyltrichlorosilane was excluded because its gas phase decomposition was almost instantaneous. Reactions of either molecular hydrogen or hydrogen radicals with the solid deposit were not included in the mechanism since hydrogen has been found to etch silicon carbide only at temperatures higher than 1700 K [18,19], and molecular hydrogen is not adsorbed on silicon surfaces [21].

Gupta et al. [20] found $SiCl_2$ to be the stable silicon-bearing surface species, and Langlais et al. [23] theoretically predicted that it should be the main species involved in silicon deposition. Hence, its reduction by hydrogen is treated as the silicon depositing reaction, and that of abstraction of hydrogen from adsorbed methylene is assumed to lead to carbon deposition. Equal amounts of deposited carbon and silicon are assumed to instantaneously react producing silicon carbide, the excess remaining as free carbon or silicon. Direct formation of silicon carbide occurs through a double desorption step (reaction SR11) involving adsorbed $SiCl_2$ and CH_2. Since reactions SR10 and SR11 involve formation of HCl, accumulation of HCl in the gas phase can suppress the deposition of Si and SiC if the two reactions are reversible.

In order to be able to use the kinetic model to describe the deposition process, we need rate constants for the forward and backward steps of each reaction. Values for the rate constants for the forward steps of the adsorption reactions (SR1, SR2, SR3, SR5, and SR7) and reactions involving a gas phase species and surface species (SR4, SR6, and SR8) can be obtained by assuming that the forward rate is equal to the collision frequency of the gaseous species with the surface multiplied by its sticking coefficient and the fractional surface coverages of the surface species involved in the reaction. For the backward rate constants of the above reactions and the rate constants of the solid deposition reactions (SR9, SR10, and SR11), the only available alternatives are to treat them as model parameters and determine their values by fitting the model predictions to the experimental data or try to estimate their values from first principles.

TRANSPORT AND REACTION MODEL

The kinetic mechanism was incorporated into a steady-state transport and reaction model for a plug flow reactor. The mass balance equations for the gaseous species, expressed in terms of partial pressures, p_i, are written:

$$\frac{\partial(up_i/RT)}{\partial z} = \sum_\rho \nu_{i\rho} R_{v\rho} \qquad (1)$$

u is the velocity of the gaseous mixture, R is the ideal gas constant, T is the temperature, and $\nu_{i\rho}$ is the stoichiometric coefficient of species i in reaction ρ.

For the surface reactions, the volumetric rate of reaction is given by $R_{v\rho} = SR_{s\rho}$, with $R_{s\rho}$ being the intrinsic rate of the heterogeneous reaction (per unit area) and S the reactor surface area per unit of reactor volume (=4 / reactor diameter). The mass balances for the surface species have the form:

$$\sum_\rho \nu_{i\rho} R_{s\rho} = 0 \qquad (2)$$

Heat effects associated with the occurrence of the reactions are neglected. Assuming that there is negligible pressure drop in the reactor, i.e., $\sum_i p_i = p \equiv$ constant, eq. (1) may be summed over i (i varying over the gas phase species) to obtain the equation giving the variation of the velocity of the gas mixture in the reactor:

$$p\frac{\partial(u/RT)}{\partial z} = \sum_i \sum_\rho \nu_{i\rho} R_{v\rho} \qquad (3)$$

Eqs. (1)-(3) are solved simultaneously along the length of the reactor to obtain the variation of the concentrations of gas phase and surface species in the reactors. Any solver designed for algebraic-differential equation systems suffices for this purpose.

DISCUSSION OF SOME RESULTS

All results presented and discussed in this section are for isothermal reaction conditions, operation at 1300 K, and with the feed containing H_2 and MTS only at a 10:1 ratio. We will first present and discuss results for the case in which only the gas phase reactions are allowed to take place. Such results help us understand the dynamics of the gas phase reactions independently of the reactions occurring on solid surfaces and thus obtain an order of magnitude estimate of the distance or equivalently space velocity or residence time needed to reach thermodynamic equilibrium in the gas phase.

Gas Phase Reactions Only

Figs. 1 and 2 present simulation results for the variation with the residence time in the reactor of the concentrations of the gas phase species for 1 atm total pressure and with only the homogeneous reactions taking place. The results for HCl are displayed in Fig. 1 along with those of the silicon species, and Fig. 2 shows the variation of the concentrations of the carbon species. The concentration vs. time curve for H_2 is shown in both figures for comparison.

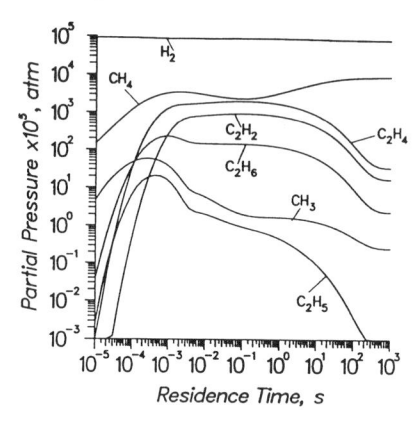

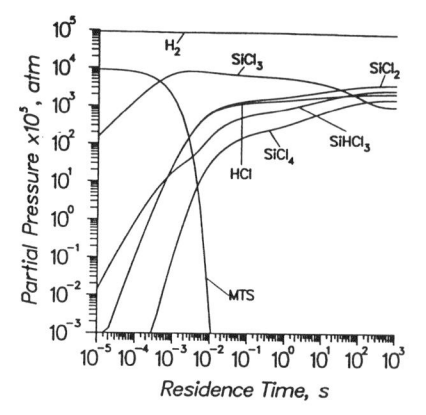

Figure 1. Concentration vs. residence time for the carbon species. $p = 1$ atm; $T = 1300$ K; $MTS:H_2$=1:10.

Figure 2. Concentration vs. residence time for the silicon and chlorine species. $p = 1$ atm; $T = 1300$ K; $MTS:H_2$=1:10.

It is seen from the results of Figs. 1 and 2 that while the decomposition of methyltrichlorosilane is almost instantaneous – its concentration drops by almost seven orders of magnitude in about 0.01 s – all the other species require residence times of more than 1000 s to reach their equilibrium concentrations. In view of the fact that much smaller residence times are encountered in typical CVD or CVI reactors, this result suggests that even in the absence of heterogeneous reactions, it is the kinetics of the problem that determine its behavior and not its thermodynamic equilibrium. The shapes of the various concentration vs. time curves indicate that the times constants encountered in the process cover a relatively broad range of values.

With Surface Reactions

Only a few preliminary results that underscore the capacity of the mathematical model to reproduce and explain most of the experimental observations made on the deposition of SiC from MTS are presented and discussed here. They are are for a cylindrical geometry reactor (1.5 cm in diameter) with the reactive mixture entering the reactor with

600 ml/min flow rate at standard conditions. Since there are no data available for sticking coefficients on SiC surfaces, we used data obtained from studies on silicon surfaces. The sticking coefficients of all radical species were assumed to be unity. The reactive sticking coefficient of acetylene (2×10^{-2}) was obtained from the work of Mogab and Leamy [16], and that ethylene (2×10^{-3}) from the study of Stinespring and Wormhoudt [17].

Computations were carried out for several sets of values for the unknown rate constants of the heterogeneous reactions. For the results discussed here, reactions SR4, SR6, SR8, SR10, and SR11 were treated as irreversible, and the rate of reaction SR9 was set equal to zero. The backward rate constants of SR1 and SR7 (expressed in kmol/m^2·s) were obtained by multiplying the forward rate constants (expressed in kmol/m^2·s·atm) by 10^{-2}. Similarly, the backward rate constants of reactions SR2, SR3, and SR5 were set equal to the forward constants times 10^{-3}. Estimates for the forward rates of the deposition reactions (SR10 and SR11) were obtained by requiring that the deposition rates predicted by the model be of the same order of magnitude as the experimental rates seen in our experiments [22] and in other studies of the literature. Thus, the forward rate constant of reaction SR10 (expressed in kmol/m^2·s·atm) was set equal to the constant obtained from the collision frequency of H_2 multiplied by 10^{-6}, while the rate constant of SR11 (expressed in kmol/m^2·s) was taken as an order of magnitude higher than that of SR10.

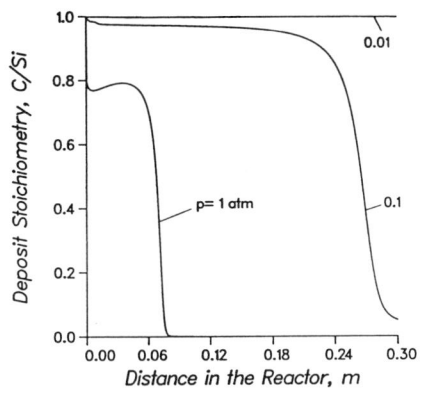

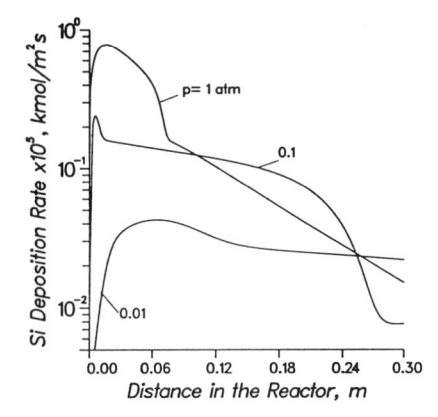

Figure 3. Deposit stoichiometry vs. distance. $T = 1300$ K; $MTS:H_2$=1:10.

Figure 4. Si deposition rate vs. distance. $T = 1300$ K; $MTS:H_2$=1:10.

Figs. 3 and 4 present the variation of the stoichiometry of the deposit and of the rate of Si (as SiC or Si) deposition, respectively, with the distance from the entrance of the reactor for three values of partial pressure (1, 0.1, and 0.01 atm). It is seen that the composition of the deposit depends strongly on the gas phase pressure in the reactor, a result seen in most experimental studies. The stoichiometry of the deposit improves with decreasing pressure because at high pressures the dominant stable carbon species is CH_4, while low H_2 partial pressures favor the formation of more reactive stable species (namely, C_2H_2 and C_2H_4). It should be noted that it is the partial pressures of the reactive species that influence the behavior of the system and not the total pressure itself. The total pressure in the reactor appears explicitly in the mathematical model only in the kinetics of reaction R12 (through the concentration of species M), and as a result, its effects on the behavior of the system are rather weak. The effects of total pressure on the deposit stoichiometry and the other variables of the process may be reproduced by the model by using an inert species to dilute the feed and thus lower the partial pressures in the gas phase.

The deposit stoichiometry remains almost constant over some distance from the entrance of the reactor, but then it goes through a relatively fast change, for 0.1 and 1 atm pressure, with silicon practically becoming the only deposition product. This also happens for 0.01 atm pressure but for much larger residence times, outside the observation window of Fig. 3. At the point of stoichiometry change, the deposition rate also undergoes a significant change (see Fig. 4). This phenomenon is not due to depletion of silicon and carbon since more than 50% of carbon and silicon contained in the feed is still present in the reactor at 1 atm when the carbon to silicon ratio starts to decrease, and more than 70% at 0.1 atm.

The results of Fig. 4 indicate that the dependence of the deposition rate on pressure and distance in the reactor is rather complex. Depending on where measurements are taken, the deposition rate for the case considered in Fig. 4 may vary monotonically with the reaction pressure (increase or decrease) or present an extremum (maximum or minimum) between 0.01 and 1 atm. It is not surprising, therefore, that there is considerable disagreement in the literature on the effects of the various operational parameters on the deposition rate from MTS. (Notice that the effects of pressure on deposit stoichiometry and deposition rate may differ quantitatively from those of Figs. 3 and 4 if one allows for third body collisions in the unimolecular steps of the reactions of Table 1, the decomposition of MTS, for instance.) The results of Fig. 4 indicate that in order to make experimental data for the MTS-SiC system useful to other workers in the area, the person reporting the data should make available not only the temperature, pressure, and composition of operation, but the reactor configuration, substrate dimensions, position of measurement, and temperature field in the reactor as well.

REFERENCES

1. J. Schlichting, Powder Metall. Int. 12 (3,4), 141 and 196 (1980).
2. T.M. Besmann and M.L. Johnson, in Proc. 3rd Int. Symp. on Ceramic Materials and Components for Engines, (Las Vegas, NE, 1988), p. 443.
3. T.M. Besmann, B.W. Sheldon, and M.D. Kaster, to appear in J. Am. Ceram. Soc. (1991).
4. M.D. Allendorf and R.J. Kee, J. Electrochem. Soc. 138, 841 (1991).
5. K.D. Annen, C.D. Stinespring, M.A. Kuczmarski, and J.A. Powell, J. Vac. Sci. and Technol. A 8, 2970 (1990).
6. J.H. Koh and S.I. Woo, J. Electrochem. Soc. 137, 2215 (1990).
7. J.N. Burgess and T.J. Lewis, Chemistry and Industry, January 1974, p. 76.
8. D.J. Ashen, G.C. Bromberger, and T.J. Lewis, J. Appl. Chem. 18, 348 (1968).
9. D.T. Clark and J.M. Tedder, Trans. Faraday Soc. 62, 393 (1966).
10. C.K. Westbrook and W.J. Pitz, Comb. Sci. Tech. 37, 117 (1984).
11. J.R. Cao and M.H. Back, Can. J. Chem. 60, 3039 (1982).
12. A.I. Kingon, L.J. Lutz, P. Liaw, and R.F. Davis, J. Am. Ceram. Soc. 66, 558 (1983).
13. M.L. Ivanova and A.A. Pletyushkin, Inorganic Materials 4, 957 (1968).
14. G.L. Harris et al., Mater. Letters 4, 77 (1986).
15. N.B. Orechkina, G.M. Tomashevskaya, and I.V. Friedman, J. Appl. Chem. USSR 60, 1447 (1987).
16. C.J. Mogab and H.J Leamy, J. Appl. Phys. 45, 1075 (1974).
17. C.D. Stinespring and J.C. Wormhoudt, J. Appl. Phys. 65, 1733 (1989).
18. T.L. Chu and R.B. Campbell, J. Electrochem. Soc. 112, 955 (1965).
19. J.M. Harris, H.C. Gatos, and A.F. Witt, J. Electrochem. Soc. 116, 380 (1969).
20. P. Gupta, P.A. Coon, B.G. Koehler, and S.M. George, in Proc. MRS Symp., edited by M.E. Gross et al., (Boston, MA, 1989), 131, p. 197.
21. T. Sakurai et al., J. Vac. Sci. and Technol. A 8, 259 (1990).
22. G. Papasouliotis and S.V. Sotirchos, ACerS Meeting, Cincinnati (1991).
23. F. Langlais, C. Prebende, B. Tarride, and R. Naslain, J. de Physique 50, C5-93 (1989).
24. E. Sirtl, L.P. Hunt, and D.H. Sawyer, J. Electrochem. Soc. 121, 919 (1974).
25. T. Aoyama, Y. Inoue, and T. Suzuki, J. Electrochem. Soc. 130, 203 (1983).
26. JANAF Thermochemical Tables, edited by M.W. Chase, Jr., et al., Supplement to the J. of Physical and Chemical Ref. Data, 3rd ed. (1985).
27. Thermochemical Data for Pure Substances, edited by I. Barin, VCH Publishers (1989).

KINETICS OF CHEMICAL VAPOR DEPOSITION OF SIC BETWEEN 750 AND 850°C AT 1 BAR TOTAL PRESSURE

DIETER NEUSCHÜTZ AND FARZIN SALEHOMOUM
Prof. Dr.-Ing. D. Neuschütz and Dipl.-Ing. F. Salehomoum, Lehrstuhl für Theoretische Hüttenkunde, Rheinisch-Westfälische Technische Hochschule (RWTH) Aachen, D-5100 Aachen, Germany

ABSTRACT

The deposition rate from mixtures of methyltrichlorosilane (MTS), hydrogen and methane was measured thermogravimetrically using a hot wall vertical reactor and planar SiC substrates. Below 850°C and at sufficiently high gas velocities, the rate of the phase boundary reaction could be determined. In the absence of CH_4 and at H_2:MTS=55, Si was deposited together with SiC. Addition of CH_4 lowered the Si content, pure SiC being deposited at CH_4:MTS above 10. The deposition rate j in the range 750 to 850°C follows the equation

$$j = j(Si) + j(SiC) = k(Si) \cdot \exp(-E(Si)/RT) \cdot [MTS]^1 + k(SiC) \cdot \exp(-E(SiC)/RT) \cdot [MTS]^{0.5}$$

with $E(Si) = 160$ and $E(SiC) = 300$ kJ/mol. Reaction mechanisms are presented to account for the observed reaction orders with respect to MTS. Between 900 and 970°C, the reaction rate decreased with temperature indicating a change in the deposition mechanism.

INTRODUCTION

Chemical vapor deposition of SiC was first investigated in the 1960s to prepare diffusion barriers on nuclear fuel particles [1]. Later, technical interest in SiC deposition focussed on fiber-reinforced ceramic materials for high-temperature applications [2,3]. Improved oxidation resistance of carbon fibers coated with SiC and improved mechanical strength of porous ceramic-matrix composites vapor infiltrated with SiC are potential benefits obtained from vapor deposited SiC.

In CVD, homogeneous gas reactions, transport processes in the gas phase, and heterogeneous reactions on the substrate surface have an influence on the deposition rate and on the composition of the deposit. For SiC formation, the starting gas mixture is typically methyltrichlorosilane (MTS) plus hydrogen. It is known that MTS decomposes readily before reaching the substrate [8], and that the Si and C activities in the gas phase may be shifted resulting in a deposit of Si or C together with SiC. At temperatures below 1100°C, MTS-H_2 mixtures are reported to lead to codeposition of Si+SiC, with a rising Si/C ratio at lower temperatures [4-7]. For a uniform deposition on a given substrate and for good penetration during chemical vapor infiltration, the gas diffusion rate should be higher than the rate of the deposition reaction.

It was the aim of the present investigation to determine, at 1 bar total pressure, the corresponding temperature and partial pressure ranges to deposit pure SiC. Previous work [1,4,8,9] was taken into account.

EXPERIMENTAL

The deposition kinetics were measured thermogravimetrically using a vertical, externally heated Al$_2$O$_3$ tube as reaction chamber with 27 mm inner tube diameter and 100 mm vertical length of the isothermal zone. The planar substrate, 10x50x1.6 mm^3, was connected to the thermoscale (Sartorius) via a magnetic coupling device, which kept the corrosive gases away from the scale. As a substrate, α-SiC (ESK Kempten) was used. Before starting the measurements, a uniform layer of 50μm of β-SiC was deposited at 1100°C. Each substrate was used for a number of measurements, until the mass increase was 3.5 g and the layer reached 1.2 mm thickness. Methyltrichlorosilane (CH$_3$SiCl$_3$) from Wacker Chemie was evaporated in a thermostat into a stream of hydrogen gas. The mixture was then cooled to a slightly lower temperature to recondensate some MTS and establish MTS saturation in the gas stream. Hydrogen and methane purities were 5.0 and 2.5, respectively. All experiments were carried out at 1 bar total pressure with partial pressures ranging from 0.01 to 0.045 bar for MTS and from 0 to 0.5 bar for CH$_4$. The Si/C ratio in the deposit was determined by electron beam microprobe analysis.

RESULTS FOR TEMPERATURES BETWEEN 750 AND 850°C

In order to measure reaction-controlled deposition rates, the influence of the total gas flow rate on the reaction rate was determined, <u>Figure 1</u>. Below 850°C, the rate became independent of the gas velocity at $\dot{V}_{tot} \geq 120$ l/h (standard conditions). It was in this range of temperatures and gas flow rates that the reaction rates were determined versus partial pressure and temperature. The deposition resulted in smooth, uniform layers over the entire substrate. The composition of the layer as determined by EBMA was 75 at% Si and 25% C. X-ray analysis showed poorly crystallized Si and β-SiC. Above 850°C, gas diffusion had a major influence on the overall deposition rate.

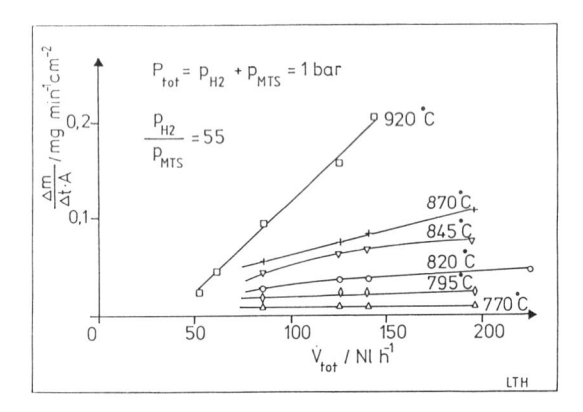

<u>Fig. 1:</u> Deposition rate from MTS+H$_2$ as a function of temperature and total gas flow rate

The addition of methane to the MTS-H$_2$ mixture lowered the Si/C ratio of the product until pure SiC was deposited. <u>Figure 2</u> shows the silicon and carbon contents of deposits formed at 870°C and increasing CH$_4$ contents. A ratio of CH$_4$/MTS ≥ 15 was required at 870°C to completely suppress the formation of elemental silicon. At lower temperature, this ratio is even higher. Tests to measure the deposition kinetics of pure SiC were run at 15 < CH$_4$/MTS < 30.

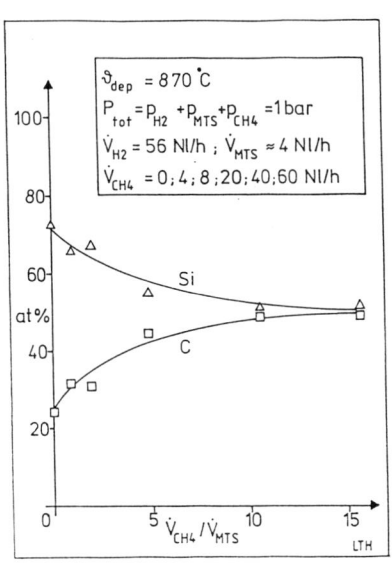

<u>Fig. 2</u>: Effect of the CH$_4$/MTS ratio on the composition of the deposit as determined by microprobe analysis

With MTS-hydrogen mixtures, the kinetics were measured at total flow rates of 130 l/h and MTS flow rates between 1 and 5 l/h. The experimental results are summarized in <u>Figure 3</u>. The different symbols represent measured values at four temperatures between 770 and 845°C. From the slope of the straight lines the reaction order with respect to MTS is seen to increase from 0.60 at 845°C to 0.74 at 770°C. At the two highest temperatures, the measured values fall below the straight lines at the highest MTS partial pressure, because gas diffusion starts influencing the overall rate. The deviation at 795°C may be caused by a slight misplacement of the thermocouple. A mixture of Si+SiC was deposited in this series.

The results obtained with MTS-hydrogen-methane mixtures are presented in <u>Figure 4</u>. The flow rates were 56 l/h H$_2$, 50 l/h CH$_4$ and 1.4 to 4.5 l/h MTS. The reaction order for MTS was 0.5 at all temperatures. Pure SiC was deposited in this series of tests.

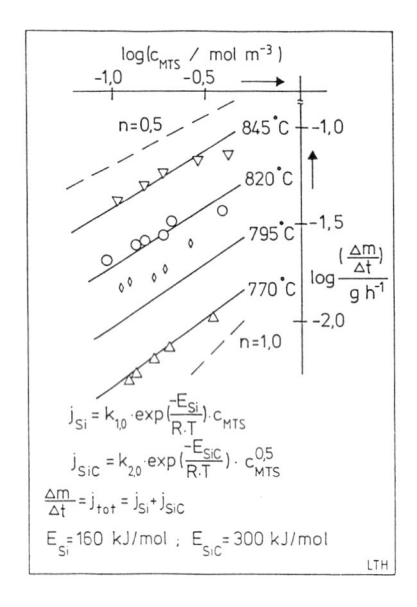

Fig. 3: Deposition rates of Si+SiC from MTS+H$_2$. Flow rates were 126 l/h for H$_2$, 1 to 5 for MTS. Symbols represent experimental values, straight lines were calculated from eq.(1).

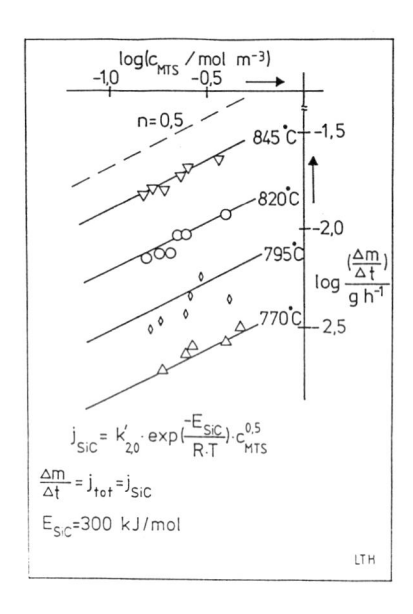

Fig. 4: Deposition rates of SiC from MTS+H$_2$+CH$_4$. Flow rates were 56 l/h for H$_2$, 50 for MTS, the ratio CH$_4$/MTS was 11 to 35. Symbols represent experimental values, straight lines were calculated from eq.(1).

RATE EQUATION AND REACTION MECHANISM

The experimental results indicate that the deposition rate of pure SiC is proportional to [MTS]$^{0.5}$, while for the mixed deposition of Si+SiC, the reaction order with respect to MTS is between 0.5 and 1, increasing with falling temperature. Since the Si/C ratio in the layer deposited from MTS-hydrogen mixtures increases with lower temperature [4-7], it may be concluded that the Si deposition is a first order reaction and that the measured rate for Si-SiC codeposition is the sum of the two individual deposition rates for Si and for SiC assumed to be independent of each other,

$$j = j(Si) + j(SiC) = k(Si) \cdot \exp(-E(Si)/RT) \cdot [MTS]^1 + k(SiC) \cdot \exp(-E(SiC)/RT) \cdot [MTS]^{0.5}. \quad (1)$$

j, k, and E are rate, rate constant, and activation energy, respectively, and the brackets denote concentrations.

Eq.(1) was used to evaluate the experimental data for Si/SiC codeposition, Figure 3. With E(Si) = 160 and E(SiC) = 300 kJ/mol and adapted values for k(Si) and k(SiC), the straight lines in Figure 3 were calculated from eq.(1), showing good agreement between experiments and proposed rate equation.

The second term j(SiC) of eq.(1) was applied to the results from MTS-H_2-CH_4 mixtures. The absolute measured rates were lower by a factor of 2.5 than the values from eq.(1) at all temperatures. The straight lines in Figure 4 were calculated from eq.(1) with this correction, then showing excellent agreement.

A tentative mechanism was developed to explain the different reaction orders for Si and SiC deposition. The reaction sequence leading to Si formation is suggested to be

$$CH_3SiCl_3 = C^* + Si^*, \quad \text{(slow)} \tag{2}$$

$$Si^* + nH_2 = Si + 2n\ H^*. \quad \text{(fast)} \tag{3}$$

The species C^* and Si^* are intermediate decomposition products of MTS, containing C and Si, respectively, with no further information on their analyses and properties.

Hence, the rate of Si deposition is

$$j(Si) \sim [Si^*] \sim [CH_3SiCl_3]^1. \tag{4}$$

The corresponding reaction sequence for SiC formation may be

$$CH_3SiCl_3 = C^* + Si^*, \quad \text{(fast)} \tag{5}$$

$$C^* + Si^* = SiC + \ldots \quad \text{(slow)} \tag{6}$$

Since (5) is assumed to be fast, this reaction will reach equilibrium (K = equilibrium constant),

$$K = [C^*][Si^*]/[CH_3SiCl_3]. \tag{7}$$

If the species C^* and Si^* are only formed by MTS decomposition, their concentrations must be equal, thus

$$K = [C^*]^2/[CH_3SiCl_3]. \tag{8}$$

According to eq.(6) the rate of SiC deposition is

$$j(SiC) \sim [C^*][Si^*]. \tag{9}$$

Since silicon species are reported to be more strongly adsorbed than carbon species [11,12], surface saturation in Si^* may be assumed. Then,

$$j(SiC) \sim [C^*] \sim [CH_3SiCl_3]^{0.5}. \tag{10}$$

RESULTS FOR TEMPERATURES ABOVE 850°C

A few tests carried out at higher temperatures revealed an anomalous behavior of the deposition rate as a function of temperature and gas flow rate: At relatively low total flow rate (70 l/h), the rate decreased with temperature above 870°C. Rising flow rates accelerated the deposition rate strongly and did not, as below 870°C, lead to a saturating value. Possibly, a change in the reaction mechanism takes place around 900°C. Assuming that one of the intermediate species essential for the deposition reaction is eventually consumed at the higher temperatures by a homogeneous gas reaction, higher gas velocity will shorten the time available for this side reaction and therefore increase the deposition rate, see Figure 1, 920°C. Methane seems to be effective in retarding this side reaction, while at lower temperatures, it solely raises the carbon activity in the system thus suppressing the deposition of free Si.

The rate anomaly of SiC deposition above 900°C has been reported previously [4,10]. More informations are required to quantitatively describe the process in this range.

ACKNOWLEDGEMENT

Financial support granted by Dornier GmbH, Friedrichshafen, and by the German Federal Ministry for Research and Technology, Bonn, is gratefully acknowledged.

REFERENCES

1. L. Aggor, W. Fritz: Chem.-Ing.-Tech. 43,472(1971).
2. E. Fitzer et al: Proc. 3rd Int. Conf. SiC, Miami Beach 1973; High Tem.-High Press. 8,187(1976); Z. Werkstofftechn. 11,330(1980); Rev. Int. Hautes Temp. Refr. 17,23(1980).
3. R. Naslain, in: Avanced Structural and Functional Materials, edited by W.G.J. Bunk, (Springer-Verlag Berlin, 1991), pp. 51-90.
4. F. Langlais, C. Prébende, B. Tarride, R. Naslain: J. de Physique, Coll. C5, Suppl. au no. 5, tome 50,93(1989).
5. S. Motojima, M. Hasegawa: J. Vac. Sci. Technol. a8(5), 3763(1990).
6. J. Chin, P.K. Gantzel, R.G. Hudson: Thin Solid Films 40, 57(1977).
7. W. Schintlmeister, W. Wallgram, K. Gigl: High Temp.-High Press. 18,211(1986).
8. J.N. Burgess, T.J. Lewis: Chem. and Ind. 19,76(1974).
9. T.M. Besmann, B.W. Sheldon, M.D. Kaster: Surf. and Coat. Technol. 1990.
10. C. Prébende: Dissertation, L'Université de Bordeaux, 1989.
11. G.S. Fischman, W.T. Petuskey: J. Amer. Ceram. Soc. 68, 185(1985).
12. M.D. Allendorf, R.J. Kee: J. Electrochem. Soc. 138,841 (1991).

KINETIC CHARACTERISTICS OF Si$_3$N$_4$ CVD

W. Y. LEE, J. R. STRIFE, and R. D. VELTRI
United Technologies Research Center, East Hartford, CT 06108.

ABSTRACT

The CVD of Si$_3$N$_4$ from SiF$_4$ and NH$_3$ gaseous precursors was studied using a hot-wall reactor in the temperature range of 1340 to 1490°C. The effects of temperature, time, flow rate, and SiF$_4$/NH$_3$ molar ratio on deposition rate and axial and radial deposition profiles were identified. The decomposition characteristics of pure NH$_3$ and SiF$_4$ were studied utilizing mass spectroscopy and compared to thermodynamic predictions.

INTRODUCTION

Si$_3$N$_4$ coatings have been successfully fabricated by CVD (chemical vapor deposition) for use in various microelectronic and structural applications. Si$_3$N$_4$ can be deposited from a wide variety of silicon containing precursors such as SiCl$_4$, SiF$_4$, SiH$_2$Cl$_2$, SiH$_4$, etc. NH$_3$ is the most widely used nitrogen-containing precursor for the Si$_3$N$_4$ deposition although N$_2$H$_6$ and N$_2$ have been occasionally used. In general, the CVD of crystalline Si$_3$N$_4$ has been achieved at temperatures above ~1300°C using SiCl$_4$-NH$_3$ and SiF$_4$-NH$_3$ precursor mixtures. Kingon et al. [1] studied the thermodynamics of the SiH$_4$-NH$_3$, SiCl$_4$-NH$_3$, and SiF$_4$-NH$_3$ precursor systems in the temperature range of 500 to 1700°C. Interestingly, the thermodynamic analysis predicted that Si$_3$N$_4$ would not be deposited above ~1300°C from the SiF$_4$-NH$_3$ system. Galasso and his co-workers [2-5] were able to experimentally deposit Si$_3$N$_4$ at these high temperatures indicating that deposition was controlled by kinetics and/or transport phenomena. This investigation was initiated to identify critical kinetic issues concerning the SiF$_4$-NH$_3$ CVD system.

EXPERIMENTAL

SiF$_4$ (99.99%) and NH$_3$ (99.9%) were separately introduced into a cylindrical graphite retort using a water-cooled, co-axial dual-path injector constructed of stainless steel. The graphite retort was placed vertically, and the gas flow direction was upward. The retort was heated by graphite heating elements which enclosed the retort. The inside diameter and length of the graphite retort were 13.3 cm and 56.9 cm, respectively. A quadrupole mass spectrometer located in the downstream of the reactor was used to monitor effluent gas composition. An orifice probe technique was used to sample the exhaust gas by differentially pumping the mass spectrometer chamber. Graphite rods of 0.63 cm diameter and 30 cm length were fixed vertically to probe deposition rate as a function of axial location in the retort. Following deposition of Si$_3$N$_4$, the graphite rods were sectioned into 1 cm length cylinders at specific axial locations and the graphite was removed by oxidation at ~650°C. The remaining coating shells were then analyzed. The utilization of a water cooled injector at the reactor inlet caused significant temperature gradients in the axial direction. The temperature profiles shown in Figure 1 were measured along the reactor centerline using a platinum-rhodium thermocouple. Peak temperatures were measured at approximately 35 cm from the reactor inlet. The total temperature gradient from the reactor inlet to the peak temperature location was about 70°C. The "nominal" temperatures listed in Figure 1 were temperatures measured using an optical pyrometer focused on the outer wall of the graphite retort at ~28 cm from the retort inlet.

RESULTS

Changes in deposition rate were measured as a function of radial position by placing graphite rods at various radial locations. As shown in Figure 2, the deposition rates

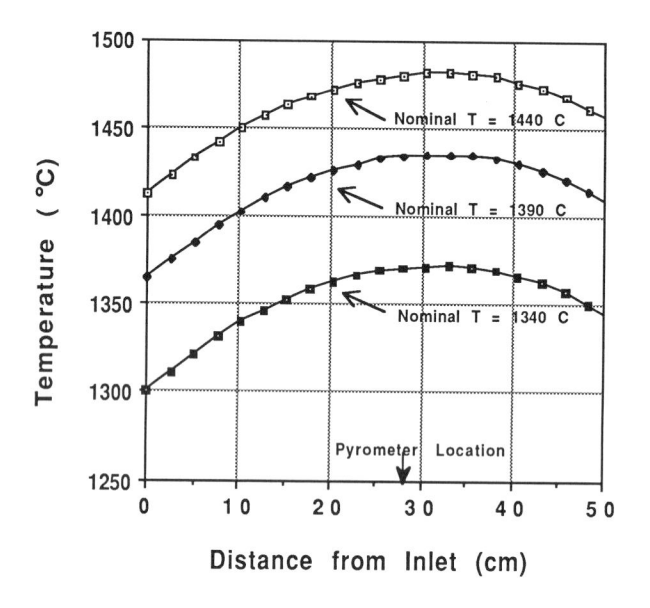

Figure 1. Axial temperature profiles measured by a thermocouple along the reactor center line while flowing 688 cm^3/min of Ar at 1.8 torr. The nominal temperature was measured using an optical pyrometer focused on the outer wall of the deposition chamber ~28 cm from the reactor inlet.

decreased in the outward radial direction at the 2.5 cm and 7.6 cm axial locations whereas the radial deposition profile became relatively uniform at more downstream locations. The significant radial divergence near the reactor entrance region was attributed to inadequate gas mixing. Similar deposition rates were observed for the rods placed radially on the opposite side of each other at 1.9 cm from the reactor center suggesting that the flow characteristics in the reactor were basically axisymmetric. The rest of data reported in this paper was obtained by placing one rod at the 1.9 cm location.

Deposition temperature was found to be the most dominant and important variable for the Si$_3$N$_4$ process. Both deposition rate and axial deposition profiles were strongly influenced by the deposition temperature. As shown in Figure 3, the deposition rate measured at 2.5 cm from the reactor inlet increased exponentially with temperature as the total flow rate was held constant at 688 cm^3/min at STP. An apparent activation energy value of 62 kcal/mole was calculated at the 2.5 cm location. Some experiments were repeated to assess the data reproducibility. Although the observed trends were clearly reproducible, measured values varied by as much as 30% in some cases. This uncertainty level could be attributed to temperature control. Temperature fluctuations of ±10°C were routinely monitored by the optical pyrometer during deposition. In comparison to the strong temperature effects shown in Figure 3, the ±10°C fluctuations were sufficient to cause the observed variation in the reproducibility data.

The curvature of the deposition rate profiles became much steeper as the nominal temperature increased. This behavior was most likely caused by faster depletion of the reagents with increased temperature. At downstream locations, the Arrhenius relationship between the deposition rate and temperature was not clearly observed because of the depletion phenomenon. As the temperature increased, "etch lines" were observed closer to the injector. The term "etch line" was used to describe the axial location which divided the reaction chamber into two zones; the deposition zone and the etching zone where Si$_3$N$_4$ deposition ceased.

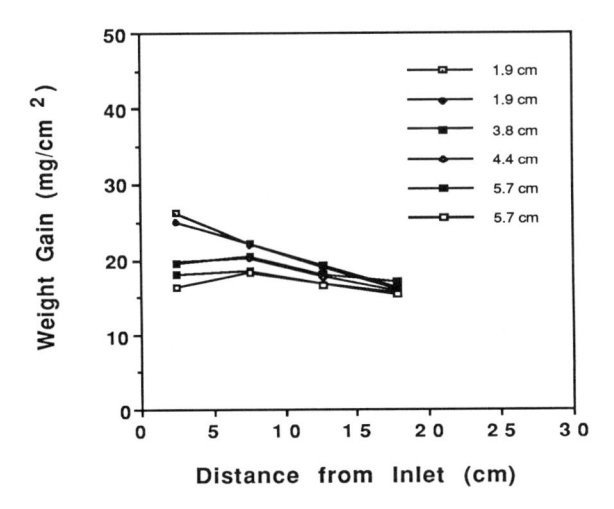

Figure 2. Axial deposition profiles obtained as a function of radial location; nominal temperature = 1440°C, pressure = 1.8 torr, time = 4 h. The radial distance was measured from the reactor center.

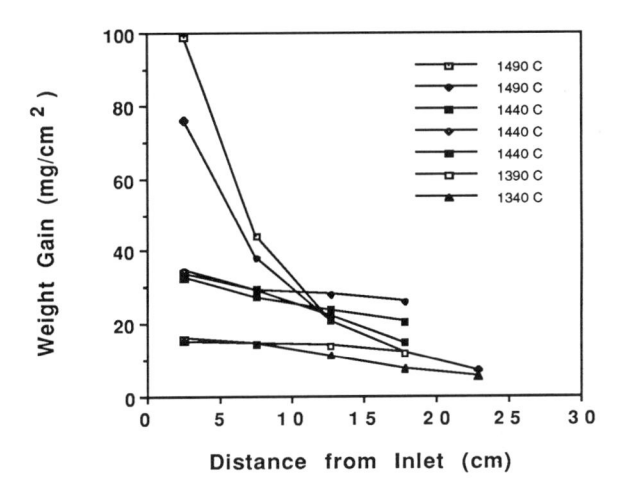

Figure 3. Effect of nominal temperature on axial deposition profile: pressure = 1.8 torr, NH_3/SiF_4 ratio = 6, total flow rate = 688 cm^3/min, and time = 4 h.

Decreasing the total flow rate from 688 to 413 cm^3/min resulted in a slight decrease in deposition rate. In addition, the exponential increase in the deposition rate with temperature was no longer evident at the 2.5 cm rod location. These results indicated that the apparent activation energy values were dependent on transport factors. For comparison, the apparent activation energy value of ~47 kcal/mole was calculated using the data reported by Galasso et al. [4] who used a cold-wall reactor to deposit Si$_3$N$_4$ from the SiF$_4$-NH$_3$ reagent system. Niihara and Hirai [6] observed that the apparent activation energy value for the amorphous growth from the SiCl$_4$-NH$_3$ reagent system was ~30 kcal/mole while that of the crystalline growth was ~53 kcal/mole.

In general, decreasing the NH$_3$/SiF$_4$ ratio from 6 to 3 increased deposition rate. Also, coating thickness increased linearly with increased deposition time indicating that the role of the initial graphite substrate surface was probably insignificant. It should be mentioned that the retort wall provided most surface area for the Si$_3$N$_4$ deposition, and the retort wall was fully coated with Si$_3$N$_4$ before these experiments were performed.

The decomposition characteristics of pure NH$_3$ and SiF$_4$ gases were studied using mass spectroscopy. As shown in Figures 4a and 4b, similar mass spectra were obtained while flowing 116 cm^3/min of pure SiF$_4$ at room temperature and 1450°C, respectively, indicating that SiF$_4$ did not thermally decompose at the high temperature. A standard SiF$_4$ mass spectrum from Ref. 7 was used to assign the SiF$_4$ peaks shown in Figures 4a and 4b. In contrast, NH$_3$ decomposed extensively into N$_2$ and H$_2$ at 1450°C as shown by comparing Figures 4c and 4d, the spectra obtained at room temperature and 1450°C. When the nominal temperature was increased to 1450°C, the intensity of the NH$_3$ peaks around 17 amu decreased about two orders of magnitude while the N$_2$ and H$_2$ signals increased by more than one order of magnitude. The N$_2$ and O$_2$ peaks in Figure 4c were probably due to an air leak. The presence of Ar peaks at 40 and 20 amu in these figures was attributed to the purge gas flowing in the cavity between the graphite reactor retort and outer stainless steel jacket. These experimental observations were consistent with SOLGASMIX-PV [8] calculations performed using thermochemical data from the JANAF Tables [9]. As listed in Table I, the calculations predicted that pure SiF$_4$ would not decompose appreciably at 1450°C while NH$_3$ was unstable at this temperature.

Table I. Stable gas species expected from the SiF$_4$+NH$_3$, pure NH$_3$, and pure SiF$_4$ systems at 1450°C and 240 Pa (1.8 torr). Equilibrium partial pressures were calculated using the SOLGASMIX-PV code [8].

Reagents	Equilibrium Partial Pressures (Pa)	
3 moles SiF$_4$ +	H$_2$	109.0
4 moles NH$_3$	N$_2$	41.1
	SiF$_4$	38.0
	HF	27.7
	SiF$_3$	19.4
	SiF$_2$	4.0
	H	0.5
	SiHF$_3$	0.3
pure NH$_3$	H$_2$	179.4
	N$_2$	59.9
	H	0.7
pure SiF$_4$	SiF$_4$	239.9
	SiF$_3$	0.05
	F	0.05

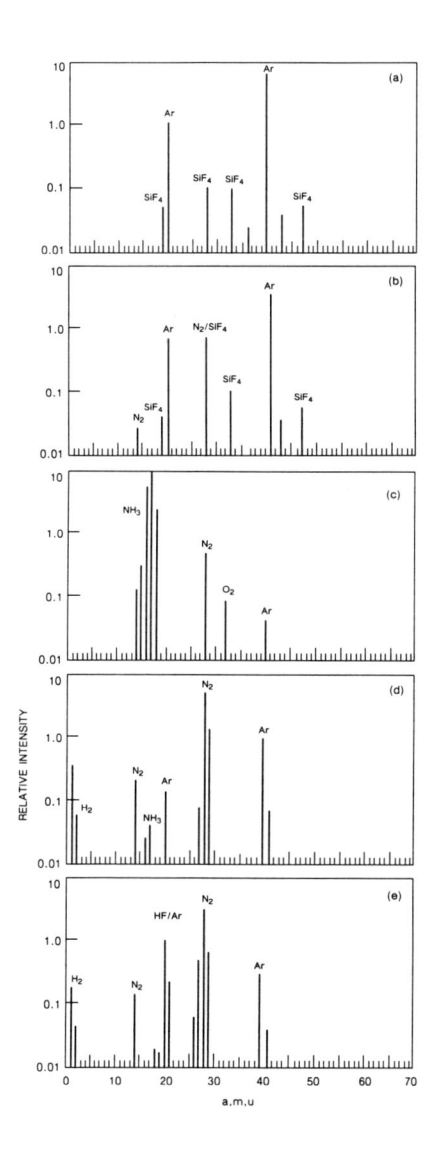

Figure 4. Mass spectra obtained using the following conditions: (a) 120 cm^3/min of SiF$_4$ at room temperature, (b) 120 cm^3/min of SiF$_4$ at 1450°C, (c) 710 cm^3/min of NH$_3$ at room temperature, (d) 710 cm^3/min of NH$_3$ at 1450°C, and (e) 120 cm^3/min of SiF$_4$ and 710 cm^3/min of NH$_3$ at 1450°C. A reactor pressure of 1.8 torr was used for these experiments.

When NH_3 and SiF_4 were simultaneously introduced at 1450°C, Figure 4e, the presence of HF in the exit stream along with N_2 and H_2 was clearly noticeable as the intensity of the 20 amu peak became stronger than that of the peak at 40 amu. The mass spectrometer probe was located downstream of a condenser which was used to trap particulates formed by low temperature reactions between unreacted SiF_4 and NH_3. Because of the condenser location, mass peak patterns corresponding to the unreacted SiF_4 and NH_3 were not visible in Figure 4e. Based on weight gain measurements on the condenser, it was estimated that approximately 94% NH_3 and 82% SiF_4 were decomposed and/or reacted before they reached the condenser stage. Thus, the mass spectroscopy and material balance data were in qualitative agreement and indicated that NH_3 was significantly depleted along the reactor length by thermal decomposition as well as deposition reaction(s) with SiF_4.

As tabulated in Table I, the thermodynamic analysis of the SiF_4-NH_3 reagent system suggested that: (1) most NH_3 would decompose and the rest would react with fluorine containing species to form HF and (2) SiF_4 was relatively stable and some would become sub-fluoride species to provide fluorine atoms necessary to form HF. Also, it was previously determined by Kingon *et al* that Si_3N_4 was not thermodynamically stable enough to be deposited as a solid phase at these conditions. In a mathematical sense, the SOLGASMIX-PV calculations would not "see" the presence of NH_3 as a starting reagent because of its thermochemical instability in reference to N_2 and H_2. Therefore, it should be interpreted that: the SOLGASMIX-PV calculations predicted that Si_3N_4 would not be deposited from the N_2-H_2-SiF_4 reagent system. Also, this implied that the Si_3N_4 deposition would not take place once NH_3 was fully decomposed.

SUMMARY

The key process parameters which governed deposition kinetics were identified. Deposition temperature was found to be the most important process parameter in affecting deposition rate and axial coating uniformity. The thermodynamic and mass spectroscopy studies indicated that the Si_3N_4 process was kinetically controlled. The depletion of NH_3 by both decomposition and deposition reactions was rapid in the reactor and was important to understanding the Si_3N_4 deposition process.

ACKNOWLEDGEMENTS

This research was supported by DARPA-DSO through WRDC Contract F33615-89-C-5628. Maj. Joseph Hager was the WRDC monitor, and Mr. William Barker was the DARPA sponsor.

REFERENCES

1. A.I. Kingon, L.J. Lutz, and R.F. Davis, J. Am. Ceram. Soc., 66, 551 (1983).
2. F.S. Galasso, U. Kuntz, and W.J. Croft, J. Am. Ceram. Soc., 55, 431 (1972).
3. F.S. Galasso, R.D. Veltri, and W.J. Croft, Ceram. Bull., 57, 453 (1978).
4. F.S. Galasso, Powder Met. Int., 11, 7 (1979).
5. R.D. Veltri and F.S. Galasso, J. Am. Ceram. Soc., 73, 2137 (1990).
6. K. Niihara and T. Hirai, J. Mater. Sci., 12, 1233 (1977).
7. F.W. McLafferty and D.B. Stauffer, The Wiley/NBS Registry of Mass Spectral Data: Volume I, (John Wiley and Sons, New York, 1989), p. 1.
8. T.M. Besmann, ORNL/TM-5775, Oak Ridge National Laboratory, Oak Ridge, TN, April 1977.
9. JANAF Thermochemical Tables, 3rd ed., Vols I and II, ed. D.R. Lide, Jr., (American Chemical Society and American Institute of Physics for Nat. Bur. Stand., U.S., 1985).

ONSET CONDITIONS FOR GAS PHASE REACTION AND NUCLEATION IN THE CVD OF TRANSITION METAL OXIDES

J. Collins[a], D.E. Rosner[a] and J. Castillo[b]
[a]Yale Univ,. Chem. Engrg. Dept., HTCRE Lab., New Haven CT 06520-2159, USA
[b]U.N.E.D., Dept. Fisica Fundamental, Apdo 60141, Madrid 28080, Spain

ABSTRACT

A combined experimental/theoretical study is presented of the onset conditions for gas phase reaction and particle nucleation in hot substrate/cold gas CVD of transition metal oxides. Homogeneous reaction onset conditions are predicted using a simple high activation energy reacting gas film theory. Experimental tests of the basic theory are underway using an axisymmetric impinging jet CVD reactor. No "vapor phase ignition" has yet been observed in the $TiCl_4/O_2$ system under accessible operating conditions (below substrate temperature $T_w=1700$ K) and further experiments are planned using more reactive feed materials. The goal of this research is to provide CVD reactor design and operation guidelines for achieving acceptable deposit microstructures at the maximum deposition rate while simultaneously avoiding homogeneous reaction/nucleation and diffusional limitations.

INTRODUCTION

The onset of gas phase reaction and particle nucleation is a common problem in the CVD of transition metal oxides, often resulting in decreased deposition rates and reduced film quality [1-3]. This is particularly true in cold gas/hot substrate CVD systems in which homogeneous reactions in the thermal boundary layer adjacent to the hot substrate produce particles which are thermophoretically repelled from the surface and, for the most part, do not deposit [4]. We call the sudden onset of significant reagent consumption by homogeneous reactions which result in non-depositing products (*e.g.* TiO_2 particles) "vapor phase ignition" (VPI). Since the onset of homogeneous reactions can effectively starve the growing surface of reagent, it is often possible to detect vapor phase ignition by a sharp drop in deposition rates with increasing surface temperature. On the other hand, in hot gas/cool substrate systems the onset of homogeneous particle nucleation can lead to increased film growth rates due to interface roughening associated with thermophoretically driven particle/vapor co-deposition [5,6]. It must also be mentioned that high temperature deposition rate decreases may be due to other causes, such as reaction product thermodynamic instability [7]. Dramatic rate decreases attributed to VPI were first reported by Ghoshtagore [4] for titania deposition from $TiCl_4$ in O_2, and his early experimental studies may still be the most thorough. Unfortunately, it difficult to use his data to predict conditions under which vapor phase ignition will occur in other reactors (let alone other chemical systems) because the transport conditions in his experiments were not well characterized.

We developed and are using an impinging-jet stagnation point reactor to experimentally study vapor phase ignition under well-defined transport conditions. In particular, we are inferring the onset of VPI in the thermal boundary adjacent to a hot substrate from observable decreases in deposition rates and changes in deposit microstructures. Future experiments will also include light scattering from particles nucleated in the boundary layer and non-intrusive measurements of local vapor phase species concentration. Experimental results are being used to assist in the development and eventual verification of a quantitative theory to predict the onset of vapor phase ignition in systems where high activation energy confines homogeneous reactions to a thin chemically reacting sublayer embedded within the thermal boundary layer adjacent to the hot deposition surface. Our objective is to develop a general theory which can be used with available homogeneous and heterogeneous chemical kinetics data to establish reactor design criteria and select optimal operating conditions which maximize deposition rates while just avoiding both VPI and vapor reagent diffusional limitations. A potentially useful byproduct of the theory will be the ability to extract global homogeneous reaction kinetic parameters from deposition rate data taken under well-defined transport conditions - *i.e.* just as it is now common practice to extract heterogeneous kinetic parameters from exponentially increasing deposition rate data using an Arrhenius plot, with a more complete analysis of the type outlined here it should also be possible to extract useful *homogeneous kinetic parameters* using deposition rate data in the fall-off region beyond VPI.

HIGH ACTIVATION ENERGY REACTING FILM THEORY

The basic theory is intended to predict the effect of homogeneous reaction on vapor deposition rates when homogeneous reactions result in non-depositing products for the typical case of cold gas/hot substrate CVD systems [8]. If the homogeneous reactions are controlled by some high activation energy process with an apparent overall activation energy E_{hom} and $(E_{hom}/RT_w) \gg 1$ (where R is the gas constant), then at sufficiently high wall temperatures there should be a chemically "frozen" outer region and a thin chemically reacting sublayer adjacent to the hot substrate. By exploiting the thinness of this reacting sublayer (to which all homogeneous reactions are confined), it is possible to generate a simple asymptotic solution for one-dimensional species mass transport and obtain expressions for the reagent mass fraction profile $\omega(y)$ normal to the substrate and the deposition flux which account for Soret as well as Fick diffusion everywhere, and reagent losses in the reacting sublayer.

The key assumptions in our present simplified model are: 1) a cold reagent gas stream at inlet temperature T_e impinging on a hot deposition surface at T_w 2) transport to the deposition surface through a one-dimensional stagnant gas film of thickness δ representing a stagnation point boundary layer 3) high activation energy homogeneous chemical kinetics $(E_{hom}/RT_w \gg 1)$ and 4) simple global homogeneous and heterogeneous chemical reaction rates $\dot{r}'''_{hom}$ and $\dot{r}''_{het}$ for a single limiting reagent, of the form:

$$\dot{r}'''_{hom} = A \cdot (\rho \cdot \omega(y))^m \cdot e^{(-E_{hom}/RT)}$$
$$\dot{r}''_{het} = B \cdot (\rho \cdot \omega(y))^n \cdot e^{(-E_{het}/RT)}$$

where ρ is the gas density, A and B are the homogeneous and heterogeneous reaction pre-exponential factors, and E_{hom} and E_{het} are their activation energies. Besides T_w/T_e, our solution involves the following key dimensionless parameters:

$$Dam_{hom} \equiv \frac{A\delta^2}{D} \cdot (\rho\omega_e)^{m-1} \qquad \left(\frac{E_{hom}}{RT_w}\right)$$

$$Dam_{het} \equiv \frac{B\delta}{D} \cdot (\rho\omega_e)^{n-1} \qquad \left(\frac{E_{het}}{RT_w}\right)$$

where D is the limiting species Fick diffusivity and Dam_{hom} and Dam_{het} are Damkohler numbers for the homogeneous and heterogeneous reactions. For the case of first order homogeneous and heterogeneous reactions (m=n=1), we obtain a simple closed form expression for the surface flux of the limiting reagent which displays the expected high temperature fall-off when $E_{hom} > E_{het}$, as shown in figure (1). Note that when the heterogeneous reaction is sufficiently fast (i.e. when Dam_{het} is sufficiently large) there is a transition from heterogeneous kinetic control to diffusion control and finally to VPI and the high temperature deposition rate fall-off region. Slower surface kinetics result in a transition from heterogeneous kinetic control directly to VPI and the fall-off region (presumably the sequence of events in Ghoshtagore's experiments). It is also evident that to maximize deposition rates, optimal operating conditions would result in incipient vapor phase ignition. However, to obtain acceptable deposit microstructures it may be necessary to operate at surface temperatures below both VPI and the onset of appreciable diffusional limitations. Since a primary goal of this work is to guide CVD reactor design and operation, experimental correlation of deposit microstructures with deposition conditions in the vicinity of VPI is currently underway to complement our CVD-rate measurements.

EXPERIMENTAL

To verify or guide improvements of our theoretical model we are now investigating the onset of VPI under well-defined transport conditions using an axisymmetric impinging jet CVD reactor. The reactor is shown schematically in figure (2). Cold reagent and carrier gas enter the reactor from the top, flow through a converging cast alumina nozzle and impinge on a polished quartz disk substrate which rests on an RF-heated graphite susceptor ($r_{substrate} = r_{jet}$). Preliminary experiments have been for titania deposition from $TiCl_4/O_2/Ar$ and $TiCl_4/N_2O/Ar$ at .1 MPa pressure using excess oxidizer. The $TiCl_4$ source is a constant temperature liquid bubbler and all lines carrying the $TiCl_4$ are teflon with stainless steel fittings and valves. To avoid water contamination the reactor is pumped down to 10 Pa for several hours before each run and only dry gasses are used (ultra high purity Ar, H_2O < .5 ppm). Our intention was to measure deposition rates with effectively no water, since it reacts violently with $TiCl_4$ even at room temperature. In some of our experiments water contamination was inferred from the presence of TiO_2 particles in the cold (300 K) reagent mixing chamber above the nozzle. Deposition rate data from these experiments were thrown out and gas handling was improved to eliminate the problem, although future experiments with controlled water addition are planned. Deposition rate measurements are by *in situ* interferometry (at a wavelength of 633 nm) and *ex situ* weight gain.

Our current range of accessible operating conditions (with well defined fluid flow) are: T_w up to 1600 K, pressure P between 0.01 and 0.1 MPa, and impinging jet Reynolds number Re_{jet} between roughly 100 and 600 (based on nozzle radius r_{jet}). Under these conditions natural convection is not expected in the primary reagent jet since $Re_{jet}^2/Gr >> 1$, (Gr is the Grashof number for the impinging jet - there may be buoyancy-driven flows away from the jet). Indeed, flow visualization of the jet using argon seeded with fine titania particles showed no flow distortions when the susceptor was hot or cold. Standard operating conditions used in our preliminary experiments are:

T_w:	900 - 1700 K	Ar	99%
P:	0.1 MPa	O_2 or N_2O	1%
Re_{jet}:	200 (nozzle velocity ≈ 1 m/s)	Oxidizer/$TiCl_4$	20/1 to 200/1

PRELIMINARY RESULTS AND DISCUSSION

The most striking preliminary result we have obtained is that under the abovementioned reactor conditions, no VPI has been observed for substrate temperatures up to 1700 K in the $TiCl_4/O_2/Ar$ system. This is very different from Ghoshtagore's experimental results, which indicated VPI at only 1125 K for the same chemical system but under rather different flow conditions [4]. As shown in figure (3), our results for titania deposition from $TiCl_4/O_2$ in Ar show a transition from heterogeneous kinetic control to diffusion control at roughly 1300 K with no high temperature rate fall-off. The relatively temperature insensitive region of our deposition rate data above 1300 K approaches the $TiCl_4$ convection-diffusion limit calculated assuming local thermochemical equilibrium at the gas/solid interface and Fick and Soret diffusion through a variable property mass transfer boundary layer [9,10].

The absence of VPI in our reactor is consistent with preliminary light scattering experiments in which a 10mW He-Ne laser was used to produce a 12 mm laser sheet which was passed through the reactor between the nozzle exit and the substrate planes. Visual inspection revealed a particle-free primary jet impinging on the substrate at all temperatures[†]. The results of our basic reacting film theory are also not inconsistent with our experimental results when rate calculations are made using heterogeneous kinetic parameters from Ghoshtagore [4] and homogeneous kinetic parameters from a recent independent (isothermal flow reactor) study of global $TiCl_4$ oxidation kinetics [11], as can be seen in figure (3). However, reacting film theory deposition rate calculations using these kinetic parameters and the approximate flow conditions in Ghoshtagore's experiments do not show a fall-off in deposition rate due to VPI until 2100 K (off scale on figure 3), and then it very gradual. In order to explain Ghoshtagore's observed VPI at 1125 K using our basic theory, it would be necessary to assume a homogeneous activation energy

---

[†] Particles were seen in the corners of the reactor where the reagent gases recirculate, but a negligible fraction of these gases can reach the substrate.

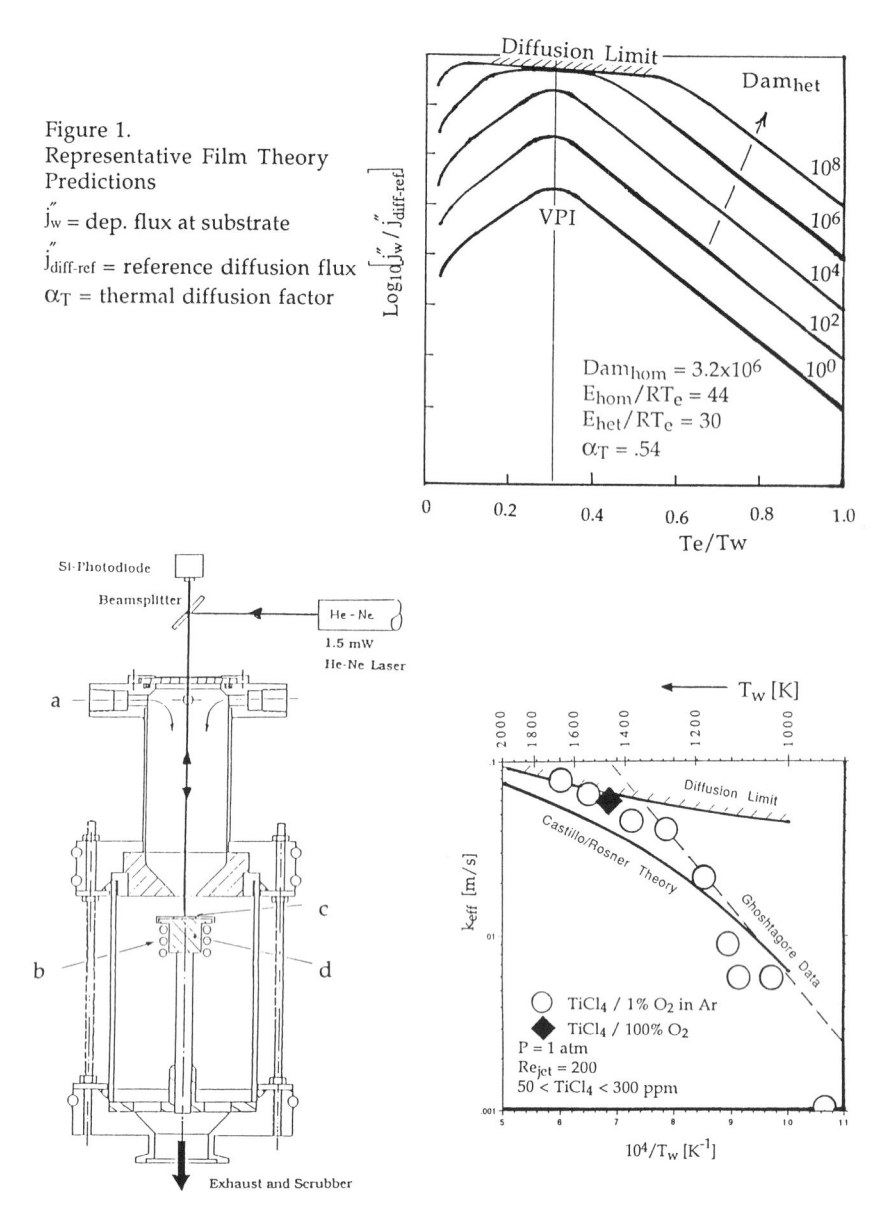

Figure 1.
Representative Film Theory Predictions

$\dot{j}''_w$ = dep. flux at substrate

$\dot{j}''_{diff-ref}$ = reference diffusion flux

α_T = thermal diffusion factor

$Dam_{hom} = 3.2 \times 10^6$
$E_{hom}/RT_e = 44$
$E_{het}/RT_e = 30$
$\alpha_T = .54$

Figure 2.
Axisymmetric Impinging Jet CVD Reactor
a) reagent inlet, b) RF coils c) substrate d) graphite susceptor

Figure 3.
Arrhenius Plot of Normalized TiO$_2$ Deposition Rates from TiCl$_4$ + O$_2$

$k_{eff} \equiv \dot{r}''_{het} / [P_{TiCl4}/RT_w]$

of roughly 345 kJ/mole as opposed to the published 89 kJ/mole. Furthermore, using the high activation energy estimated in this way from Ghoshtagore's data, our basic theory would predict VPI at only 1300 K under our reactor conditions, which was not observed. These inconsistencies have not yet been resolved. To do so may require a more sophisticated treatment of homogeneous chemical reactions than is now incorporated in our theory. However, there are uncertainties in the transport conditions in Ghoshtagore's experiments, including the possibility for buoyancy-driven recirculation around his deposition substrate. Recall that in our experiments no particles are observed in the primary reagent jet, but they are plainly visible in recirculating zones in the reactor. Trace water contamination in Ghoshtagore's (or our) experiments may also contribute to the discrepancy in VPI conditions.

Avoiding VPI would be good news if our goal was simply to produce high quality TiO_2 films, but our goal is to study VPI and its sensitivity to reactor operating parameters. To force VPI to occur at lower temperatures in our jet impingement reactor we have tried halving the total gas flow rate (*i.e.* doubling the reagent residence time in the thermal boundary layer), but this did not yield a clear VPI. Increasing the O_2 partial pressure from .001 to .1 MPa (as in the Ghoshtagore experiments) had no measurable effect on deposition rates at 1500 K (with $O_2/TiCl_4$ always > 30). Another approach we considered was using N_2O as an oxidizer, since N_2O begins to dissociate to N_2 and atomic O at roughly 1300 - 1400 K under our standard operating conditions (residence time in thermal boundary layer ≈ 17 ms). O atom attack of $TiCl_4(g)$ should be much faster than O_2 attack, and should lower the VPI temperature [12]. Preliminary experiments for $N_2O/TiCl_4$ = 20, 40 and 200 have not yet given conclusive evidence of VPI at accessible surface temperatures, but still higher N_2O concentrations may finally give the anticipated effects. Future experiments may also include more reactive metal sources and organometallics known to react in the gas phase at lower temperatures [3].

CONCLUSIONS AND FUTURE WORK

While we have not yet been able to observe $TiCl_4$ vapor phase ignition in our impinging jet reactor (fig. 2), we are confident that our basic model of high activation energy reacting gas films will prove to be useful for many CVD applications once the theory has been experimentally verified for some prototypical chemical system. The most obvious reason for our difficulty in achieving VPI below 1700 K in our reactor at atmospheric pressure is the very short reagent residence time in the thermal boundary layer adjacent to the substrate. The sensitivity of the system to trace amounts of water has not yet been determined by experiments with controlled water addition.

We must also acknowledge that the current version of our theory is unable to reconcile recent isothermal flow reactor data on global homogeneous $TiCl_4$ oxidation kinetics [11] with the deposition rate data of Ghoshtagore - nor are the homogeneous oxidation kinetics estimated using the present theory and Ghoshtagore's CVD rate data consistent with our own preliminary deposition experiments. However, it would be premature to conclude that any useful theory for VPI in CVD reactors requires a more complex description of the relevant homogeneous reaction kinetics. Other more straight-forward enhancements to the basic theory under consideration are treating real viscous flow, as opposed to an equivalent stagnant gas film (including Stefan flow for non-dilute systems), and accounting for variable thermophysical properties.

Once our high activation energy reacting gas film theory, with any necessary modifications, is experimentally verified it will provide rational criteria to reactor designers and process engineers on how to achieve the maximum possible deposition rates while still avoiding VPI and diffusional limitations. Coupling the prediction of VPI onset conditions with semi-empirical correlations of deposit microstructures in the vicinity of VPI will then help find the operating conditions needed to produce films of desired microstructure at the maximum rate.

ACKNOWLEDGEMENTS

This work (carried out in the Yale High Temperature Chemical Reaction Engineering (HTCRE) Laboratory) was supported by NASA Lewis Research Center under Grants NTG-5037 and NAG 3-884, the US Air Force under Grant 89-0223 and the Yale HTCRE Lab. Industrial Affiliates (Shell, GE, DuPont, and Union Carbide). Thanks are also due to Tony Macini of the student machine shop and Dick Downing from Gibbs machine shop for their help in the design and construction of our CVD reactor.

REFERENCES

1. Y. Takashi *et al.*, J. Cryst. Growth, **74**, 409 (1986).
2. H. Yamane and T. Hirai, J. Mat. Sci. Let., **6**(10), 1229, (1987).
3. M. Balog *et al.*, J. Electrochem. Soc., **126**(7), 1203 (1979).
4. R.N Ghoshtagore, J. Electrochem. Soc., **117**(4), 529 (1970); R.N. Ghoshtagore and A.J. Noreika, *ibid*, **117**(10), 1310 (1970).
5. H. Komiyama *et al.*, in Chemical Vapor Deposition X, edited by G.W. Cullen (Electrochem. Soc. Proc. **87**-8, Pennington NJ 1987) pp. 1119-1128.
6. H. Komiyama and T. Osawa, Jap. J. Appl. Phys., **24**(10), L795 (1985).
7. M. Tirtowidjojo and R. Pollard, J. Cryst. Growth, **98**, 420 (1989).
8. J. Castillo and D.E. Rosner, presented at the 1990 AIChE Meeting, paper no. 55d; J. Electrochem. Soc. publication in preparation.
9. D.E. Rosner and J. Collins in Chemical Vapor Deposition XI, edited by K.E. Spear and G.W. Cullen (Electrochem. Soc. Proc. **90**-12, Pennington NJ 1990) pp. 49-60.
10. D.E Rosner, Transport Processes in Chemically Reacting Flow Systems, (Butterworths, Boston, 1986), pp. 307-404 (3rd Printing 1990).
11. S.E. Pratsinis *et al.*, J. Am. Ceram. Soc., **73**(7), 2158 (1990).
12. J.D. Chapple-Sokol *et al.*, J. Electrochem. Soc., **136**(10), 2993 (1989).

RAPID GROWTH OF CERAMIC FILMS BY PARTICLE-VAPOR CODEPOSITION*

ROBERT H. HURT AND MARK D. ALLENDORF
Combustion Research Facility, Sandia National Laboratories, 8361
Livermore, California 94551-0969

* This work is supported by the Support of the Department of Energy Office of Conservation Advanced Industrial Materials Program.

ABSTRACT

Particle-enhanced chemical vapor deposition (PECVD) is capable of producing ceramic films at high deposition rates. A mathematical model of the particle-vapor codeposition process has been developed and has been applied to PECVD processes to predict deposition rate enhancements and deposit properties.

I. INTRODUCTION

Recently, substantial increases in the rate of chemical vapor deposition of TiO_2, ZrO_2, and AlN have been reported by introducing particles into the CVD process [1,2]. Improvements in AlN deposition rates of up to two orders of magnitude have been obtained with this technique. In general, particles may be introduced in the form of an aerosol, generated *in situ* by a controlled amount of gas-to-particle conversion (homogeneous nucleation), or in the form of an entrained powder, independently charged to the reactor [3]. In this article, processes in which the rate of CVD is enhanced by the codeposition of particles, achieved either by seeding the gas phase or by in situ aerosol formation, will be referred to generically as "Particle-Enhanced Chemical Vapor Deposition" or PECVD processes.

The structure of deposits formed by this technique is depicted in Figure 1. It is believed [1] that reactions of gas-phase species occur on the enhanced surface area of a porous region at the interface between the growing deposit and the gas phase. There is a close relation between this process and chemical vapor infiltration (CVI) processes [4], in which porous bodies (formed, for example, from compressed powders) are densified by internal CVD growth, the difference being that in PECVD the porous body is formed gradually with CVI-type densification occurring simultaneously in a restricted region at the porous interface of the growing deposit.

In the existing literature there have been numerous studies of particle deposition and of vapor deposition processes and their applications. There has not, however, to the authors' knowledge, been a theoretical treatment of the particle-vapor *codeposition* process and, as a result, the factors determining the growth rate and properties of such deposits are not well established. In the present article, a random sphere formulation is developed and used in a unified treatment of both PECVD subprocesses: particle deposition and densification of porous media by vapor deposition.

II. SUMMARY OF THE RASSPVDN MODEL DEVELOPMENT

In this section, the development of the "<u>R</u>andom <u>S</u>phere Model of <u>S</u>imultaneous <u>P</u>article and <u>V</u>apor <u>D</u>eposition" or "RASSPVDN" is summarized. The complete development can be found elsewhere [5]. The model considers the local particle-vapor codeposition process, isolated from processes occurring prior to deposition, such as nucleation and growth of seed particles in the gas phase (which are highly system-specific). Seed particles, which are assumed to be spheres of radius R_p, approach the growing film and deposit in the porous interface at a rate F_o (particles/cm^2-sec). At the same time, vapor deposition occurs on all exposed surfaces at a uniform rate, G (μm/min).

Formulated in this way, the model describes CVD processes whose rates are limited by surface chemical kinetics. The volumetric deposition rate can be defined as dV_t/dt, where V_t (cm^3/cm^2) is the total solid volume in the deposit normalized by area of the bare substrate. There are two contributions to the deposition rate, the first being vapor deposition on the enhanced surface area at a rate $(A/A_o)G$, where A is the total surface area of the deposit, and the second the direct addition of solid volume through seed particle deposition $F_o(4/3)\pi R_p^3$. The total volumetric deposition rate is then given by

$$\frac{dV_t}{dt} = (A/A_o)G + F_o \frac{4}{3}\pi R_p^3 \qquad (1)$$

The deposition rate *enhancement* is defined as the volumetric deposition rate dV_t/dt divided by G, the volumetric deposition rate for vapor deposition (CVD) in the absence of particles. In order to solve Equation 1, an accurate description of the complex geometry of the porous deposit is required to relate the amount of internal surface area to the process parameters F_o, R_p, G, and the deposition time, t. A mean-field approach is adopted here that allows a unified treatment of both ballistic particle deposition and the densification of porous media by vapor deposition.

The Random sphere formulation: Gavalas [6] has treated the growth and coalescence of cylindrical pores during carbon gasification by a random capillary model. In a similar fashion, the growing film can be considered to consist of the original substrate surface plus the collection of all seed particles (assumed to be spheres), which have grown steadily by surface growth from the vapor phase since the time of their deposition (see Figure 1). These enlarged spheres overlap with their neighbors, producing a complex, three-dimensional porous body, whose geometry is completely determined if the sizes of all the spheres and the locations of their centers are specified.

Figure 1. The particle-vapor codeposition process.

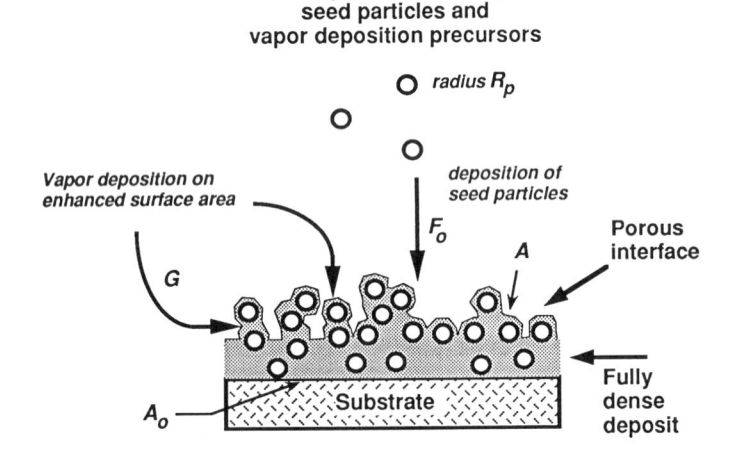

It is assumed that the sphere centers are distributed randomly in space with a mean number density per unit volume γdr, which is dependent upon height, z above the substrate and upon deposition time, t. Since simultaneous deposition and growth lead to a deposit comprising, at any instant, spheres of many different sizes, γ will be a function of both z and r, the sphere radius. The problem is thus reduced to finding the evolution of the distribution $\gamma(r,z)$ with time and deriving expressions for the deposit properties of interest from $\gamma(r,z)$ at any instant. We seek an expression for $V_s(Z_e)$, the solid volume fraction of the deposit at a height, Z_e, above the substrate. To calculate $V_s(Z_e)$ or, equivalently, $1 - \theta(Z_e)$, consider a random volume element at height Z_e. The solid volume fraction in the plane at Z_e is equal to the probability, P, that this volume element lies within the domain of one or more of the growing spheres that constitute the deposit. If the probability is dP_s that the volume element at Z_e lies within the volume of a *single* given sphere, then by the properties of the Poisson distribution, the probability that the element lies within *none* of the spheres is

$$1 - P(Z_e) = e^{\left(-\int dP_s\right)} \quad (2)$$

where the individual probabilities have been integrated over all spheres. Performing this integration over three spatial dimensions and across the spectrum of particle sizes present in the deposit yields an expression for the porosity or solid volume fraction at distance Z_e from the substrate:

$$V_s(Z_e) = 1 - \theta(Z_e) = 1 - e^{\left[-\int_0^\infty \int_{|Z_e-z|}^\infty \gamma(r,z)\pi\{r^2 - (Z_e-z)^2\} \, dr \, dz\right]} \quad (3)$$

Equation 3 is the fundamental equation describing the geometry of the porous deposit. The total surface area of the deposit can now be obtained by considering a differential amount of film growth on existing spheres:

$$A/A_o = \frac{1}{G}\frac{dV_t}{dt} \quad (\text{for } F_o = 0) \quad (4)$$

where V_t is the total solid volume in the deposit, obtained by integrating V_s (Equation 3) over all z. Both A/A_o (cm^2/cm^2) and V_t (cm^3/cm^2) are normalized by the area of the bare substrate A_o.

Surface growth by vapor deposition: Deposition from the gas phase also affects the distribution $\gamma(r,z)$. In the absence of diffusion limitations, there will be uniform growth on all exposed surfaces by vapor deposition (G μm/min), resulting in a steady increase in the radius of all particles and shifting the distribution $\gamma(r,z)$ toward larger particle sizes:

$$\frac{\partial \gamma(r,z)}{\partial t} = \frac{\partial \gamma(r,z)}{\partial r} \cdot G \quad (5)$$

Simultaneously, a flat deposit grows on the original substrate surface by vapor deposition at a uniform rate, G (μm/min).

Particle Deposition: The mechanism of particle deposition determines the nature of the porous interface, which in turn determines the PECVD growth rate. Two important particle deposition mechanisms will be considered: 1. deposition at unity sticking coefficient. Surface forces are dominant in this mechanism and particles are brought to rest at the point of first impact with any surface in the interfacial region of the growing deposit, and 2. gravitational settling. In this mechanism surface forces between the particles and the deposit are negligible in comparison to gravitational or inertial forces and seed particles may undergo multiple interactions with solid surfaces, coming to rest only at positions in the bed, at which further downward motion is impossible. The first deposition mechanism produces loose particle packings with low particle volume fraction, while the second produces dense packings with particle volume fractions that can exceed 50%. Which deposition mechanism is applicable in a given case is a function of the particle properties, chiefly size, density, surface roughness, or the presence of small amounts of a liquid phase. Any combination of large particles, high density, or smooth dry surfaces tends to promote deposition by gravitational settling. A detailed description of orthogonal ballistic deposition with unity sticking coefficient is obtained from within the random sphere formulation, as described by Hurt and Allendorf, [5]. The complex process of gravitational settling, in contrast, is treated using an empirical relation that mimics random particle packing.

III. RESULTS

From a nondimensionalization of the complete set of model equations, it becomes clear that the evolution of deposit properties in dimensionless time, $\tau = t\, G/R_p$, is a function only of the one parameter $F_o R_p^3/G$. A FORTRAN code has been developed to solve the model equations and the numerical results are presented below.

There are two components to the deposition rate enhancement: 1) the enhancement of the surface area for vapor deposition and 2) the direct deposition of solid contained in the seed particles. At time = 0, the substrate surface is bare and A/A_o is unity for all values of $F_o R_p^3/G$. As particles deposit, the porous interface develops and A/A_o rises, eventually reaching an asymptotic value corresponding to the establishment of a self-preserving interface. A steady-state condition has been achieved with respect to the interface structure and the film growth rate. The growth rate enhancement in this steady-state condition is plotted in Figure 2, along with the seed particle volume fraction in the steady-state deposit, as a function of the dimensionless parameter $F_o R_p^3/G$. Rate enhancements remain less than a factor of two up to a dimensionless particle deposition rate ($F_o R_p^3/G$) of 0.01, thereafter increasing rapidly with increasing $F_o R_p^3/G$. Deposition with unity sticking coefficient gives higher surface area enhancements by factors of up to 8. The seed particle volume fraction increases with $F_o R_p^3/G$ to a steady-state value, which depends on the deposition mechanism. The limiting seed particle volume fraction is equivalent to the solid volume fraction in a bed of particles, deposited in the absence of vapor deposition: 0.65 for gravitational settling, approximately 0.1 for deposition at unity sticking coefficient. This asymptote indicates that high PECVD rates are theoretically possible (with rate enhancements over two orders of magnitude), while maintaining relatively low seed particle volume fractions (ca. 10% for deposition at unity sticking coefficient).

Steady-state, self-preserving profiles of the solid volume fraction at the interface are plotted in Figure 3 for deposition at unity sticking coefficient at various values of $F_o R_p^3/G$. Increasing $F_o R_p^3/G$ increases the width of the interfacial region.

Figure 2. PECVD rate enhancement factor and seed particle volume fraction at steady-state vs. $F_o R_p^3/G$. Dashed lines: deposition by gravitational settling; dotted lines: deposition with unity sticking coefficient.

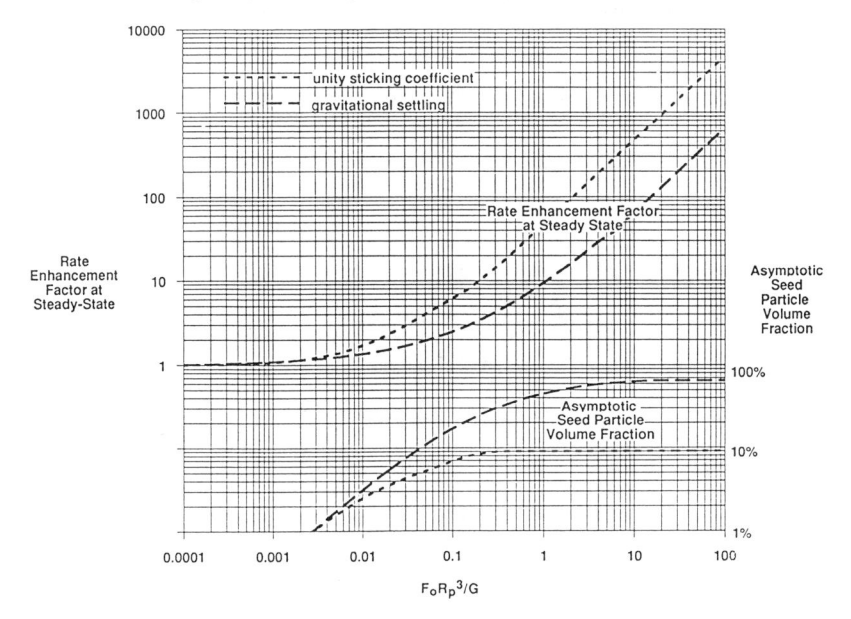

Figure 3. Self-preserving profiles of solid volume fraction at the interface of the growing deposit for various $F_o R_p^3/G$. Particle deposition with unity sticking coefficient.

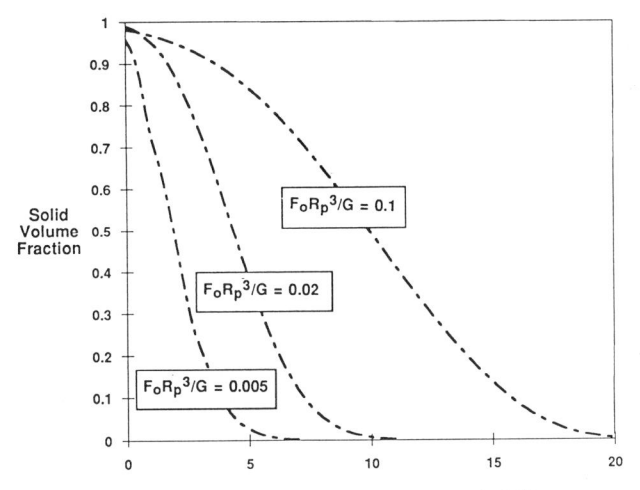

IV. DISCUSSION

Results of the model illustrate that particle-vapor codeposition produces deposits having characteristic physical features, consisting of a fully dense region and a porous region at the gas/deposit interface. The model predicts the existence of a unique, direct relationship between the growth rate enhancement in the kinetically controlled regime and the single dimensionless parameter $F_o R_p^3/G$. The extent of the rate enhancement at a given value of $F_o R_p^3/G$ is a function of the particle deposition mechanism, which in turn is dependent on the surface properties of the materials involved. Under most conditions, the enhanced surface area is primarily responsible for the growth rate enhancement, the contribution of the seed particles being much smaller. Seed particle volume fractions are predicted to increase with increasing rate enhancement up to an asymptotic value that is dependent on the deposition mechanism. It is, therefore, theoretically possible to achieve high rate enhancements while maintaining relatively low seed particle fractions in the final deposit. The use of very fine particles, in particular, which deposit with sticking coefficients near unity should limit the volume fraction of seed material required to approximately 10%.

The RASSPVDN model provides a number of other valuable predictions regarding the role of seed particle size. First, the model results indicate that the steady-state rate enhancement depends only upon the dimensionless parameter $F_o R_p^3/G$ or, for a given vapor deposition (CVD) rate, only upon the rate of addition of particle volume, $F_o(4/3)\pi R_p^3$. The steady state rate enhancement is therefore *independent* of particle size at constant mass or volumetric particle feed rate and constant deposition mechanism. In practice, large changes in particle size will often be accompanied by a shift in the deposition mechanism, leading to a more complex dependence on size overall. The model also predicts that high rate enhancements will be accompanied by porous interfacial zones of significant thickness, which could lead to rough and friable surfaces of PECVD deposits. Improved surface properties may be obtained by densifying the interface by CVD (in the absence of particles) at the end of the coating process. The time required for "post-densification" can, according to the model, be a significant fraction of the total production time. Since the densification time scales as $t \sim G/R_p$, the time required for post-densification of the porous interface (using CVD in the absence of particles) can be reduced by the use of finer particles.

In the kinetically controlled regime, arbitrarily large rate enhancements are possible by operating at very high dimensionless particle deposition rates $F_o R_p^3/G$. In practice, the finite rate of transport or of homogeneous gas-phase reactions may limit the rate at which *fully dense material* can be deposited. Transport limitations may arise from the bulk gas to the deposit edge, or within the porous interfacial zone in the deposit itself. Diffusion limitations within the interface will reduce the concentration of the reactive species and thus the densification rate at points far from the deposit surface, resulting in porous, friable, low-quality coatings. Scaling laws derived from the model indicate that the use of high reactant concentration and/or small particle size will promote efficient mass transfer through the interface and the production of fully dense material at high growth rates. It is hoped that the model results presented here will guide the development of PECVD processes into parameter ranges producing high-quality material at high deposition rates.

REFERENCES

[1]. H. Komiyama, T. Osawa, H. Kazi, T. Konno *Mat. Sci. Monographs* **38A** 667 (1986).
[2]. Y. Shimogaki and H. Komiyama *Chem. Lett.* 267 (1986)
[3]. A. N. Scoville and P. Reagan. presented at the *14th Annual Conference on Composite Materials and Structure*, Cocoa Beach, Florida, January (1990)
[4]. Besmann T.M., Sheldon B.W., Lowden R.A., and Stinton D.P. *Science* 253, 1104 (1991)
[5]. R. H. Hurt, and M. D. Allendorf. *AIChE Journal*, **37**:10 1485 (1991).
[6]. G. R. Gavalas *AIChE J.* **26**:4 577 (1980)

STEP COVERAGE MODELING OF THIN FILMS DEPOSITED BY CVD USING FINITE ELEMENT METHOD

CHING-YI TSAI* and SESHU B. DESU**
*Department of Engineering Science and Mechanics
**Department of Materials Engineering
Virginia Polytechnic Institute and State University, Blacksburg, VA, 24061

ABSTRACT

A two-dimensional finite element model was developed to study the step coverage of submicron trenches with arbitrary shape under chemical vapor deposition processes. Parameters that characterize the step coverage were found to be the surface Damkohler number, ratio of diffusion coefficients inside and outside of the trench, and aspect ratio of the trench geometry. Efforts were concentrated on studying the step coverage of SiO_2 film deposited from SiH_4/O_2 precursors within rectangular shape trenches. The model predictions were found to be in good agreement with reported experimental results.

INTRODUCTION

Although chemical vapor deposition processes (CVD) are known to have better film conformity than those of physical vapor deposition processes (PVD), as aspect ratio (width/depth) of the trenches becomes smaller, conformity of the deposition materials may be lost, resulting in creation of voids inside the trenches. This problem is known as the step coverage problem. Poor step coverage leads to non-planar surfaces and thus non-uniform electrical resistance. Poor step coverage also leads to significant problems in lithographic processes.

Some efforts have been spent in analyzing the deposition profiles of the CVD processes inside the trenches using the fundamental Boltzmann equations. Due to the difficulties associated with obtaining numerical solutions of the Boltzmann equations directly, simulation techniques, such as Monte Carlo Simulations, have been widely used to obtain solutions of the Boltzmann equations statistically [1-2]. Although Monte Carlo simulations were successful in simulating the film step coverage of the CVD processes, they suffer from expensive computational costs [3].

In 1939, Thiele [4] proposed a one-dimensional (1-D) model for simulating simultaneous diffusion and chemical reactions in a single pore. This 1-D model was recently used by McConica et al. (1990) [5] to study the step coverage of CVD tungsten process in cylindrical contact holes. Using this 1-D model, McConica et al. [5] obtained qualitative agreements between the model's predictions and experimental results.

In general, 1-D step coverage models require lumping the trench geometry and, thus, necessitate the assumption that the film thickness is uniform over the bottom of the trench. Furthermore, the film thickness at the entrance side wall is also assumed to be identical with the film thickness along the flat surface adjacent to the trench mouth in these 1-D models. In other words, obtaining quantitative predictions of film thickness profiles over the trench surface is very difficult using these 1-D models [6-7]. A two-dimensional (2-D) model would alleviate the problem of geometric lumping for simulating the film step coverage.

In this paper, a 2-D diffusion-reaction model was developed to study the step coverage of the CVD processes within submicron trenches or contact holes. Parameters that control the step coverage were obtained through the dimensional analysis of the governing equations and boundary conditions. A Fortran programming code based on nonlinear finite element method (FEM) was then developed to simulate the processes.

This step coverage finite element model can also be used to study the chemical vapor infiltration (CVI) process within the fibrous preform.

THEORY

Consider either a long rectangular trench of width W and depth L or a cylindrical tube of radius R and depth L as an idealized model of a via cut into the surface of a patterned wafer. Uniform deposition thickness within the trench is obtained if concentration of the reactive species is uniform around the trench geometry.

The following assumptions were made in order to model the step coverage problem for the chemical vapor deposition processes (CVD).

1. In order to calculate the concentration pattern along the flat surface adjacent to the trench mouth, the shaded area, as shown in Fig.1, is selected as domain for the analysis of the diffusion—reaction process. Because the size of the domain is much smaller as compared to the size of the boundary layer adjacent to the surface of the wafer, it is assumed that the domain is located inside the boundary layer. Therefore the convection flow of the gas inside the domain is neglected.

2. The diffusion constants either inside or outside the trench can be calculated as, $1/D_t = 1/D_{mol} + 1/D_k$, where D_{mol} and D_k are the diffusion coefficients for the molecular and Knudsen diffusion respectively [8]. The diffusion constants of the gas species outside the trench is assumed to be independent of the trench geometry.

3. The dominant chemical reaction inside the domain is the sticking of one major reactive species onto the substrate surface.

4. The geometry change of the trench due to the film deposition is neglected.

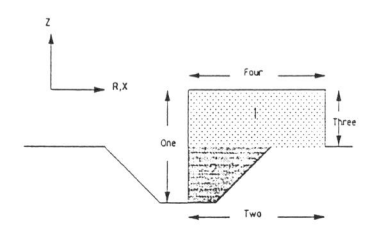

Fig.1. Illustrations of the domain and associated boundaries considered in the finite element modeling.

Analysis of the step coverage problem

With the coordinates depicted in Fig.1, the concentration of the reactive species within the domain can be analyzed by solving the following diffusion equation,

Rectangular Coordinates
$$D_t\left(\frac{\partial^2 C}{\partial X^2} + \frac{\partial^2 C}{\partial Z^2}\right) = 0 \tag{1a}$$

Cylindrical Coordinates
$$D_t\left[\left(\frac{\partial}{R\partial R}\left(\frac{R\partial C}{\partial R}\right) + \frac{\partial^2 C}{\partial Z^2}\right)\right] = 0 \tag{1b}$$

where D_t and C are the diffusion coefficient (cm^2/sec) and molar concentration (moles/cm^3) of the reactive species respectively.

In order to obtain the concentration distribution of the reactive species over the entire domain, conditions of the species along the boundaries of the domain must be specified.

Symmetric condition is specified along the boundaries one and three. Along the boundary two, the assumption that the rate of mass transfer at the surface is equal to the rate of surface reaction is used. Along the boundary four, the concentration of the reactive species is assumed to be constant. The value of this constant (C_{i0}) can be determined from the information of the deposition rate on the flat surface adjacent to the trench mouth. For a first order chemical reaction, the value of this constant will not affect the step coverage of the trenches.

FINITE ELEMENT MODELING

First, the following group of parameters were used to non–dimensionalize the governing equation and boundary conditions of the step coverage problem,

$$X/L = X^*, \ R/L = R^*, \ Z/L = Z^*, \ C/C_{i0} = C^*, \ D_t/D_0 = D_t^*$$

where D_0 is the diffusion coefficient of the reactive species outside the trench. This constant D_0 is independent of the trench geometry.

Then, the finite element formulations of the problem were obtained by multiplying the diffusion equation, (1a) or (1b), with the weighting function, $\Psi(X,Z)$ or $\Psi(R,Z)$, and integrating over the entire domain [9]. After implementing the natural boundary conditions along boundaries one, two, and three into the FEM model, the finite element formulations of the problem were obtained. The star symbol was dropped in the subsequent expressions for clarity.

$$\sum_{j=1}^{n} K_{ij}^{(e)} C_j^{(e)} = F_i^{(e)} \quad (2a)$$

where

$$K_{ij}^{(e)} = \int_{\Omega^e} D''Y''\left(\frac{\partial N_i}{\partial Y}\frac{\partial N_j}{\partial Y} + \frac{\partial N_i}{\partial Z}\frac{\partial N_j}{\partial Z}\right) dYdZ + \int_{\Gamma_3^e} Dam N_i N_j ds$$

$$F_i^{(e)} = 0 \quad (2b)$$

with
  $D'' = 1.0$, for elements located outside the trench
  $D'' = D_t/D_0$, for elements located inside the trench
  $Y'' = 1.0$, and $Y = X$ for rectangular coordinates
  $Y'' = R$, and $Y = R$ for cylindrical coordinates
  $Dam = K_s L/D_0 = $ Surface Damkohler Number,
  where K_s is the rate constant (cm/sec)

Four–node linear rectangular elements, as shown in Fig.2, were used to discretize the domain. Fortran programming codes "FEM2D", included in the appendix of reference [9] were modified to solve the finite element formulations above.

Three parameters, that is, surface Damkohler number (Dam), ratio of the diffusion constants (D_t/D_0), and aspect ratio (W/L or R/L) of the trench, were found to characterize to this step coverage finite element model.

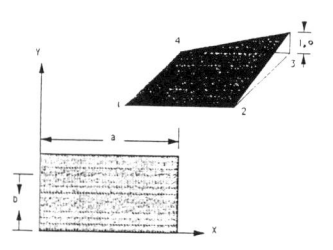

Fig.2 Illustrations of a typical four-node rectangular element
and one of its interpolation functions.

RESULTS AND DISCUSSION

SiO_2 film deposited from SiH_4/O_2 precursors

Kawahara et al. (1991) studied the step coverage of SiO_2 film deposited from SiH_4/O_2 precursors with trenches of 2.5 μm depth and 0.5 to 2.5 μm width [10]. Step coverage of the trenches was measured for the deposition temperatures of 450°K, 520°K, 580°K, and 640°K. The sticking coefficient of the reactive species within this temperature range is around 0.05 to 0.70 as reported by the same authors.

Parameters used in the finite element analysis of this problem are presented in Table I. The experimental results are well simulated by the proposed FEM model, as shown in Fig.3. Predictions from FEM model are also compatible to those from Monte Carlo simulations [10].

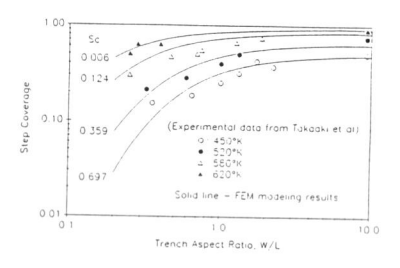

Fig.3. Step coverage dependence on the trench aspect ratio
and deposition temperature for SiO_2 thin films deposited
from SiH_4/O_2 precursors.

Parametric study of the FEM step coverage model

Fig.4 and Fig.5 present the dependence of the step coverage upon the three parameters with the following combinations,

Table I Simulation parameters for the SiH_4/O_2 LPCVD process

T		W/L D_t/D_0 (1.0E−2)							
°K	Dam	0.2	0.4	0.6	0.8	1.2	1.6	2.0	10
450	0.034	1.6	2.4	2.8	3.1	3.6	3.9	4.1	4.8
520	0.015	1.3	2.0	2.3	2.6	3.0	3.2	3.4	4.0
580	.0045	1.2	1.7	2.0	2.3	2.6	2.8	3.0	3.5
640	.0020	1.0	1.5	1.8	2.0	2.3	2.5	2.6	3.1

W/L = 0.1, 0.5, 1.0, 10.0
D_t/D_0 = 0.001 or 0.01 when the aspect ratio of the trench is 0.1
Dam = 0.0001, 0.0005, 0.001

For the CVD processes with large value of diffusion constant ratio (D_t/D_0 = 0.01), as shown in Fig.4, the step coverage is insensitive to the surface Damkohler number and the aspect ratio of the trench if the aspect ratio is above a critical value. This critical value is about 1.0 for the simulation conditions used above. Below the critical value, the step coverage is significantly affected by the trench aspect ratio.

For the CVD processes with small value of diffusion constant ratio (D_t/D_0 = 0.001), as shown in Fig.5, changes of the step coverage with respect to the surface Damkohler number are significant for all values of trench aspect ratios.

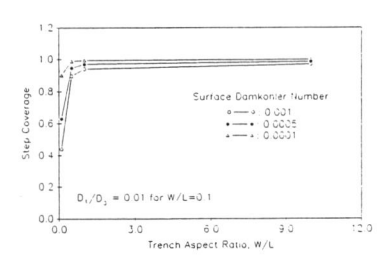

Fig.4. Parametric study of the step coverage dependence upon the surface Damkohler number, trench aspect ratio, and ratio of diffusion coefficients inside and outside of the trench.
D_t/D_0 is 0.01 for W/L=0.1 in this study.

CONCLUSIONS

A 2–D nonlinear finite element model, which simulates simultaneous diffusion and surface reaction processing over the trench geometry, has been successfully developed to study the step coverage for trenches with any arbitrary shape. Good agreement between the experimental results of SiO_2 thin films deposited from the SiH_4/O_2 precursors and the model's predictions was found. The numerical results are also similar to those obtained by Monte Carlo Simulations which generally require more computational time and cost.

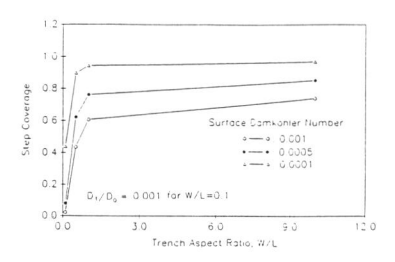

Fig.5. Parametric study of the step coverage dependence upon the surface Damkohler number, trench aspect ratio, and ratio of diffusion coefficients inside and outside of the trench. D_t/D_0 is 0.001 for $W/L=0.1$ in this study.

The proposed FEM model can be used to study the step coverage of trenches with any CVD processes as long as the kinetic data are available. This model assumes that the deposition process within a trench is dominated by only one major reactive species. However it is not difficult to extend this FEM model to processes where more than one species govern the deposition.

ACKNOWLEDGEMENT

The authors are indebted to Professor J.N. Reddy for his help concerning the Finite Element Method. Thanks also to the Center for Composite Materials and Structures of Virginia Polytechnic institute and State University for providing financial support to one of the authors (Ching Yi Tsai).

REFERENCES

1. M.J. Cooke and G. Harris, J. Vac. Sci. Technol., A, 7, 3217 (1989).
2. A. Yuuki, Y. Matsui, and K. Tachibana, Japanese J. of App. Phys., 28, 212 (1989).
3. T.S. Cale, G.B. Raupp and T.H. Gandy, J. Appl. Phys., 68, 3645 (1990).
4. Eugene E. Petersen, Chemical Reaction Analysis, (Englewood Cliffs, New Jersey, 1965).
5. S. Chatterjee and C.M. McConica, J. Electrochem. Soc., 137, 328 (1990).
6. Kazunori Watanabe, and Hiroshi Komiyama, J. Electrochem. Soc., 137, 1222 (1990).
7. J. Schlote, K.W. Schroder, and K. Drescher, ibid., 138, 2393 (1991).
8. Nyan–Hwa Tai and Tsu–Wei Chou, J. of the Amer. Cera. Soc., 72, 414 (1989).
9. J.N. Reddy, An Introduction to the Finite Element Method, (McGraw–Hill, New York, 1984).
10. Takaaki Kawahara, Akimasa Yuuki, and Yasuji Matsui, Japanese J. of App. Phys., 30, 431 (1991)

PYROLYTIC CARBON DEPOSITION ON GRAPHITIC SURFACES

Ismail M. K. Ismail
University of Dayton Research Institute
c/o Phillips Laboratory/ RKFC, Edwards Air Force Base, CA 93523-5000, USA

ABSTRACT

The present study outlines the importance of carbon active sites in controlling the kinetics of pyrolytic carbon deposition on a graphitized pitch fiber at 1025°C. Blockage of active sites with chemisorbed hydrogen retards deposition rates substantially. Removal of the chemisorbed hydrogen from the occupied sites raises deposition rates to the normal values noted on a fresh "clean" surface. The effect of surface activation prior to deposition is discussed. Activating the surface generates additional active sites and enhances the rates. However, not all the newly developed sites contribute to the kinetics. After the deposition of the first carbon layer, a fraction of those newly developed sites is instantaneously blocked and does not further contribute to subsequent deposition. The remaining fraction, along with the original sites available before activation, keeps replicating during the remaining course of deposition. That is, the disappearance of one active site after carbon deposition is associated with the generation of a new site. This trend continues up to the deposition of 60 carbon layers.

INTRODUCTION

The chemical vapor deposition (CVD) of pyrolytic carbon (pyro c) on carbon substrates is an important process for fabricating carbon composites which are used in many aerospace applications. The CVD process has two main advantages. First, by changing the experimental deposition conditions, one can obtain carbon composites having a wide spectrum of properties. Second, compared to other methods used for fabricating carbon composites, such as coal-tar or pitch impregnation, CVD is simpler, cleaner, and less expensive. In addition, it takes place in one step which operates at lower temperatures and pressures than the other industrial processes that are currently involved in fabricating carbon composites.

Extensive studies on the CVD of pyro c have appeared in the literature, including the structure and properties of deposits [1-4], the relations between deposit properties and processing conditions [3,4], the kinetics and mechanism of deposition [1,3,5,6], and the characteristics of deposits [1,3,5,7-9]. The kinetics of deposition depends on the type of substrate, provided that other experimental parameters are fixed [10]. For example, at 1025°C and 1 atm, the kinetics under a flow of 10% CH_4 in Ar was not the same for all substrates [10]. The results indicated that at least four categories of different substrates are possible: non-porous carbons with small surface areas, microporous carbons with large internal surface areas, non-porous carbons with large external surface areas, and activated carbons with considerable external and internal surface areas [10]. Thus, the total surface area (TSA) of substrates played a major role in CVD kinetics. The importance of carbon active surface area (ASA) to CVD kinetics, however, was briefly outlined, and it was speculated that the deposition started on carbon active sites. This was supported by the results of propylene pyrolysis on the active sites of a graphitized carbon black [11,12]. At 600-800°C and 0.012 Torr starting pressure, the carbon deposit replicated the original active surface area of the substrate during the deposition of the first few layers [11]. It was clearly ascertained that the pyrolysis and deposition reactions under those experimental conditions were only occurring on the carbon active sites [11,12].

In the present article, the study focuses on one substrate; a graphitized pitch-based carbon fiber subjected to different surface treatments prior to pyro c deposition. The treatments either enhanced the active surface area or blocked it and, as a result, the deposition kinetics changed considerably. The objective of this work is to address the importance of active sites to the CVD process, and to illustrate that regardless of the number of pyro c layers deposited on the surface, certain types of the starting sites keep replicating. Disappearance of one site during deposition is concurrently associated with the generation of a new active site.

EXPERIMENTAL

A graphitized unsized pitch carbon fiber (VSB-32), manufactured by Amoco Performance Products was used in this study. The fiber had a low level of metallic impurities, an active surface area of 0.029 m^2/g, a TSA of 0.54 m^2/g, and a geometric area of 0.18 m^2/g. The fiber diameter was 10.5 nm and its density was 2.02-2.11 g/cc.

The CVD experiments were executed using two different apparatus. The deposition was performed in a thermogravimetric analyzer (TGA) connected to a high vacuum system. A known weight of the fiber (~40 mg) was suspended on the balance beam, the system was evacuated to 10^{-5} Torr, backfilled with Ar to ambient pressure, heated to 1025°C in Ar flowing at 150 cc/min, and finally held under those conditions for 1 h. Unless otherwise specified, a mixture of 10% CH$_4$ in Ar was passed over the sample; the increase in fiber weight was monitored with time for 15 h. In some experiments prior to deposition, the fiber was subjected first to a specific treatment which will be outlined in the text.

The volumetric system displayed in Figure 1 served three purposes: to investigate deposition kinetics under static conditions at subambient pressures; to deposit a pre-calculated exact number of pyro c layers (1-60 layers) on the substrate; and to measure the active surface area of the samples before and after pyro c deposition. Details on measurements of active surface area have been outlined elsewhere [13]. In the case of successive deposition of pyro c layers, the fiber samples were evacuated at 1000°C for 2 h and a pre-calculated dose of pure CH$_4$ was injected into the reactor. The change in system pressure was monitored with time until the required number of carbon layers deposited. The excess gas was pumped away and an active surface area measurement, using O$_2$ at 350°C, was made.

Fig1: Static reactor used for CVD and ASA measurements.

RESULTS AND DISCUSSION

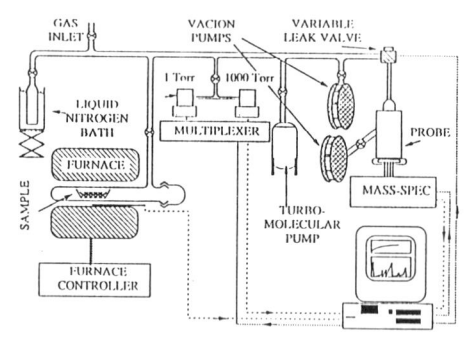

Enhancement of Deposition Rates by Surface Activation

Figure 2 illustrates the effect of surface activation on CVD rates as measured by the TGA under a flow of 30% CH$_4$ in Ar. The relationship between weight gain (due to pyro c deposition) and exposure time was linear, and the slope of each line gave the average deposition rate (mg pyro c/g fiber. min). The bottom line represents deposition on the as-received fiber when subjected to the normal pretreatment conditions described in the previous section. The top line was taken on a sample that had been pre-activated *in-situ* before the deposition reaction. While the fiber was heated isothermally at 450°C, air was admitted, and the sample lost weight due to carbon gasification. When 44% level of burn-off (BO) was achieved, air was interrupted, the system was outgassed, backfilled with Ar, and heated to 1025°C. Activating the sample in air enhanced the starting active surface area by a factor of 5 but hardly increased the TSA [14]. Yet, the CVD rate on the activated sample was only twice that on the original fiber (2.94 vs 1.52 mg/g.min). This means that not all the newly developed active surface area was contributing to CVD kinetics. Further linearity was obtained with each sample up to 120%

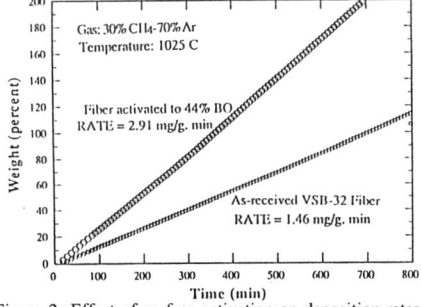

Figure 2: Effect of surface activation on deposition rates.

weight gain for the as-received fiber, or 200% weight gain for the activated fiber. This suggests that a fraction of the starting active surface area, which was contributing to CVD kinetics, either remained unchanged or kept replicating during the course of deposition.

Retardation of CVD Rates after Blockage of Active Sites

Figure 3 illustrates the change in sample weight gain as a function of time after subjecting the surface to different treatments. Line I represents the reference state when the as-received (untreated) fiber was exposed to 10% CH_4 in Ar at 1025°C. The plot is linear and a 39 % weight gain was achieved after 15 h exposure. When a different mixture with 10% CH_4-10% H_2-80% Ar was passed over another fresh fiber sample, the rate of deposition was smaller by more than one order of magnitude, and the weight gain was insignificant, as shown by line II. The presence of H_2 in the flowing mixture retarded pyro c deposition because H_2 chemisorbed on active sites and formed C-H bonds. To confirm this, three additional experiments were made (lines III, IV and V). Another fresh as-received fiber sample was held in Ar at 1025°C and then exposed for 20 h to H_2 flowing at 150 cc/min. This treatment allowed most of the active sites of the fiber to chemisorb H atoms such that the active sites were blocked with hydrogen. At the end of the 20 h hold, the sample was exposed to the standard mixture of 10% CH_4 in Ar and line III was obtained. It is noted that during the first 5 h, there was an induction period, since the weight of pyro c deposit was insignificant. The rate then gradually increased with time and finally (after 12 h) attained a constant value close to that of line I, i.e., line I and III finally became parallel. The induction period noted during the first 5 h is attributed to the beginning of gradual desorption of H_2 with concurrent liberation of a few original active sites. Between 5 and 12 h

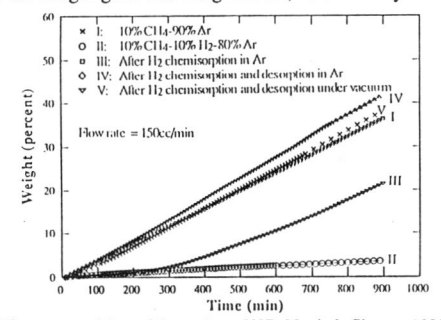

Fig 3: Deposition of Pyro C on VSB-32 pitch fiber at 1025C.

exposure, the deposition continued at an increasing rate due to the continual liberation of additional sites. In this region, the sample was gaining weight due to pyro c deposition, but losing weight as a result of hydrogen desorption. Due to the small molecular weight of hydrogen and larger atomic weight of carbon, the net result was a weight gain. After 12 h exposure, most of the effective active sites contributing to the kinetics were recovered and their contribution to the kinetics was stabilized.

The next experiment was made to support this finding. A new as-received sample was exposed to H_2 in the same manner as the previous one. Then after the 20 h hold in H_2, an ultra high pure Ar was passed over the sample replacing Ar for an additional 20 h period. The sample was finally exposed to the mixture of 10% CH_4 in Ar and line IV was obtained. Since lines I and IV are close, the deposition kinetics on both samples were probably the same. However, line IV had a slightly higher slope than I because the fiber was slightly activated in Ar. Analysis of the gas showed that it had trace amounts of oxygen (< 3ppm). With long exposure to Ar for 20 h at 1025°C, the sample was activated to a small level of burn-off. As discussed in the previous section, activating the fiber prior to deposition enhanced the CVD rate. It was then speculated that the activation process in Ar could have been eliminated if the desorption step at 1025°C was performed under vacuum. This was confirmed by performing the last run of this series of experiments. A new sample was treated in H_2 in the same manner as the last one. However, instead of subsequent treating in Ar, the sample was held at 1025°C under high vacuum (better than 10^{-5} Torr) for 20 h. The sample was quickly flushed with Ar to ambient pressure and exposed to the 10% CH_4 in Ar mixture, line V was obtained. Since lines I and V almost superimpose on each other, the kinetics of deposition and role of active sites in both cases were identical. In other words, whether the deposition is made on a fresh untreated fiber, or on a sample whose active sites are first blocked then regenerated, the CVD rates are practically the same, provided that the fiber surface was not subjected to significant activation.

Development of Active Sites by Oxidation

The population of active sites at the surface of the fiber was enhanced in two different ways, by treatment in an oxygen plasma or by high temperature oxidation (HTO) in air at 600, 700, or 925°C. The plasma treatment was performed using a Branson/IPC unit (S3000 Series) with a RF power=100 W and O_2 flow rate of 100 cc/min. While Figure 4 (A and B) illustrates the variation of active surface area with burn-off using the two treatments, Figure 5 correlates the term 100xASA/TSA (which is the % of surface active) to burn-off. Activating the fiber to 2% BO enhanced the active surface area by a factor of 5 and raised the % of surface active from 4.6 to ~25, regardless of the type of surface treatment. At higher levels of burn-off, the increase in active surface area with burn-off was linear with both types of activation (Figure 4, A and B). Yet, at comparable levels of burn-off, the activation in plasma was more effective in generating a larger number of active sites than HTO; as shown by lines A and B in Figure 4. It may be added that the plasma treatment was also more effective than HTO in enhancing the total Surface Area. The net result was that at comparable levels of burn-off, the % of surface active

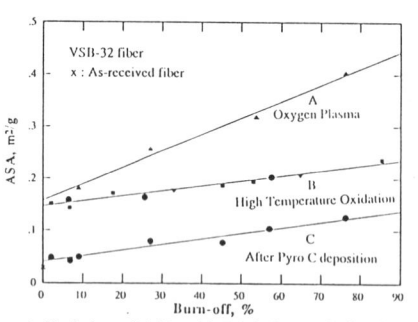

Figure 4: Variation of ASA with BO before and after Pyro C deposition.

with oxygen plasma and HTO was the same as shown in Figure 5; all the data points conformed to one plot. Thus, regardless of the level of burn-off achieved above 2 %, or the method of activation, the maximum % of surface active, which may be utilized in the chemical vapor deposition process, is constant at ~ 25.

Replication of Specific Types of Active Sites after Active Surface Area Deposition

When the first layer of pyro c deposited on the surface, a fraction of original active sites was instantaneously blocked and did not contribute any further to deposition kinetics. For the as-received fiber and the samples activated by HTO or in the plasma, the variation of active surface area with burn-off after deposition of the first layer of pyro c conformed to one plot; line C of Figure 4 which is practically parallel to line B. The drop in active surface area from line B, which belongs to the samples activated by HTO, to C was constant between burn-off of 2 and 85 %. This means that for a given burn-off above 2%, there are three types of active sites: Type I whose area is given by the original active surface area of the as-received fiber before activation, Type II which is equivalent to the additional (new) active surface area developed up to 2% BO, and Type III which develops above 2% BO. The original Type I sites are exclusively present at the external surface because the as-received fiber is nonporous. These sites kept replicating during the course of deposition.

Next we consider the deposition of successive pyro c layers. The active surface area of the as-received fiber after deposition of 1, 2, and 85 layers of pyro c were 0.02, 0.02, and 0.04 m²/g, respectively. Their average value is 0.03 m²/g which is equal to the original active surface area of the as-received fiber. Therefore, the area of the active sites of Type I is approximately 0.03 m²/g. For the sample activated by HTO to 2% BO, the starting active surface area dropped from 0.15 to 0.04 m²/g after deposition of the first layer. The active surface area then remained constant at 0.04-0.03 m²/g after further deposition of pyro c up to 62 layers. Similarly, for the sample with 9% BO, whose starting active surface area was 0.18 m²/g, the active surface area dropped to 0.04-0.05 m²/g after deposition of additional pyro c up to 52 layers. Thus, the drop from line B to C is equivalent, on the average, to 0.12-0.13 m²/g, which is the additional area of active sites developed at 2% BO. Therefore, with the samples subjected to HTO, the constant area of Type II sites is approximately 0.12-0.13 m²/g and is independent of the level of burn-off. However, the area of Type III sites, which is the additional active surface area developed above 2% BO, increases with increasing level of activation.

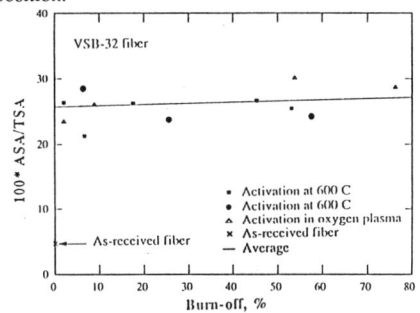

Figure 5: Dependence of percent surface active on burn-off.

As for the samples activated in the oxygen plasma, neither the area of Type II nor Type III sites is constant because lines A and C are not parallel. The number of Type II and III sites increases with increasing the level of burn-off. This outlines the main difference between the activation by HTO or inside the plasma. In spite of this difference, there is some commonality between the two treatment methods; the active surface area of Type III sites at a given burn-off is the same. For example, after 27% BO in the plasma, the active surface area was enhanced to 0.243 m²/g, which is higher than the active surface area developed by HTO at the same level of burn-off. After the deposition of the first pyro c layer on the sample, the active surface area dropped to 0.07 m²/g. This value remained constant at 0.08-0.07 m²/g after the deposition of additional pyro c up to 56 layers. This area represents the combined areas of Type I and III sites at 27% BO. It is worth adding that with this particular plasma activated sample, the active surface area of Type II sites is 0.16-0.17 m²/g. Knowing that after pyro c deposition, the data points for all the samples activated by both methods conformed to one line, we conclude that the area of Type III sites is only dependent on level of burn-off but not on type of treatment. The area of Type II sites is either constant after activation by HTO, or increases with burn-off after treatment in the plasma.

SUMMARY AND CONCLUSIONS

1. For pyro c deposition on the surface of the graphitized pitch carbon fiber at 1025°C and under a flow of CH_4/Ar mixture (ratio 1 : 9), the presence of active sites is needed for the reaction to proceed. If these sites are blocked, the deposition rate becomes very small.

2. Under the present experimental CVD conditions, the heterogeneous gas/solid reaction appears predominant, the CVD mechanism on the graphitized substrate is probably controlled by collision between reactive gaseous species and the active sites of the carbon. The contribution of the gas phase reactions to deposition rates is insignificant at 1025°C.

3. Depending on the method of surface treatment, the fiber surface has different types of active sites. The as-received fiber has only Type I sites. The samples activated by different methods to different levels of burn-off have Type II and III sites, in addition to Type I.

4. The area of Type II sites, which develops at the early stages of activation, depends on the method of activation and level of burn-off. This area, which is probably located inside pores, does not contribute to the CVD mechanism after deposition of the first pyro c layer.

5. The active surface area of Type III sites is dependent on the level of burn-off but independent of the method of activation. Apparently this area contributes along with Type I sites to CVD kinetics.

ACKNOWLEDGMENT:*The financial support of Air Force Office of Scientific Research under contract F06411-83-C-0046 with Phillips Laboratory is appreciated. I would like to thank Dr. Wesley P. Hoffman for helpful technical discussion.*

REFERENCES

1. J. C. Bokros, Chemistry and Physics of Carbon, Volume 5, Edited by P. L. Walker, Jr, (Marcel Dekker, Inc., New York, 1969), p. 1.
2. J. H. Je and Jai Young Lee, Carbon 22, 563 (1984).
3. W. V. Kotlensky, Chemistry and Physics of Carbon, Volume 9, Edited by P. L. Walker, Jr and P. A. Thrower (Marcel Dekker, Inc., New York, 1973), p. 173.
4. S. Marinkovic and S. Dimitruevic, Carbon 23, 691 (1985).
5. J. L. Kaae, Carbon 23, 665 (1985).
6. P. A. Tenser, Chemistry and Physics of Carbon, Volume 19, Edited by P. A. Thrower (Marcel Dekker, Inc., New York, 1975), p. 19.
7. J. Goma and A. Oberlin, Carbon 23, 85 (1985).
8. J. Goma and A. Oberlin, Carbon 24, 135 (1986).
9. I. M. K. Ismail, Carbon 27, 958 (1989).
10. I. M. K. Ismail, M. M. Rose and M. A. Mahowald, Carbon 29, 575 (1990).
11. W. P. Hoffman, F. J. Vastola and P. L Walker, Jr., Carbon 23, 151 (1985).
12. W. P. Hoffman, F. J. Vastola, P. L. Walker, Jr., Carbon 26, 485 (1988).
13. I. M. K. Ismail, Carbon 25, 653 (1987).
14. I. M. K. Ismail, Carbon 26, 749 (1988).

PART II

In Situ Diagnostics

SURFACE CHEMISTRY OF FLUORINE-CONTAINING MOLECULES RELATED TO CVD PROCESS ON SILICON NITRIDE: SiF$_4$, XeF$_2$, AND HF*

DUANE A. OUTKA
Sandia National Laboratories, Livermore, CA 94551

ABSTRACT

The reactivity of several fluorine-containing molecules on a polycrystalline silicon nitride (Si$_3$N$_4$) surface is studied under ultrahigh vacuum (UHV) conditions using temperature programmed desorption (TPD) and Auger electron spectroscopy (AES). The chemistry of fluorine on Si$_3$N$_4$ is of interest in understanding the high temperature chemical vapor deposition (CVD) of Si$_3$N$_4$, which uses SiF$_4$ as a starting material. XeF$_2$ is reacted with a Si$_3$N$_4$ surface to prepare and characterize various surface SiF$_x$ ($1 \leq x \leq 3$) species. These are identified by the chemical shift induced by the fluorine atoms in the Si (LMM) Auger peak and by changes in the TPD. Of these species, SiF$_2$ is stable to the highest temperature. SiF$_2$ is also formed by the reaction of SiF$_4$ with a Si$_3$N$_4$. Because SiF$_2$ is so stable, its decomposition is proposed as a rate-determining step in the CVD deposition of Si$_3$N$_4$ from SiF$_4$. Gaseous HF, which is a product of the CVD process, does not dissociate on Si$_3$N$_4$ and is therefore unlikely to cause the etch-like marks on the Si$_3$N$_4$ coating that are observed under certain conditions.

INTRODUCTION

Silicon nitride (Si$_3$N$_4$) is being developed as a high-temperature oxidation barrier for lightweight, carbon-composite parts in turbine engines [1]. Currently, such parts can be coated in a small-scale CVD reactor from starting materials of silicon tetrafluoride (SiF$_4$) and ammonia (NH$_3$). The CVD reaction is conducted at 1700 K and the overall reaction is:

$$3 \text{ SiF}_4 \text{ (g)} + 4 \text{ NH}_3 \text{ (g)} \rightarrow \text{Si}_3\text{N}_4 \text{ (s)} + 12 \text{ HF (g)}. \tag{1}$$

The silicon-fluorine chemistry is of particular interest in this CVD process because SiF$_4$ is a starting material and the Si-F bond is the strongest bond that must be broken, up to 167 kcal mol^{-1} [2]. This study examines the reactivity of a variety of fluorine-containing molecules on the Si$_3$N$_4$ surface in order to gain insight into this CVD process. No previous UHV surface studies involving bulk Si$_3$N$_4$ have been reported.

EXPERIMENTAL

The sample was a polycrystalline β-Si$_3$N$_4$ obtained from Allied Signal Aerospace company. It was 1.3 cm in diameter and 1.5 mm thick. Groves (approx. 0.38 mm wide) were cut into the edge of the sample with a diamond saw in order to hold the sample and thermocouple. For heating, a 0.3-mm tungsten wire was formed into a flat zig-zag filament and placed against the back of the sample. This was stiff enough to hold the sample when the ends of the filament were bent to lie in the groves cut into the sample. For temperature measurement, a chromel-alumel thermocouple (0.125 mm diameter wire) was spot welded to a 0.125 mm thick Ta foil that was folded and pressed into the grove cut into the sample.

The Si$_3$N$_4$ sample was cleaned by bombardment with 5 keV argon ions. Then the sample was heated to 1300 K or higher. Si$_3$N$_4$ begins to thermally decompose at these temperatures, and heating provided a consistent thermal treatment and reproducible Si to N AES ratio. Si$_3$N$_4$ is an insulator, but AES could be performed by lowering the primary beam voltage to 1 keV. The Auger spectrum of the clean Si$_3$N$_4$ surface is shown in Fig. 1.

---

* This work is supported by DARPA and the U. S. Department of Energy under contract number DE-AC0476DP00789.

In the TPD setup, a conical shield was used to suppress interference from atoms and molecules desorbing from sample supports. The heating rate was approximately 4.5° C sec^{-1}. During TPD, several values of m/e (mass to charge ratio) were monitored simultaneously by controlling the mass spectrometer with a computer.

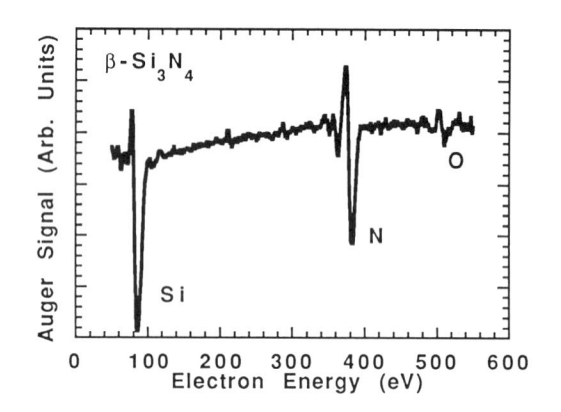

Fig. 1

Auger spectra of clean β-Si$_3$N$_4$. The principal elements on the surface was Si (87 eV) and N (381). Trace amounts of O (506), Ar (213), and C (272) were also present.

XeF$_2$ RESULTS

Although gaseous XeF$_2$ is not involved in the CVD process, it is useful in the study of fluorine chemistry on Si$_3$N$_4$ because it allows various SiF$_x$ ($1 \leq x \leq 3$) surface species to be prepared and their stability studied. XeF$_2$ is a fluorinating agent which decomposes on Si$_3$N$_4$ to deposit fluorine. The Xe is desorbed under the conditions of exposure and does not interfere. The exposure to XeF$_2$ was performed with the gas at room temperature and with the surface heated to 400 K and above.

Following exposure to XeF$_2$, fluorine was adsorbed on the Si$_3$N$_4$ surface at Si atom sites as demonstrated by AES and TPD. The fluorine AES peak at 690 eV could not be observed because of the low primary beam energy of 1 keV, which was used to minimize sample charging. The presence of fluorine could be inferred, however, from a shift to higher kinetic energy of the Si (LVV) AES peak due to the proximity of the fluorine atoms (Fig. 2). This shift is due either to the electronegativity of the fluorine or surface-induced perturbation of the Si (LVV) spectrum [3]. No such shift was observed in the nitrogen AES peak, indicating that fluorine bonds to the surface only via Si. The presence of fluorine on the surface was confirmed by TPD and the desorption of fluorine containing molecules.

The AES spectra indicate the presence of three SiF$_x$ ($x \leq 3$) fragments. SiF$_3$ and SiF are most apparent at low temperatures, while SiF$_2$ is most apparent at high temperatures and therefore is the most stable. Fig. 2 shows the Si AES peak as a function of surface temperature during XeF$_2$ exposure. Depending upon exposure conditions, several Si AES peaks are observed from 87 to 96 eV in 3 eV increments. The results are interpreted by assuming that the Si peak shifts 3 eV to higher energy for each adjacent fluorine atom. For example, the clean Si$_3$N$_4$ surface has a Si peak at 87 eV. Upon exposure to XeF$_2$ at temperatures below 800 K, the Si peak splits and shifts to 90 and 96 eV, which are attributed to SiF and SiF$_3$ surface species, respectively. In contrast, exposure to XeF$_2$ at temperatures of 800 K and above yields two Si peaks at 87 and 93 eV, which are attributed to Si$_3$N$_4$ and SiF$_2$, respectively.

These changes in the AES are paralleled by changes in the TPD spectrum following exposure to XeF$_2$ (Fig. 3). The interpretation of the TPD is complicated, however, because of rearrangement of ions in the mass spectrometer [4] and because decomposition and recombination reactions may result in desorption products that differ from the original surface species. There is evidence for a change in surface composition above 1000 K, however. Below 1000 K the dominant desorption product is SiF$_3$ (m/e = 85), which is probably a cracking fragment of SiF$_4$ (m/e = 104) [4]. The desorption of SiF$_4$ is attributed to the decomposition and recombination of SiF and SiF$_3$ surface species to form SiF$_4$, which readily desorbs. Above 1000 K SiF$_2$ (m/e = 66) appears as a new desorption product along with SiF$_3$ (m/e = 85) (Fig. 3). The appearance of SiF$_2$ in the TPD suggests that SiF$_2$ is present on the surface and that a new pathway for removal of fluorine from the surface becomes available, the direct desorption of SiF$_2$. A signal for SiF$_3$ (m/e = 85) is still observed either because the SiF$_2$ groups undergo simultaneous rearrangement to form SiF$_4$ which desorbs or because of recombination of SiF$_2$ with F on the walls of the chamber or in the mass spectrometer ionizer. Nevertheless, the TPD indicates a change in surface composition at 1000 K which is consistent with the AES results.

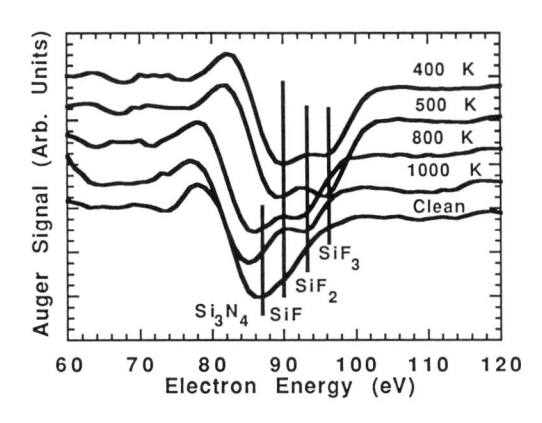

Fig. 2

Effect of XeF$_2$ exposure temperature on the Si Auger line. At temperatures below 800 K, SiF$_3$ and SiF are the predominant surface species. At 800 K and above, SiF$_2$ is the predominant surface species.

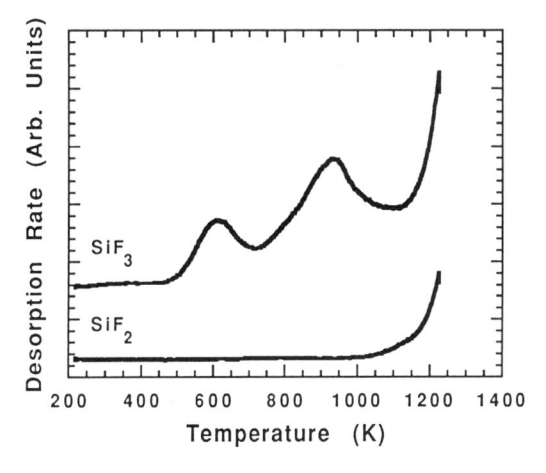

Fig. 3

TPD following exposure of Si$_3$N$_4$ to XeF$_2$. SiF$_3$ (m/e = 85) is the major fluorine-containing product desorbed below 1000 K. This is probably a cracking fragment of SiF$_4$. SiF$_2$ (m/e = 66) is observed as a new desorption product above 1000 K.

The transition to SiF$_2$ occurs as somewhat different temperatures in AES and TPD, because the heating time for the two experiments differ. That is, in the AES experiments, the sample was annealed for several minutes at each temperature while during TPD the sample was quickly flashed. The important conclusion from these results is that SiF$_2$ is the most stable surface species on Si$_3$N$_4$.

SiF₄ RESULTS

SiF$_4$ dissociatively adsorbs on Si$_3$N$_4$ to deposit Si and F containing species, which can be identified based upon the XeF$_2$ results. There is no evidence for molecular adsorption of SiF$_4$ on Si$_3$N$_4$ down to surface temperatures of 100 K. Instead, the fluorine that is deposited on Si$_3$N$_4$ by SiF$_4$ is stable to over 1000 K, which indicates that SiF$_4$ dissociates on Si$_3$N$_4$.

Fig. 4 shows the changes in the AES following exposure of Si$_3$N$_4$ to SiF$_4$. Again, only the Si peak is affected, which indicates that the nitrogen is not directly participating in the bonding of the F to the surface. Following exposure to SiF$_4$, the Si peak shifts from 87 eV to 93 eV, which indicates, by comparison to the XeF$_2$ results above, that SiF$_2$ is the principal species on the surface. The Si peak also grows with respect to the N peak indicating that Si is deposited on the surface as well as F. Upon heating to 1500 K, the Si AES peak returns to 87 eV indicating that the fluorine has been removed from the surface.

The removal of fluorine from the Si$_3$N$_4$ surface is indicated in the TPD by a desorption peak centered at about 1400 K (Fig. 5). This peak corresponds to the desorption that just begins to rise at 1000 K in Fig. 3 and is associated with decomposition and desorption of surface SiF$_2$. A simple Arrhenius analysis of this peak, which assumes coverage-independent, first-order desorption and a pre-exponential factor of 1 x 10^{13} s^{-1}, yields an activation energy of 92 kcal mol^{-1}. The experiments in Fig 5 were carried to a higher temperature than those in Fig. 3 to show the entire desorption peak at the expense of burning-out the heater.

The reaction of SiF$_4$ differs from that of XeF$_2$, however, in that only SiF$_2$ is formed on the surface, regardless of temperature. Exposure of Si$_3$N$_4$ to SiF$_4$ for surface temperatures from 500 to 1000 K yielded only results similar to those shown in Figs. 4 and 5 with no evidence for SiF or SiF$_3$ formation such as observed with XeF$_2$ above.

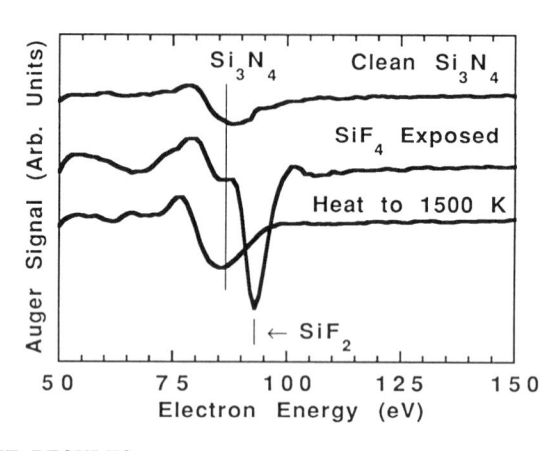

Fig. 4

Effect of SiF$_4$ exposure on the Si Auger line. The top spectrum is the clean Si$_3$N$_4$ surface, the second is following exposure to SiF$_4$, and the bottom is following heating to 1500 K. SiF$_2$ is the principal surface species formed upon exposure to SiF$_4$.

HF RESULTS

The reaction of HF with Si$_3$N$_4$ is of interest because it is a product of the CVD reaction and has been proposed to account for "etch" lines sometimes observed on parts near the exit of the reactor. Therefore, the adsorption of HF was examined to determine whether HF would react with Si$_3$N$_4$.

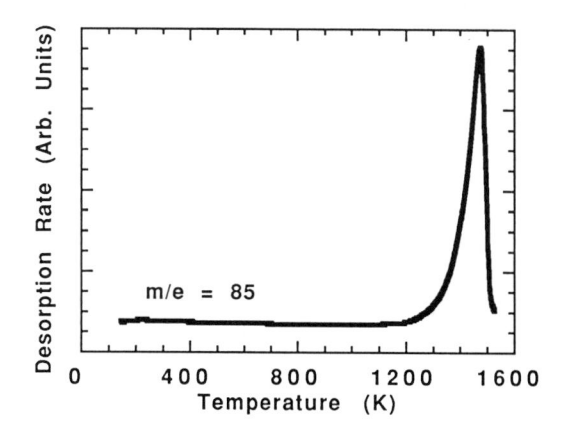

Fig. 5

TPD following exposure of Si_3N_4 surface to SiF_4. The peak is identified with the decomposition of SiF_2 surface species. Desorption of m/e=85 (shown) and m/e=66 (not shown) are both observed with the same shape.

HF does not dissociate on Si_3N_4 for surface temperatures up to 1000 K. Instead, only molecular adsorption of HF is observed on Si_3N_4 if the surface is cooled to 100 K. HF desorbs at 150 K in the TPD (Fig. 6). This indicates that HF will not react with Si_3N_4 under CVD conditions and is not responsible for etch lines observed on parts coated in the CVD reactor under certain conditions.

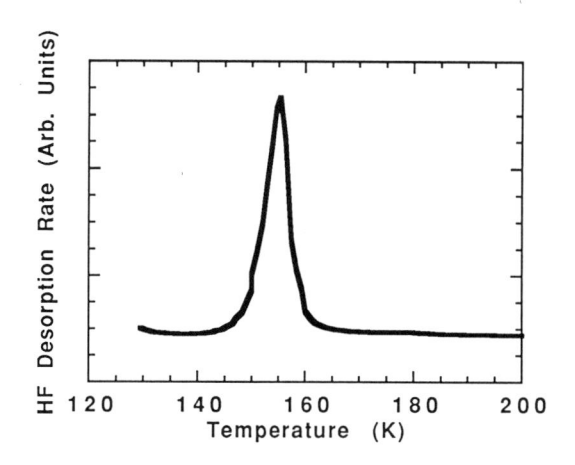

Fig. 6

TPD following exposure of Si_3N_4 surface to HF. The only desorption product observed is HF (m/e=20), which indicates that HF adsorption is molecular.

DISCUSSION

The study of fluorine-containing molecules on Si_3N_4 was undertaken to provide insight into the high temperature CVD deposition of Si_3N_4 from SiF_4. Based upon these UHV results, several predictions can be made about the mechanism of the CVD process.

The Si-F bond is the strongest bond involved in the CVD reaction, and breaking of this bond is likely to be the rate-determining step of the CVD process. Despite the strength of the Si-F bond, the dissociative adsorption of SiF_4 on Si_3N_4 proceeds at modest temperatures, 500 K.

This indicates that the dissociative adsorption of SiF$_4$ on Si$_3$N$_4$ is not a rate-determining step in the CVD process.

The present results suggest instead that the decomposition of surface SiF$_2$ is the rate determining step of the CVD process since this is the most stable intermediate that has been observed on Si$_3$N$_4$. SiF$_2$ has an activation energy for desorption of approximately 92 kcal mol^{-1}. If decomposition of surface SiF$_2$ is rate-determining, then the kinetics for the CVD process would be independent of the gas-phase concentrations because no gas phase species are involved in the rate-determining step. This is provided the pressure is not so low that the impingement rate of gases on the surface becomes rate-determining. That is, at low pressures, < 10^{-4} Torr, the CVD rate will be linear with pressure because it is limited by the impingement rate, but at higher pressures, the rate will plateau as the decomposition of surface SiF$_2$ become rate-determining. The CVD is typically performed at 1 Torr and 1700 K, which is calculated to be within the plateau region, assuming a sticking probability of 10^{-3} and the desorption parameters discussed earlier.

The adsorption of SiF$_4$ on Si$_3$N$_4$ is similar in several respects to the adsorption of SiF$_4$ on silicon. Only two papers have previously studied SiF$_4$ adsorption and these are on silicon [5, 6]. In both cases, SiF$_4$ dissociates at room temperature or below. Furthermore, in the study on Si(111)(7x7) [6], molecular adsorption of SiF$_4$ was observed only below 90 K, which is consistent with the absence of molecular adsorption on Si$_3$N$_4$ at surface temperatures of 100 K and above. Chuang [5] also reports that SiF$_2$ is formed following the adsorption of SiF$_4$ on an unspecified silicon surface, which is the same result observed here on Si$_3$N$_4$. On Si(111)(7x7), however, SiF and SiF$_3$ are the observed dissociation products. This difference may be due to different experimental conditions. For example, the experiments on Si(111)(7x7) were performed at low temperature, 30 K, whereas the results of Chuang and the present results on Si$_3$N$_4$ were performed at 300 K and above. Of course, it may simply be the case that different intermediates are formed on Si$_3$N$_4$ and on different silicon surfaces.

The reactivity of SiF$_x$ groups observed here on Si$_3$N$_4$ also resembles in some respects the behavior of SiF$_x$ on silicon surfaces. In particular, the TPD spectrum of Fig. 3 with two SiF$_3$ peaks and a higher temperature SiF$_2$ peak is quite similar to the TPD observed on Si(100) following F atom exposure [7]. The peaks occur at 200-300 K higher temperature on Si$_3$N$_4$, however, indicating that the surface species are more stable on Si$_3$N$_4$. It is also interesting to note that a study of XeF$_2$ adsorption on a variety of silicon surfaces at room temperature found that SiF and SiF$_3$ are the predominant surface species [8]. This is the same result observed here following XeF$_2$ exposure below 800 K. The behavior of fluorine on silicon at high temperatures, above 800 K, has not been investigated in detail so a comparison of whether SiF$_2$ becomes the predominant surface species on silicon at high temperature such as occurs here on Si$_3$N$_4$ cannot be made at this time.

REFERENCES

1. J. R. Strife and J. E. Sheehan, Am. Ceramic Soc. Bull., 67 369 (1988).

2. P. Ho and C. F. Melius, J. Phys. Chem., 94 5120 (1990).

3. S. M. Durbin and T. Gog, Phys. Rev. Lett., 63 1304 (1989).

4. M. J. Vasile and F. A. Stevie, J. Appl. Phys., 53 3799 (1982).

5. T. J. Chuang, J. Appl. Phys., 51 2614 (1980).

6. C.-R. Wen, S. P. Frigo, and R. A. Rosenberg, Surface Sci., 249 117 (1991).

7. J. R. Engstrom, M. M. Nelson, and T. Engel, Phys. Rev. B, 37 6563 (1988).

8. F. R. McFeely, *et al.*, Phys. Rev. B, 30 764 (1984).

DETERMINATION OF THE ROUGHNESS OF CVD SURFACES BY LASER SCATTERING

MAX KLEIN AND BERNARD GALLOIS

Department of Materials Science and Engineering
Stevens Institute of Technology, Hoboken, NJ 07030

ABSTRACT

A laser scattering apparatus was developed for the determination of surface roughness and other surface statistical parameters of chemically vapor-deposited coatings. Visual examination of HeNe laser scattering patterns reflected from polished sapphire and CVD titanium nitride surfaces showed a sensitivity to roughness differences of tens of nanometers. The scattering apparatus was integrated with a cold-wall CVD reactor. The root mean square roughness of silicon carbide deposits on silicon in the early stages of growth was determined from the intensity of the specularly reflected beam. Changes in roughness and the spatial arrangement of depositing crystallites were monitored *in situ* by angular resolution of the scattered light spectra. Both *ex situ* and *in situ* results were in good agreement with profilometric examinations of the rough surfaces.

INTRODUCTION

An understanding of the microstructural development of coatings deposited by chemical vapor deposition (CVD) is essential to the control of the properties of such coatings. For instance, the steady growth of columnar grains typical of deposits from the vapor has been shown not to originate at the substrate/coating interface in certain cases of titanium carbide and titanium nitride growth on vitreous graphite [1]. This phenomenon has been exploited to grow fine-grained thick coatings [2]. A method for observing changes in the growing surface is desirable to provide the capacity for in-process modification of the deposit and hence, its properties.

Previously, laser light-scattering techniques have been used to inspect the surface finish of manufactured parts and the results related to stylus profilometric measurements [3,4]. Therefore, laser light scattering has been proposed as a non-intrusive, *in situ* technique for monitoring the growth of CVD coatings [1,5-7] since changes in microstructural development are manifested in the surface morphology. In this work, the effectiveness of light scattering in describing the surfaces of well-characterized samples *ex situ* and of early CVD deposits *in situ* is demonstrated.

EXPERIMENTAL DETAILS

The laser scattering apparatus is shown in Figure 1, integrated with a cold-wall CVD reactor. The laser scattering apparatus consisted of a 5 mW HeNe laser whose intensity was attenuated by a combination 1/2-wave plate and beamsplitting/polarizing cube. The resulting s-polarized beam was passed through a spatial filter to remove intensity fluctuations in the beam profile, and was focused on a detector by the re-collimating lens. For preliminary studies, the beam was reflected from samples mounted *ex situ* on a rotatable stage with which the angle of beam incidence was varied.

Initially, the detector was a lensless, 35 mm single-lens reflex camera which captured

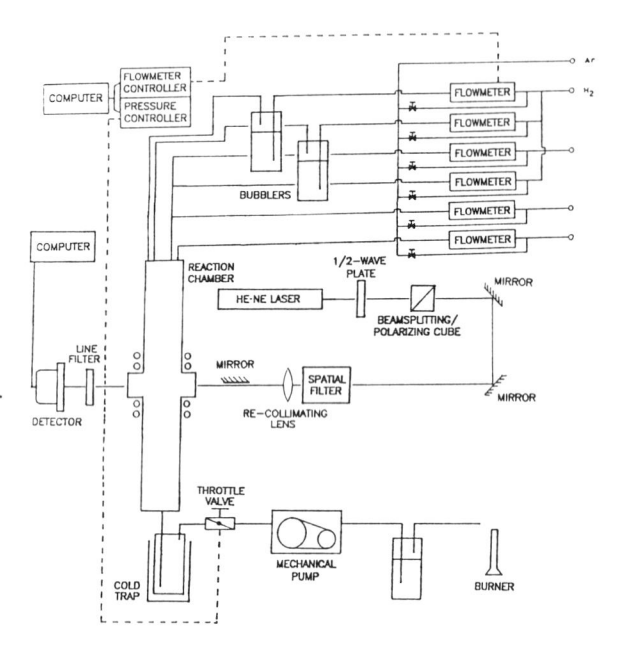

Fig. 1. Schematic diagram of the laser scattering apparatus integrated with the cold-wall CVD reactor.

laser scattering patterns on photographic film. Later, the camera was replaced by a linear 512-element photodiode array which detected the intensity of the scattered light as a function of photodiode position. A 25 μm slit was placed across the detector to create 25 μm-square detection elements on 25 μm centers. The slit was positioned such that measurements were made in the plane of incidence, that is, the plane containing both the incident and specularly reflected beams. The computer interface to the detector was capable of recording scattered spectra every 0.015 seconds, to a total of 255 spectra per experiment. A narrow band-pass filter was placed ahead of the detector to block spurious background radiation.

The laser scattering apparatus was then mounted on the cold-wall CVD reactor for *in situ* studies as shown in Figure 1. Front surface-reflecting dielectric mirrors directed the beam to the substrate surface at a predetermined angle of incidence.

The cold-wall reactor was used to deposit series of timed silicon carbide coatings on silicon for *ex situ* and *in situ* studies. Silicon wafer substrates were heated on a graphite susceptor by radio frequency induction with a 450 kHz RF generator. The reaction chamber was a quartz tube modified by the addition of quartz windows in the reaction zone to provide an entrance and exit for the laser beam. The windows were offset from the chamber walls to prevent clouding.

The reactant gases, hydrogen and methyltrichlorosilane (MTS) were supplied by a metered flow of hydrogen through a liquid bubbler of MTS which, when supplemented by additional hydrogen, yielded a total input gas ratio of H_2:MTS of 400:32.5 sccm at 26.5 kPa (200 torr). Computer control of gas flows, temperature and pressure enabled accurate commencement and termination of each experiment.

RESULTS AND DISCUSSION

The effect of a rough surface on the angularly resolved distribution of light scattered from the surface depends on the nature of the roughness. On an intuitive level, rougher surfaces give rise to more diffuse scattering of the reflected light. A demonstration of this principle is shown by the scattering patterns of Figure 2. Figures 2 (A) and 2 (C) are photographs of laser light scattered from polished sapphire substrates whose surface profiles are shown in 2 (B) and 2 (D), respectively. The peak-to-trough roughnesses of the sapphire substrates are seen to differ by less than 10 nm with the consequence of increasingly diffuse scattering by the rougher surface. Typical surface morphologies of CVD titanium nitride exhibiting mean grain sizes of 0.9 μm and 2.3 μm are shown in Figures 2 (F) and 2 (H), respectively. The corresponding scattering patterns from the surfaces are shown in Figures 2 (E) and 2 (G). A striking difference in scattering pattern is observed between the two samples, which display considerable speckle contrast. Note that the Airy rings about the central (specular) beam are the result of the spatial filter used and are not a result of laser scattering. The relative distortion of the rings, however, may be considered another indication of roughness.

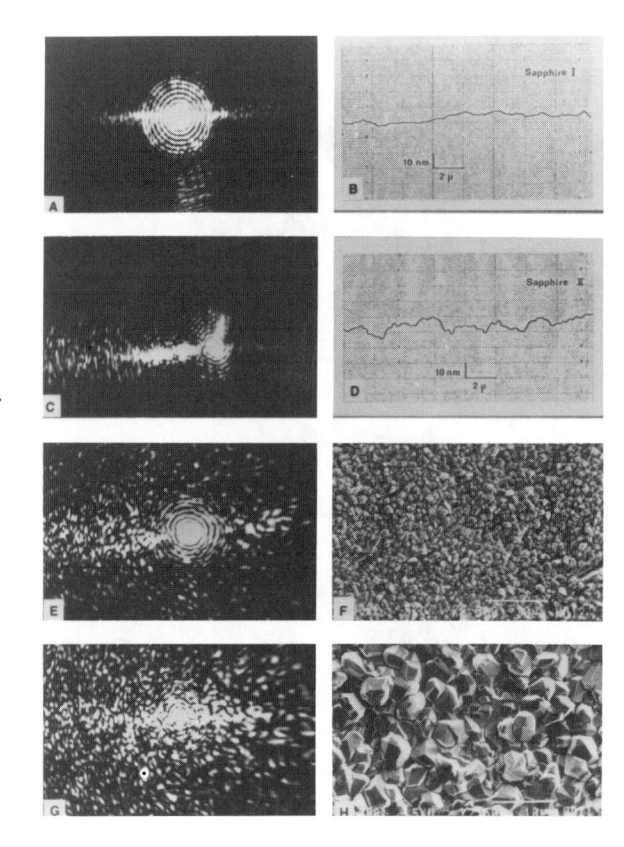

Fig. 2. (A) and (C): Photographs of scattering patterns from sapphire surfaces (B) and (D).

(E) and (G): Photographs of scattering patterns from CVD titanium nitride surfaces (F) and (H).

The relationship between the intensity of the diffuse component of the scattered light and the root mean square (rms) roughness of a surface has been quantified [8]:

$$\frac{I}{I_o} = \exp[-((4\cdot\pi\cdot\sigma_{rms}\cdot\cos\theta_i)/\lambda)^2] \qquad (1)$$

where σ_{rms} is the rms roughness and λ is 0.6328 µm, the wavelength of the incident light. Also, I is the intensity of the specular reflection from the rough surface, I_o is the total intensity of all light reflected from the surface, and θ_i is the angle of beam incidence.

This scalar approach has been applied to electromagnetic scattering from surfaces with a Gaussian height distribution [9] and σ_{rms} of the order of 1 µm [10,11]. To be exact, I_o should be measured by a Coblentz or collecting sphere. For practical purposes, however, I_o can be taken to be the intensity of the specular reflection from a 'perfectly' smooth surface of the same material as the rough surface [10]. In this case, the smooth surface was a clean silicon wafer substrate.

Root mean square roughness measurements were taken on the series of silicon carbide-on-silicon deposits made at 1273 K, shown in Figure 3. The micrographs show the progress of the coating from individual submicrometer-sized clusters at 15 seconds to a

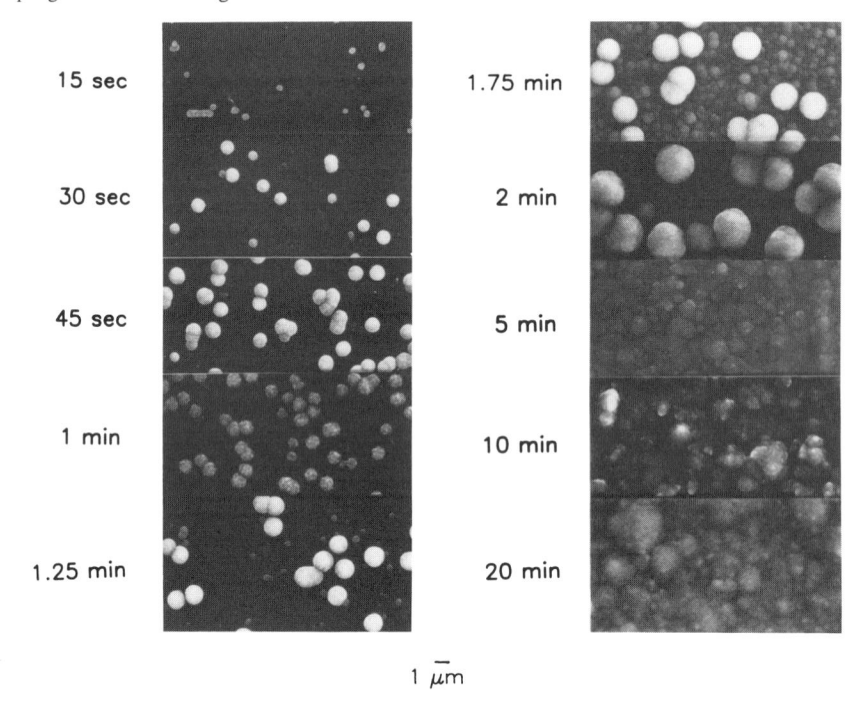

Fig. 3. Scanning Electron Micrographs showing development of the morphology of silicon carbide deposited on silicon at 1273 K.

Fig. 4. Specular beam profiles scattered *ex situ* from silicon carbide surfaces for measurement of rms roughness.

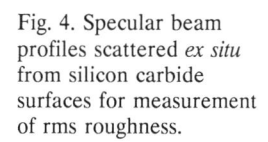

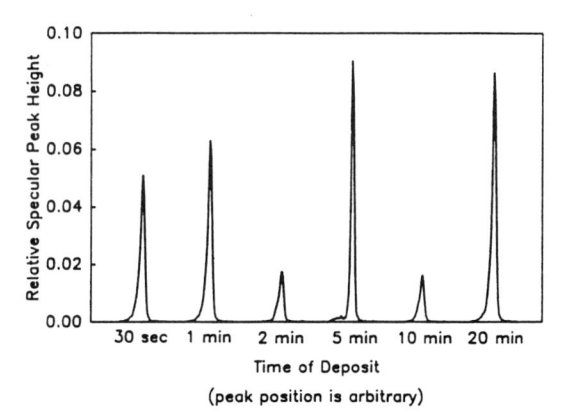

Fig. 5. Comparison of rms roughnesses obtained *ex situ* by profilometry and specular reflectance with *in situ* specular reflectance results.

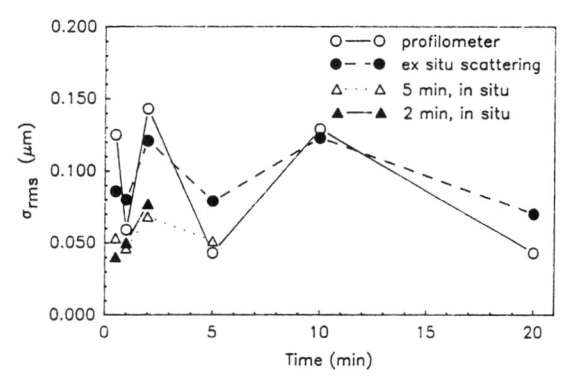

continuous deposit of nodular morphology after 2 minutes. The relative peak heights of the specular beam reflected from the 30 second, 1, 2, 5, 10, and 20 minute deposits, which are representative of a variety of surface morphologies, are shown in Figure 4. The intensity of each reflection, I, was determined from the integrated area under the peak, and σ_{rms} was calculated.

Profilometer traces covering at least 50 μm of each surface were also taken, and the standard deviation of vertical height displacement from the mean, which is the definition of σ_{rms}, was calculated. The results of these *ex situ* measurements are compared in Figure 5. Both methods indicate a similar trend with good agreement. The largest discrepancy is at 30 seconds, when the separation of clusters on the surface is greatest. The apparent 'bracketing' of the roughness values obtained from scattering by the profilometer measurements may be the result of a bandwidth limitation placed on the detected scattered intensities by the geometry of the scattering apparatus, which includes the size of the slit across the detector array.

A more complete description of the surface statistical parameters of a rough surface is possible through a vector scattering treatment of the angular distribution of scattered light from the surface. For a surface whose roughness falls within a 'smooth-surface' limit, that is $\sigma_{rms} \ll \lambda$, the intensity distribution of scattered light can be directly related to the Power Spectral Density (PSD) of the surface. The PSD, being the square magnitude of the Fourier transformation of the surface profile, contains information about both the vertical and horizontal character of the surface.

If the surface features are sufficiently isotropic in arrangement, the incident beam will be scattered with intensity maxima corresponding to the spatial frequencies (f_x) of surface features, since the surface can be considered to be a composite of many sinusoidal diffraction gratings. The positions of the scattered maxima are then those that would be predicted by the familiar grating equation for the plane of incidence:

$$f_x = \frac{\sin \theta_s - \sin \theta_i}{\lambda} \qquad (2)$$

where θ_s is the angle to the scattered peak. The vector scattering analysis also allows the intensities of the maxima to yield information about the rms roughness variance and rms slope variance of the scattering features by evaluating the zeroth and second moments of the PSD, respectively [3,12].

In situ experiments to explore both the scalar and vector approaches to the scattering problem were performed on silicon carbide deposits, using the scattering apparatus integrated with the cold-wall reactor. A sequence of scattering patterns from the sample surface was collected during the early stages of growth of each coating. The specularly reflected beam was captured in these studies to determine surface roughness from the change in intensity of the specular beam. An incident beam of higher intensity was also used to explore the generation of peaks far from the specular beam due to the spacing of the rough surface features.

Excerpts from the series of spectra taken near the specular beam during a 5-minute *in situ* experiment are shown in Figure 6. The spectra are presented in perspective view with scattered intensity plotted against position on the photodiode array and increasing time. The specular beam, seen to the right of the frame, is truncated because its intensity exceeded the capacity of the photodiodes. The change in rms roughness was calculated from the variation in intensity of the specular beam by fitting a Gaussian intensity profile to each truncated peak, and integrating the intensities to obtain σ_{rms} according to Equation (1). *In situ* roughnesses were calculated for the same time intervals as the *ex situ* studies and are shown for the 5-minute experiment and a 2-minute experiment in Figure 5. The agreement among the *in situ* studies is good, and the trend is very similar to the *ex situ* results. The magnitudes of the roughnesses obtained by the two sets of experiments differ only by a scaling factor, which may be related to an error in the initial calculation of I_o for the *in situ* studies.

Fig. 6. Series of spectra obtained *in situ* from silicon carbide during a 5 minute experiment.

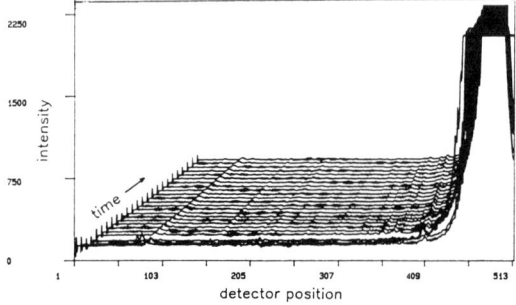

Fig. 7. Spatial separations (µm) observed during the first 21 seconds of a 2 minute *in situ* experiment.

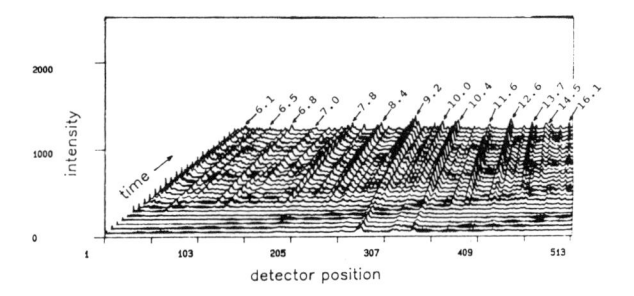

In another set of *in situ* experiments, spectra were taken farther from the specular beam where intensity maxima would be expected due to the spacing of surface features such as those displayed in Figure 3. With the detector position well-established, inter-feature spacings may be assigned to the spatial frequencies corresponding to the various maximum peaks according to Equation (2). The results of such an analysis are shown in Figure 7 for spectra taken during the first 21 seconds of a 2-minute experiment. The specular beam is located a known distance to the right of the frame allowing a high intensity incident beam to be used. Note that the initial spectra are relatively free from any distinguishing features. At 13 seconds, the first peaks appear at spatial frequencies corresponding to larger spacings of 11 to 16 µm. With increasing deposition time, a variety of spacings down to 6.7 µm become more prominent. Smaller peaks between 6 and 7 µm appear at 13 seconds but gradually fade by 20 seconds.

It is not clear which of the peaks can be justifiably assigned spatial frequencies based on the coherent isotropic scattering discussed earlier. Also, the sizes of the initial clusters rapidly exceed the smooth-surface limit necessary for an exact vector scattering analysis of the spectra. It is important to note, however, that significant changes in the scattered spectra occurred with progress of the deposition.

CONCLUSIONS

The laser scattering technique has been shown by several methods to be sensitive to subtle changes in surface morphology. While actual CVD surfaces seldom display the ideal periodicity and smoothness requisite for an exact solution to the scattering problem, laser scattering is still effective as a general method for monitoring the rms roughness of deposits. It can also be used as an *in situ* diagnostic for the appearance of certain features of a deposition process as has been demonstrated elsewhere for the epitaxial growth of silicon [13]. If the surface features possess some degree of periodicity, the spatial frequencies of the features can be obtained from the scattered spectrum. Therefore, with some knowledge of the nature of the depositing material, laser scattering is a promising technique for monitoring the growth of CVD coatings, which may find the greatest application in manufacturing situations requiring process control.

ACKNOWLEDGMENTS

The authors gratefully acknowledge the support of the Army Research Office, Division of Materials Science, under contract DAAG29-85-K-0214. Max Klein was supported by an Army Research Office fellowship, DAAL03-86G-0070.

REFERENCES

1. M. Klein and B. Gallois in Chemical Vapor Deposition of Refractory Metals and Ceramics, edited by T.M. Besmann and B.M. Gallois (Mater. Res. Soc. Proc. 168, Pittsburgh, PA 1990) pp. 93-99.
2. J.S. Paik, PhD thesis, Stevens Institute of Technology, 1991.
3. E.L. Church, H.A. Jenkinson and J.M. Zavada, Opt. Eng.16 (4), 360-374 (1977).
4. E. Marx and T.V. Vorburger, Appl. Opt. 29 (25), 3613-3626 (1990).
5. M.M. Klein, PhD thesis, Stevens Institute of Technology, 1991.
6. B.W. Sheldon and T.M. Besmann in Chemical Vapor Deposition of Refractory Metals and Ceramics, edited by T.M. Besmann and B.M. Gallois (Mater. Res. Soc. Proc. 168, Pittsburgh, PA 1990) pp. 99-106.
7. B.W. Sheldon and T.M. Besmann in Evolution of Thin-Film and Surface Microstructures, edited by C.V. Thompson, J.Y. Tsao and D.J. Srolovitz (Mater. Res. Soc. Proc. 202, Pittsburgh, PA 1991) pp. 161-166.
8. H. Davies, Proc. IEEE, Pt. III 101, 118 (1954).
9. P. Beckmann and A. Spizzichino, The Scattering of Electromagnetic Waves from Rough Surfaces (Macmillan, New York, 1963), p.93.
10. H.C. Bennett and J.D. Porteus, J. Opt. Soc. Am. 51 (2), 123-129 (1961).
11. L.H. Tanner and M. Fahoum, Wear 36, 299-316 (1976).
12. J.C. Stover, S.A. Serati and C.H. Gillespie, Opt. Eng.23 (4), 406-412 (1984).
13. D.J. Robbins, A.J. Pidduck, A.G. Cullis, N.G. Chew, R.W. Hardeman, D.B. Gasson, C. Pickering, A.C. Daw, M. Johnson, and R. Jones, J. Cryst. Growth 81 (1-4), 421-427 (1987).

IN-SITU LIGHT-SCATTERING MEASUREMENTS DURING THE CVD OF POLYCRYSTALLINE SILICON CARBIDE

Brian W. Sheldon[1,2], Philip A. Reichle[1], and Theodore M. Besmann[1]
1 - Metals and Ceramics Division, Oak Ridge National Laboratory, Oak Ridge, TN 37831.
2 - Division of Engineering, Brown University, Providence, RI 02912.

ABSTRACT

Light-scattering was used to monitor the chemical vapor deposition of silicon carbide from methyltrichlorosilane. The nucleation and growth of the SiC features caused changes in the surface topography that altered the angular scattering spectrum that was generated with a He-Ne laser. These scattering spectra were then analyzed to obtain information about the nucleation and growth processes that are occurring.

INTRODUCTION

The formation of polycrystalline materials by chemical vapor deposition (CVD) proceeds by nucleation and growth mechanisms that control the resultant microstructure. In spite of its technological importance, most of the current understanding of the relationship between the CVD process and the microstructure and properties of polycrystalline materials is qualitative and empirical. Better design and control of CVD materials will be possible with an improved understanding of the nucleation and growth processes that determine the microstructure.

The CVD of silicon carbide has been studied extensively [1]. Because of its high strength and chemical stability at elevated temperatures, polycrystalline SiC produced by CVD is potentially important for a variety of applications. Also, SiC matrix composites for high-temperature structural applications are being produced commercially by chemical vapor infiltration (CVI), which is essentially CVD in a porous structure [2]. A variety of different microstructures have been observed in CVD SiC and empirically correlated to the temperature, pressure, and gas composition that was used for deposition [1]. However, the mechanisms which govern these observed microstructural differences are not well understood.

Light-scattering is commonly used to measure surface roughness [3-5], thus it has been proposed as an in-situ method of monitoring the evolution of the surface topography during CVD [6,7]. These types of measurements were made during the CVD of SiC, and are presented here along with a brief analysis.

DESCRIPTION OF EXPERIMENTS

Silicon carbide was deposited from methyltrichlorosilane (MTS), using H_2 as a carrier gas. The substrates were prepared by depositing polycrystalline SiC onto graphite disks, and then polishing the deposited material to a mirror finish. The local smoothness of these surfaces (i.e., the absence of scratches) was verified with scanning electron microscopy.

The reactor consisted of a vertically mounted silica tube, where the reactant gases were introduced through the top flange, and the vacuum system was connected to the bottom flange (i.e., a down-flow configuration). The substrate was positioned inside of a custom-designed, cylindrical graphite heating element. The temperature of the sample was measured with an optical pyrometer that was sighted through a slit in the heating element. Figure 1 is a schematic of the light-scattering apparatus that was used as an in-situ monitor during this process. The 10 mW He-Ne laser was directed at the substrate through a port in the reactor, and the specular beam reflected from the substrate exited through this same port. The intensity of this specular beam was measured by reflecting it off of a mirror and into a single photodiode. Both the

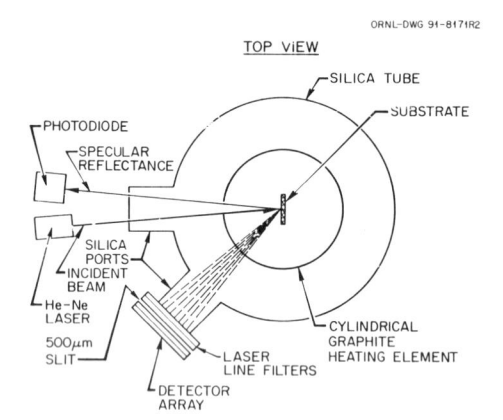

Fig. 1 Schematic of the in-situ light-scattering apparatus.

incident and reflected beams passed through slits in the heating element. Both the incident and specular intensities were measured by passing the beams through a small hole and onto the photodiode. The use of this aperture facilitated alignment and reproducibility. However, the specular reflectance underwent significantly more divergence than the incident beam, thus the aperture may have caused slightly low specular intensity measurements.

A portion of the scattered light from the substrate was collected by a linear array of 1024 photodiodes, after passing through a slit in the heating element and a second port in the reactor. The substrate temperature was between 950 and 1000°C for all of the runs that were conducted; lower temperatures were not possible because Si rather than SiC deposition has been observed [8,9]. Thus, at the temperatures that were studied there is a significant amount of thermal background radiation. This did not affect the specular reflectance measurements; however, the intensity of this background exceeded the intensity of the scattered light. To remove most of the thermal background radiation, two laser-line filters (narrow band pass filters which transmit only at a wavelength of 632.8 nm) were positioned in front of the detector array. A narrow slit was also positioned between the line filters and the detector.

For the experimental measurements to be quantitatively meaningful, the specular and scattered intensities must be compared to the incident beam intensity. To accomplish this, detailed calibration measurements prior to deposition were necessary because of the windows, mirror, and line filters that were used. The angular position of the detector array was also determined with a set of careful calibration measurements.

RESULTS

The intensity of the specular reflectance is plotted versus the deposition time in Figure 2. The top line (hollow circles) shows only a small decrease in the specular intensity as deposition proceeds, while the bottom line (filled circles) shows a significantly larger decrease. The filled circles were measured during deposition at a higher temperature and a higher MTS concentration, thus the faster decrease in the specular intensity is reasonable because deposition, and hence surface roughening occurs at a faster rate.

Figure 3 shows a sequence of scattering spectra that were collected during the same experiment where the lower line in Figure 2 was measured. The magnitude of the scattering signals initially increases rapidly, and after several minutes the rate of increase is much slower. During this CVD process individual SiC features nucleate, grow, and eventually coalesce into a polycrystalline film [1]. The initial rapid increase in the scattering intensities corresponds to the nucleation and growth of individual SiC features. After these features coalesce into a polycrystalline film the surface topography changes at a much slower rate, thus the scattering spectra apparently change at a much slower rate.

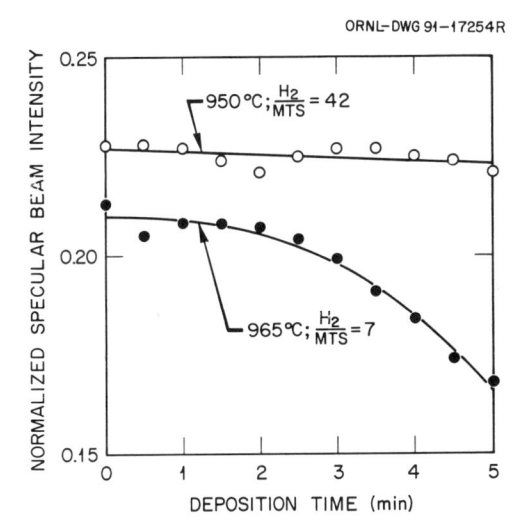

Fig. 2 The intensity of the specular beam, I_s (normalized to the incident intensity, I_i) versus deposition time. The open circles were measured with the substrate at 950°C and H_2/MTS=42. The filled circles were measured with the substrate at 965°C and H_2/MTS=7. In both cases the total pressure was 3 kPa.

ANALYSIS AND DISCUSSION

The specular intensity is the easiest optical measurement that can be made in the system that was used here. It provides an in-situ monitor of changes in the surface topography, but it can not be used to obtain detailed quantitative information. As noted above, the results in Figure 3 indicate that intensity changes in the scattering spectra can provide more precise information (e.g., monitoring the formation of a coalesced polycrystalline film). In theory, a more detailed analysis is also possible because the angular scattering spectrum contains all of the information that is necessary to quantitatively describe the surface topography [3]. To conduct this type of analysis, the angular scattering spectra in Figure 3 are first converted to the power spectral density (PSD). The one and two minute spectra in Figure 3 were converted to PSD versus r plots that are shown in Figure 4. The independent variable, r, is given by:

$$r = k [\sin\theta_s - \sin\theta_I] \quad (1)$$

where k is:

$$k = \frac{2\pi}{\lambda} \quad (2)$$

and θ_I and θ_S are the incident and scattering angles, respectively. The reference point for these angles is the surface normal (i.e., an incident beam that is perpendicular to the surface corresponds to $\theta_S = 0$). The laser wavelength, λ, is 0.6328 μm for the He-Ne laser that was used.

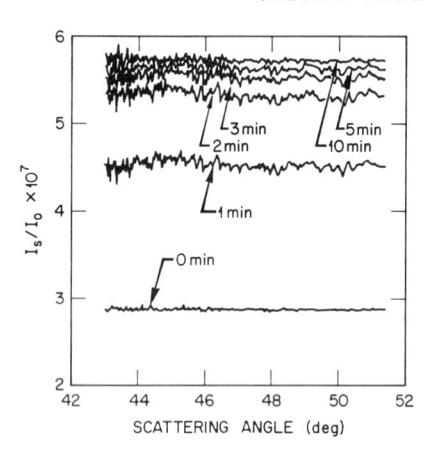

Fig. 3 A series of angular scattering spectra measured with a substate temperature of 965°C, a total pressure of 3 kPa, and $H_2/MTS=7$.

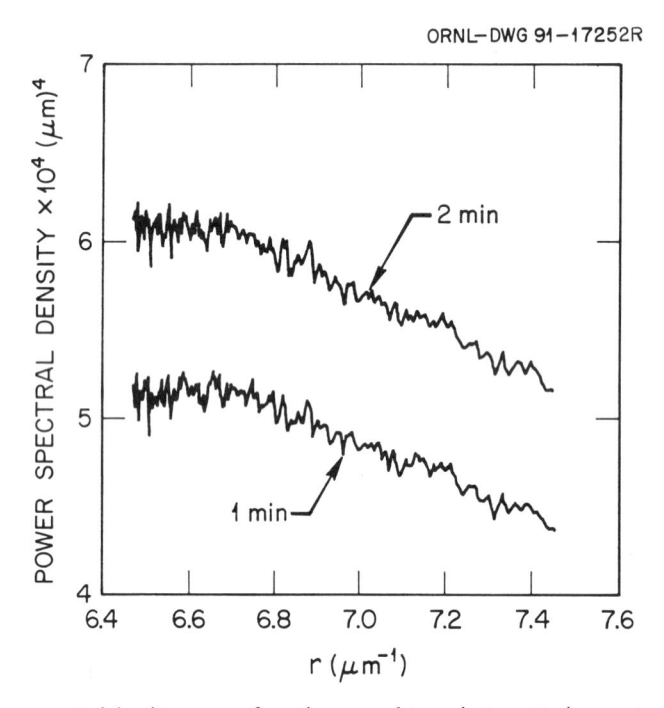

Fig. 4 Power spectral density versus r, from the one and two minute scattering spectra that are shown in Figure 3.

Experimental results such as those shown in Figure 4 can be analyzed with an appropriate expression for the PSD that depends on the mathematical description of the surface profile that is used. The standard approaches are based on Gaussian surface profiles [3], which do not accurately describe the surfaces that form during the CVD of polycrystalline materials. During the early stages of CVD, shot-models can accurately describe the nucleation and growth of individual surface features. By building an appropriate nucleation and growth model into a shot-model of the surface profile, the parameters that describe the nucleation and growth kinetics can be obtained directly from the light-scattering spectra [10]. In this previous work, these nucleation and growth based shot-models were successfully used to analyze out-of-reactor light-scattering measurements that were made on CVD surfaces.

The experimental PSD results in Figure 4 were first analyzed with a simple nucleation and growth model that assumes that nucleation occurs in a short burst at one point in time, and is followed only by growth of this collection of nuclei. This behavior has been observed experimentally during the CVD of polycrystalline silicon [11,12]. This model leads to a relatively simple expression for the power spectral density [3,10]:

$$PSD = \frac{\eta}{4} (ut)^6 \left(\frac{J_2[(ut)r]}{[(ut)r/2]^2} \right)^2 \qquad (3)$$

where η is the nucleation density (nuclei/μm^2), u is the growth rate ($\mu m/min$), t is time, and J_2 is a Bessel function. Evaluating the infinite series for J_2 gives:

$$PSD = \frac{\eta}{16} (ut)^6 \left(1 - \frac{[(ut)r]^2}{6} + \frac{7[(ut)r]^4}{576} - \frac{[(ut)r]^6}{1920} + ... \right) \qquad (4)$$

The leading term in Eq. (4) can be viewed as the intensity of the PSD, which is a function of both the nucleation and growth parameters (η and u). The polynomial factor that follows this term is an infinite series that can be viewed as the "shape" of the PSD curve, which is only a function of the growth rate and the deposition time (i.e., not the nucleation density). In Figure 4 the intensity of the scattering spectra increases with time, while the shape of the PSD shows almost no change with time. This is not consistent with Eq. (3) because the shape should also change as growth proceeds. This is evidenced by nonlinear regression results according to Eq. (4), that yield values of η = 51 nuclei/μm^2 and u = 0.25 $\mu m/min$ for the one minute PSD and η = 53 nuclei/μm^2 and u = 0.13 $\mu m/min$ for the two minute PSD. Close examination of Eq. (4) also shows that the PSD is very sensitive to small changes in u and t, and less sensitive to the value of η. The experimental PSD results in Figure 4 were also analyzed with other nucleation and growth models that are described elsewhere [10]. In general, the nucleation and growth parameters that were obtained with these models were also inconclusive.

As expected, these results are less precise than the out-of-reactor measurements that were performed previously [10]. There are several reasons for this:

1. It was possible to make scattering measurements over a much wider angular range in the out-of-reactor case.

2. It was necessary to remove most of the thermal background radiation from the heating element and the hot substrate with narrow-band pass filters. This substantially reduces the intensity of the scattered signal.

3. In addition to reducing the signal intensity, the windows and narrow-band pass filters that were used between the substrate and the detector create other problems associated with refraction and unwanted reflections. These were accounted for with calibration measurements, but some inaccuracies still exist.

In future research, efforts to reduce these problems should be explored. The angular range can be expanded with a wider detector array, or by using multiple detector arrays. Reducing the problems associated with the thermal background radiation will require a modified reactor design, or a stronger laser to generate stronger scattered signals. Some improvements might also be obtained by making better calibration measurements.

CONCLUSIONS

Laser light-scattering was used as an in-situ monitor of nucleation and growth during the CVD of polycrystalline SiC. The intensities of the specular reflection and the scattered light were used to qualitatively monitor changes in the surface topography during deposition. Previously, out-of-reactor scattering measurements have been used to quantitatively analyze nucleation and growth parameters. Similar analyses were applied less successfully to the in-situ measurements that were made here. Improved quantitative analysis will require improvements in the CVD light-scattering apparatus.

ACKNOWLEDGEMENTS

This research was supported by the U.S. Department of the Air Force, Office of Scientific Research, Bolling Air Force Base, D.C. under U.S. Department of Energy Interagency Agreement 1854-C027-A1 under contract DE-AC05-84OR21400 with Martin Marietta Energy Systems. The authors also acknowledge Jerry McLaughlin for technical support and Harry Livesey for preparing the drawings.

REFERENCES

1. J. Schlichting, *Powder Metall. Int.*, **12**, 141 and 196 (1980).

2. T.M. Besmann, B.W. Sheldon, R.A. Lowden, and D.P. Stinton, *Science*, **253**, 1104 (1991).

3. E.L. Church, H.A. Jenkinson, and J.M. Zavada, *Opt. Eng.*, **18**, 125 (1979).

4. J.C. Stover, S.A. Serati, and C.H. Gillespie, *Opt. Eng.*, **23**, 406 (1984).

5. J.M. Bennett and L. Mattson, "Introduction to Surface Roughness and Scattering", (Optical Society of America, Washington D.C., 1989).

6. M. Klein and B. Gallois in "Chemical Vapor Deposition of Refractory Metals and Ceramics", edited by T.M. Besmann and B. Gallois (Mater. Res. Soc. Proc. **168**, Pittsburgh, PA, 1990), pp. 93-98.

7. B.W. Sheldon and T.M. Besmann in "Chemical Vapor Deposition of Refractory Metals and Ceramics", edited by T.M. Besmann and B. Gallois (Mater. Res. Soc. Proc. **168**, Pittsburgh, PA, 1990), pp. 99-106.

8. F. Langlais, C. Prebende, B. Tarride, and R. Naslain, *J. de. Phys.* **50/C5**, 93 (1989).

9. T.M. Besmann, B.W. Sheldon, and M.D. Kaster, *Surf. Coat. Tech.* **43/44**, 167 (1990).

10. B.W. Sheldon and T.M. Besmann, "Evolution of Thin Film and Surface Microstructure", edited by C.V. Thompson, J.Y. Tsao, and D.R. Srolovitz (Mater. Res. Soc. Proc. 202, Pittsburgh, PA, 1991), pp. 161-66.

11. J. Bloem, *J. Crystal Growth*, **50**, 581 (1980).

12. J. Bloem and W.A.P. Claassen, *Philips Tech. Rev.*, **41**, 60 (1983).

INTERACTIVE USE OF ELECTRON MICROSCOPY AND LIGHT SCATTERING AS DIAGNOSTICS FOR PYROGENIC AGGREGATES

Richard A. Dobbins, Division of Engineering, Brown University, Providence, RI 02912

ABSTRACT

Both electron microscopy and light scattering have played an important role in elucidating the processes of inception, growth, and oxidation of carbonaceous particles in flames. The techniques developed have application to the various pyrogenic materials including the metallic oxides, carbides, etc. Thermophoretic sampling has been developed to afford efficient extraction of particle samples from hot reaction zones. The sampling procedure preserves the particle morphology for subsequent analysis by transmission electron microscopy (TEM). Studies using this technique have shown aggregate structures with fractal dimensions of 1.6 to 1.8, a result that is consistent with the computer simulations of the cluster-cluster aggregation process. Diverse morphologies, including microparticles found in the particle inception zone, reveal the evolution of these aggregates. The optical cross sections for polydisperse aggregates which are used to interpret the laser scattering/extinction tests (LSE) are described. Population averaged properties - volume fraction, volume mean diameter, monomer and aggregate number concentrations, mean-square radius of gyration - are derived. The interactive use TEM and LSE data leads to a global description of the aggregate dynamic processes which are found to be regionally partitioned within the laminar hydrocarbon diffusion flame.

I. Introduction

During the last twenty years there has been a frequent use of LSE tests to monitor the formation and growth of particulate materials. The use of LSE in combustion research to diagnose the formation of carbonaceous particles in flames has an extensive history [1-5]. It is to be noted that the experimental procedures have progressed more rapidly then has the theoretical foundations that provide the basis for the interpretation of the tests results. Thus the initial observations were analyzed on the falacious assumption of a population of monodisperse spherical particles. TEM provides entirely different information, which when rationalized with the LSE data provides important insight into the nature of the particle formation processes.

TEM observations of particles are normally made by ex situ observations that require a sampling procedure which does not alter the state of the material to be examined. While obtrusive sampling has been avoided in the past, more recently these methods have proven capable of providing the vital information on particle morphology [6,7] upon which the cross sections for scattering and extinction are based. On the other hand, TEM analysis cannot provide data on the aerosol suspension - properties such as particle number or volume concentration, surface area per unit volume, etc.

LSE has the great advantage of providing in situ observations of the population of particulate material occupying a small volume in a precisely defined space. Because of the wide dynamic range of photomultipliers, it is possible to make observations over a wide range of particle concentrations even in the presence of very strong gradients of particle properties [8]. Furthermore, multiangle observation and tomographic inversion methods permit the LSE diagnostic methods to be applied to complex nonsymetric geometries of the reactor system [9]. A weakness of the LSE observations is their inability to provide direct information on the particle morphology upon which the data interpretation is dependent.

II. Thermophoretic Sampling

The extraction of particulate material from a high temperature environment can be achieved by the rapid insertion of the microscope grid to the point of interest. The technique is referred to as thermophoretic sampling since the temperature gradient in the thermal boundary layer drives the particles to the grid surface. The grid is then directly transferred to the electron microscope for detailed examination of the captured sample. The evolution of the particle morphology is analyzed by taking a series of samples extracted at selected positions along the particle path. The details of this technique are described elsewhere [6].

The most important information that is yielded by thermophoretic sampling is the morphological character of the particle field. For example, ceramic particles at a temperature well above their melting point will normally be present in the form of polydisperse spheres. Similarly, particles formed in the early inception stage may be isolated spherules. The optical behavior of these particles is well described by Mie theory calculations modified to include a polydispersity of the particle population. A common morphology that is not accommodated by the spherical assumption is the aggregate configuration which has been encountered either as an intermediate or as a terminal state of particle development.

Aggregated (clusters or agglomerates) of primary particles (monomers or spherules) are formed by the process of cluster-cluster aggregation and constitute a major particle morphology [10-13]. The quantitative description and diagnostic analysis of aggregates is facilitated by the concept of mass fractals. Thus the number of primary particles per aggregate is related to the radius of gyration R_g and the primary diameter d_p by

$$n = k_f (R_g/d_p)^{D_f} \tag{1}$$

where D_f is the fractal dimension and k_f is the prefactor. Aggregates formed by cluster-cluster aggregation have fractal dimensions of 1.8 to 1.9 according to numerical simulations[3]. Laboratory experiments of several different types yield fractal dimensions in the range of 1.6 to 1.9 which is good agreement with the theoretical results. TEM experiments readily yield values of d_p and, with somewhat more effort, provide values of n. The R_g of an aggregate can be found from the 2-D projection provided by TEM. Stereoscopic TEM methods and computerized data acquisition provide clear advantages when these techniques are available.

Aggregates undergoing the clustering process will inevitably display polydispersity since the number of primary particles per aggregate will be highly variable. For a population of polydisperse aggregates Eq. (1) can be recast as

$$\overline{n^1} = \overline{\kappa} \ \overline{R_g^2} / d_g^2 \tag{2}$$

Here $\overline{R_g^2}$ is defined as appropriate for the interpretation of the optical observations and the quantity $\overline{\kappa}$ assumes the dependence on D_f, k_f and the size PDF. The quantity $\overline{n^1}$ is the first moment of the size distribution function. A knowledge of d_p is readily determined by TEM and becomes an important input into the full description of the aggregate field. A knowledge of the $\overline{n^1}$ and d_p yields the volume average of the volume equivalent spheres of the members of the aggregate population,

$$D_{30} = \left[\overline{n^1}\right]^{1/3} d_p \tag{3}$$

The value of the $\overline{\kappa}$ is approximately 2.6 to 3.3 [14] for size distributions that approximate the self preserving distribution. A detailed size distribution is obtainable from the TEM micrographs if a more accurate value of $\overline{\kappa}$ is required. Further details of the aggregate field require a knowledge of the particle concentration which is more readily obtainable from the scattering measurements as discussed below.

III. Laser Scattering/Extinction Measurements

The observation of the scattering and extinction of light at a well defined laser wavelength provides two additional pieces of information that are sensitive to different moments of the size distribution function in the case of aggregates consisting of an absorbing material. In this instance a measure of a mean size is possible as is illustrated elsewhere [14,15]. For dielectric particles both scattering and extinction depend on the same moment ratio and no information on aggregate size is afforded. In this event the angular distribution of scattering is of greater interest [16,17]. The differential volumetric scattering cross section $Q_{vv}(\theta)$ at the angle θ from the forward direction for randomly oriented, polydisperse aggregates [18] is given by

$$Q_{vv}(\theta) = N_a f_n (\overline{n^1})^2 x_p^6 F(m) f(X)/k^2$$

where

$$F(m) = |(m^2 - 1)/(m^2 + 2)|$$

$$f(X) = \exp(-X/3), \qquad X < 1.5 D_f$$

$$k = 2\pi/\lambda$$

$$q_i = 4\pi \sin(\theta_i/2)/\lambda$$

$$x_p = \pi d_p/\lambda$$

$$X = q_i^2 \overline{R_g^2}$$

with N_a equal to the number of aggregates per unit volume, and f_n is a moment ratio ≈ 1.7 [14, 15].

Equation (4) applies to aggregates with $X \leq 1.5\ D_f$ which is given here by way of illustration. In this case, observations of the light scattered at two angles yields the ratio $R_{ij} = Q_{vv}(\theta_i)/Q_{vv}(\theta_j)$ where $\theta_j > \theta_i$, and gives from Eq. (4) the mean square radius of gyration,

$$\overline{R_g^2} = a_{ij} \ln R_{ij}/k^2 \tag{5}$$

where a_{ij} is a constant that depends on the angles θ_j and θ_i [15]. Equation (5) is notable in that it requires no knowledge of the size distribution or refractive index [18]. With $\overline{R_g^2}$ and the value of d_p from TEM, the values $\overline{n^1}$ and D_{30} can be found.

IV. Interactive use of TEM and LSE

TEM aids LSE when used to (a) define morphology, (b) provide values of d_p for quantitative data reduction and (c) explore regions of complex morphology. An illustration of the first two levels of interactive use is now provided. Dissymetry measurements give $\overline{R_g^2}$, TEM gives values of d_p which through Eqs. 2 and 3 give $\overline{n^1}$ and D_{30}. With the refractive index of the material, the quantity $F(m)$ is known along with all quantities on the right hand side of Eq. (4) except the aggregate number concentration. The latter is found by the measurement of the absolute value of $Q_{vv}(\theta)$ using a gas of known scattering cross section and concentration as a calibration medium [19] or an aerosol of known size distribution, refractive index and concentration [17]. Once the aggregate number concentration N_a is known, the volume fraction of the material f_v and the particulate surface area per unit volume S_t can then be determined by

$$f_v = N_a \frac{\pi}{6} D_{30}^3 \tag{6}$$

$$S_t = N_a \overline{n^1} \pi d_p^2 \tag{7}$$

The values of f_v and S_t are important in the determination of heterogeneous reaction rates. The abbreviated results presented here are given in more detail in the references listed below.

Striking morphology gradients do occur in the case of the carbonaceous materials produced in diffusion flames. Isolated singlet microparticles have been first identified by thermophoretic sampling, and their discovery provides valuable information on the early formation of graphitic soot from a PAH precursor particle [20]. This discovery indicates that gradients in particle morphology are accompanied by gradients in refractive index and size distribution as well. Quantitative analysis of the particle field in the presence of these complications provides an interesting challenge for both LSE and TEM. Metal oxide particles have been observed as fractal aggregates and as polydisperse compact spheres [16,17] and both diagnostic methods are therefore of value in this instance.

V. Summary

The data yielded by the TEM and LSE experiments are of dissimilar but complementary natures. The interactive application of these techniques provides information that is beyond the capabilities of either method used alone.

Acknowledgement

This work was sponsored by the U.S. Department of Commerce, National Institute of Standards and Technology, Center for Fire Research under Grant No. NANB1D1110.

References

1. A. D'Alessio, A. DiLorenzo, A. Borhese, F. Baretta and S. Masi, The sixteenth (International) Symposium on Combustion, The Combustion Institute, 1977, p 695.

2. A. D'Alessio, in Particulate Carbon Formation During Combustion (D.C. Siegla and G. W. Smith, Eds.), Plenum Press New York, 1981.

3. J. Kent, H. Jander, and H. Gg. Wagner, The Eighteenth (International) Symposium on Combustion, The Combustion Institute, 1981, p 1117.

4. B.S. Haynes, and H. Gg. Wagner, Ber. Bunsenges. Phys. Chem. 84, 499 (1980).

5. I.J. Jagoda, G. Prado, and J. Lahaye, Combust. and Flame 37, 261 (1980).

6. R.A. Dobbins and C.M. Megaridis, Langmuir 3, 254 (1987).

7. C.M. Megaridis and R.A. Dobbins, Combust. Sci. and Tech. 71, 95 (1990).

8. R.J. Santoro, H.G. Semerjian and R.A. Dobbins, Combust. and Flame 51, 203 (1983).

9. H.G. Semerjian, R.J. Santoro, P.R. Emmerman and R. Goulard, Int. J. Heat Mass Transfer 24, 1139 (1981).

10. D.W. Schaefer, J.E. Martin, P. Wilzius and D. S. Cannell, Phys. Rev. Ltrs. 52, 2371 (1984).

11. R. Jullien and R. Botet, Aggregation and Fractal Aggregates, World Scientific Press, Singapore (1987).

12. R. Botet and R. R. Jullien, Annales de Physique 13, 153 (1988).

13. D.W. Schaefer, MRS Bulletin, XIII, 22 (1988).

14. R. Puri, T.J. Richardson, R.J. Santoro and R.A. Dobbins, Submitted for publication, 1991.

15. R.A. Dobbins, R.J. Santoro and H.G. Semerjian, Twenty-third (International) Symposium on Combustion, The Combustion Institute, 1990, p 1525.

16.   M.R. Zachariah, D. Chin, H.G. Semerjian, and J.L. Katz, Applied Optics 28, 530 (1989).

17.   H. Chang and P. Biswas, Personal Communication, November 1991.

18.   R.A. Dobbins and C.M. Megaridis, Applied Optics 30, 4747, (1991).

19.   R.R. Rudder and D.R. Bach, J. Opt. Soc. Am. 58, 1260 (1968).

20.   R.A. Dobbins and H. Subramaniasivam, In Press (1991).

TUNABLE DIODE LASER ABSORPTION SPECTROSCOPY OF THE PYROLYSIS OF METHYLSILAZANE

H.C. Sun, Y.W. Bae[*], E.A. Whittaker, and B. Gallois[*]

Department of Physics and Engineering Physics
*Department of Materials Science and Engineering
Stevens Institute of Technology, Hoboken, NJ 07030

Abstract

An understanding of the chemical processes occurring in the gas phase during metalorganic chemical vapor deposition is needed to design novel precursors and for the subsequent control of the composition and the microstructure of the solid product. Tunable diode laser absorption spectroscopy provides a means to precisely monitor specific bond rupture in the precursor during pyrolysis. Methylsilazane [CH_3SiHNH]$_n$, a precursor to silicon-based ceramic thin films, was used to investigate the potential of this technique. Below the decomposition temperature, the intensity of the absorption line at 871.6±0.1 cm^{-1} corresponding to one of the harmonics from Si-CH$_3$, increased linearly with the vapor pressure of methylsilazane up to 800 Pa and then decreased exponentially. The typical linewidths of the absorption line was approximately 0.006 cm^{-1}, orders of magnitude narrower than would be observable using conventional infrared techniques. The absorption line was detectable over a pressure range from less than 1 Pa to 10 kPa.

Introduction

There has been an increasing interest in novel synthetic routes of thin films by the pyrolysis of organometallic precursors. Metalorganic chemical vapor deposition (MOCVD) provides a means of low-temperature processing of pure metals and alloys, metal and metalloid carbides, borides, nitrides, silicides, and related thin films which has both important technical and economical implications for structural and microelectronic device applications.[1]

Despite the demonstrated potential utilities, little is known of the chemical processes by which organometallic compounds are converted to solid thin films upon pyrolysis.[2] This stems from the nature of the complex molecular structures that serve as precursors, and the even more complex and intractable structures that are intermediates in the pyrolysis process. Optimization of deposition conditions are, thus, usually made by experimentally varying the numerous process parameters until acceptable deposits are obtained. As the need to design novel organometallic precursors rises and control of composition and microstructure of pyrolytic products becomes more demanding, studies aimed at understanding of the chemical processes which arise during the film formation pose a formidable challenge.

In an attempt to understand the chemical reactions in detail we have implemented high-resolution tunable diode laser spectroscopy (TDLAS) as a chemically selective, high sensitivity diagnostic. Individual diodes,[1] commercially available with center frequencies between 300 cm^{-1} and 3000 cm^{-1},[3-5] are chosen to match the IR absorption spectrum of the species and vibrational transitions of interest. Measurement of the line integrated absorption strength of a particular transition can be correlated with the species concentration with appropriate calibration procedures. In addition, by implementing novel modulation techniques we are able to monitor species concentration with high sensitivity and in real time.

In this report, we present preliminary data on the detection of methylsilazane, a novel organometallic precursor used in the synthesis[7-10] of dielectric thin films at low deposition temperatures. This precursor has a rich infrared spectrum making it an ideal candidate for TDLAS detection.

Experimental procedures

Methylsilazane was synthesized following the published procedure.[11] Its molecular structure and properties were described elsewhere.[12] A gas cell equipped with GaAs windows was used for the studies by Fourier Transform Infrared Absorption Spectroscopy (FTIR) and TDLAS. Carbonylsulfide and ammonia were used for the calibration of the laser frequencies.

A schematic diagram of the TDLAS setup is shown in Fig. 1. The rather complex signal processing scheme reflects the implementation of two modulation techniques which afford very high sensitivity detection free from etalon fringes induced by the windows of the reaction chamber. High sensitivity is attained by frequency modulating the laser at 40 MHz, an approach known as frequency modulation spectroscopy (FMS). The etalon fringes are rejected by combining FMS with a novel type of wavelength modulation spectroscopy. The two methods result in absorption lineshapes which resemble derivatives of the actual absorption lines. Details of how FMS achieves high sensitivity[13-15] and rejects optical fringes[6] are discussed elsewhere.

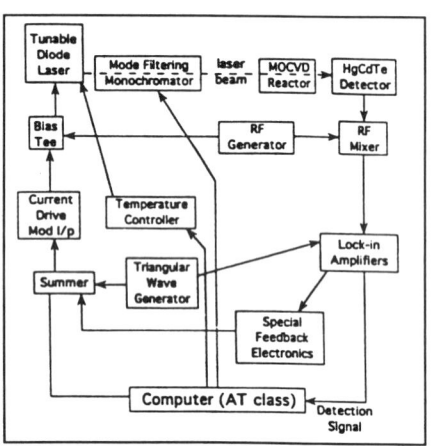

Figure 1 Schematic diagram of tunable diode laser absorption spectrometer.

For this report the laser was used in two modes of operation. Temperature tuning provides medium resolution broadband spectral scans. Current tuning with the above modulation techniques implemented provides high resolution narrow band scans. In both cases the laser was controlled by a personal computer.

Results and Discussion

The gross features of characteristic vibrational modes resulting from the groups of atoms attached to silicon in gaseous methylsilazane are shown in the conventional FTIR spectrum of Fig. 2. A strong line centered at 938 cm^{-1} (Si-N-Si asymmetric stretch) together with peaks at 3392 and 1184 cm^{-1} arising from N-H stretching and bending modes indicate the presence of Si-NH-Si structures in the precursor.[16] The symmetric -CH$_3$ deformation gives a very sharp and intense band at 1264 cm^{-1} and is the most characteristic feature for methyl groups bonded to silicon.[16] The line at 1264 cm^{-1} is accompanied by equally intense bands at 888 and 765 cm^{-1} from the -CH$_3$ rocking and the Si-C stretching vibrations.[16] Si-H band stretching at 2155 cm^{-1} is one of the most distinctive features of organosilicon compound.[16]

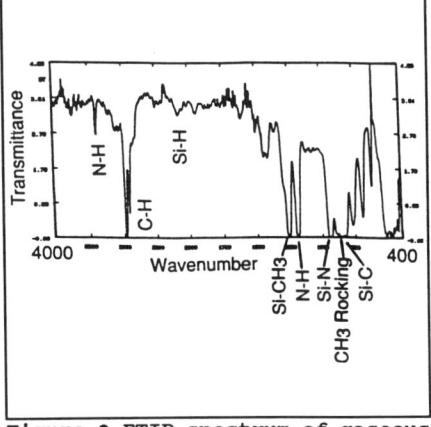

Figure 2 FTIR spectrum of gaseous methylsilazane precursor.

Almost all the absorption bands appearing in the FTIR spectrum can be detected by using TDLAS. The current laser installed in our system tuned from 855 cm^{-1} to 905 cm^{-1}. A broad scan over this entire range using temperature tuning was performed on methylsilazane as shown in Fig. 3. The strongest absorption peak at around 870 cm^{-1} was chosen to be examined in high resolution. The resultant typical absorption line is shown in figure 4. The peak was identified by comparing with peak positions of carbonylsulfide and ammonia to be at 871.6±0.1 cm^{-1}. It was found to be one of the harmonics of the rotational-vibrational mode arising from the -CH$_3$ rocking in methylsilazane.[16] The

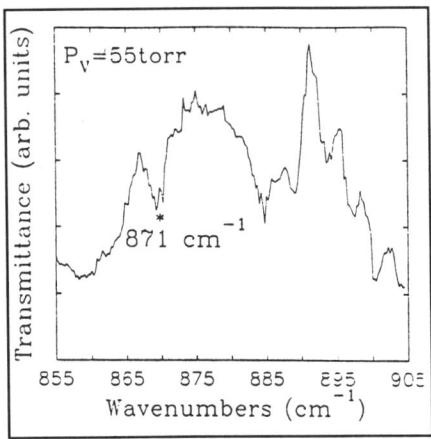

Figure 3 A broad band scan using TDLAS.

0.1 cm^{-1} uncertainty was due to thermal drift of the laser. Uncertainty down to 0.0003 cm^{-1} can be easily achieved when more than one optical detector is available to perform simultaneous measurement. Series of scans with the same experimental parameters were done at various vapor pressures of methylsilazane. The peak signal from each scan is plotted against gas pressure as shown in figure 5. The absorption signal varies as a product of an exponential decay term that is due to the absorption of the laser carrier and a linear term that is due to the differential absorption of the laser side bands.[13-15] Figure 5 shows three different regions. First, the linear region apparent below 0.5 Torr is the low pressure regime where the absorption peak is predominantly Doppler broadened. The second region is the transition region between 0.5 and 6 Torr. At 0.5 Torr, the absorption profile starts to be pressure broadened so that the curve starts to deviate from linearity. Then, the linear term competes with the exponential term. The third region starts at around 6 Torr. This is the high absorption region where the exponential term predominates.

The whole curve serves as a calibration curve for the real time monitoring experiment. In practice, the first region is the most important part as complete bond rupture means sharp decrease in absorption. If the third region is to be used, sensitivity can be improved by applying simple TDLAS without our modulation techniques. The signal to noise ratio obtained at 10 mTorr with 16 Hz detection bandwidth was in excess of 300. Since 870 cm^{-1} was not the peak lasing mode of our laser, this signal to noise ratio means submillitorr monitoring in the kHz detection bandwidth range could be possible with the right laser for different representative peaks. This also indicates the possibility of submicrotorr detection for steady state condition.

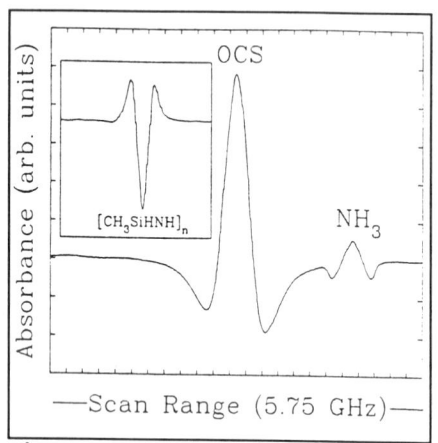

Figure 4 Identification of methylsilazane absorption peak by OCS and NH$_3$ references.

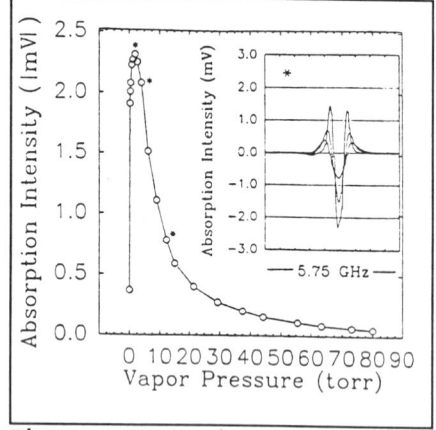

Figure 5 CWFMS signal calibration of methylsilazane as a function of vapor pressure.

A careful design of the reaction chamber will boost the sensitivity up by half or even one order of magnitude.

Summary

We investigated the possibility of using branches of tunable diode laser absorption spectroscopy to monitor reaction mechanism of MOCVD processes. The infrared absorption modes of an organometallic precursor, methylsilazane, were investigated using both medium- and high-resolution laser scans. The proven capabilities of high sensitivity, high resolution, and fast data acquisition time of this technique offer a possible diagnostic tool for the *in situ* monitoring of chemical processes taking place during the pyrolysis of organometallic precursors.

Acknowledgements

The authors gratefully acknowledge Mr. Premar for supplying the methylsilazane compound and assisting in the FTIR analysis. This work was partially supported by the New Jersey Advanced Technology Center for Surface Engineered Materials.

Reference

1. G.S. Girolami and J.E. Gozum, in *Chemical Vapor Deposition of Refractory Metals and Ceramics*, (Proc. Mater. Res. Soc. Symp.), edited by T.M. Besmann and B.M. Gallois (Materials Research Society, Pittsburgh, PA, 1990), Vol. 168, p. 319.

2. G. Pouskouleli, Ceram. Int. 15, 213 (1989).

3. J. Reid, M. El-Sherbiny, B.K. Garside, and E.A. Ballik, Appl. Opt., 19, 3349 (1980).

4. J. Wormhoudt, A.C. Stanton, A.D. Richards, and H.H. Sawan, J. Appl. Phys., 61, 142 (1987).

5. H.C. Sun and E.A. Whittaker, *IEEE LEOS'90* (Conference Proceeding), p. 577 (1990).

6. H.C. Sun and E.A. Whittaker, Appl. Opt., to be published.

7. H. Du, B. Gallois, and K.E. Gonsalves, J. Am. Ceram. Soc. 73 (3), 764 (1990).

8. H. Du, B. Gallois, and K.E. Gonsalves, Chem. Mater. 1 (6), 569 (1989).

9. H. Du, Y.W. Bae, B. Gallois, and K.E. Gonsalves, in *Chemical Vapor Deposition of Refractory Metals and Ceramics*, (Proc. Mater. Res. Soc. Symp.), edited by T.M. Besmann and B. Gallois (Materials Research Society, Pittsburgh, PA, 1990), Vol. 168, p. 331.

10. Y.W. Bae, B.J. Wilkens, H. Du, K.E. Gonsalves, and B. Gallois, *Proc. 2nd Int. Conf. Electron. Mater.*, (Materials Research Society, Pittsburgh, PA, 1990) (in press).

11. D. Seyferth, G.H. Wiseman, and C. Prud'Homme, J. Am. Ceram. Soc. 66 (1), c-13 (1983).

12. D. Seyferth and G.H. Wiseman, *Ultrastructure Processing of Ceramics, Glasses, and Composites*, edited by L.L. Hench and D.R. Ulrich (John Wiley & Sons, New York, 1984), p. 265.

13. M. Ghertz, G.C. Bjorklund, and E.A. Whittaker, J. Opt. Soc. Am. B, 2, 1510 (1985).

14. C.B. Carlisle, D.E. Cooper, and H. Preier, Appl. Opt., 28, 2567 (1989).

15. P. Werle, F. Slemr, M. Ghertz, and C. Brauchle, Appl. Phys. B, 49, 99 (1989).

16. D.R. Anderson, *Analysis of Silicones*, edited by A.L. Smith, (John Wiley & Sons, New York, 1974), p. 261.

AXIAL CONCENTRATION PROFILE OF H$_2$ PRODUCED IN THE CVD OF Si$_3$N$_4$.

STEPHEN O. HAY AND WARD C. ROMAN
United Technologies Research Center, E. Hartford, CT 06108

ABSTRACT

Silicon nitride (Si$_3$N$_4$) has been demonstrated to be an effective high temperature anti-oxidant when deposited in its α-crystalline form. The Materials Technology Laboratory at UTRC has developed a pilot-scale chemical vapor deposition (CVD) reactor capable of depositing α-Si$_3$N$_4$ from ammonia (NH$_3$) and silicon tetrafluoride (SiF$_4$) at 1.8 torr and 1440 C. Coherent anti-Stokes Raman spectroscopy (CARS) has been applied to measure H$_2$ produced in this reactor. Axial concentration measurements have been performed both in the presence and absence of SiF$_4$. Previous CARS measurements demonstrated the importance of surface (Si$_3$N$_4$) catalyzed decomposition of NH$_3$:

$$2NH_3 \rightarrow 3H_2 + N_2$$

as a competing reaction to:

$$4NH_3 + 3SiF_4 \rightarrow Si_3N_4 + 12HF$$

in the CVD reactor under deposition conditions. The observed hydrogen concentration profiles confirm these measurements and allow quantitative comparison between the competing reactions. NH$_3$ decomposition is suppressed 20% by the addition of SiF$_4$ in a 6:1 (NH$_3$:SiF$_4$) molar ratio. No decomposition is observed in the absence of Si$_3$N$_4$.

Introduction

Silicon nitride (Si$_3$N$_4$) is an effective high temperature anti-oxidant [1] when deposited in its α-crystalline form. The UTRC Materials Technology Laboratory has developed a thermal CVD process [2] for α-Si$_3$N$_4$ deposition from ammonia (NH$_3$) and silicon tetrafluoride (SiF$_4$) at 1.8 torr and 1440 C. In order to validate the modeling performed [3] on the UTRC reactor design, it is desirable to measure species concentration profiles within an operational reactor. Our objective is to apply non-intrusive optical diagnostics within the UTRC reactor environment for *in situ* species concentration and temperature measurements. To accomplish this objective, a CVD reactor with multiple optical access ports was constructed. This access allows *in situ* measurements to be made at reference operating conditions at several positions downstream of the reactant injectors.

CARS was selected as the diagnostic to be applied for several reasons. First, this technique requires only limited line of sight optical access to the reactor. Second, the resultant signal is coherent. This allows background discrimination of intense black body background by simple beam propagation through a distance, R (the background drops off as $1/R^2$ while the coherent signal is attenuated only due to diffractive and scattering losses which are small). Third, CARS is applicable to all species possessing a Raman active vibrational mode and is therefore theoretically capable of being used to observe the ground state of any molecular species. And finally, a rovibrational CARS spectra can be interpreted to yield both a species concentration and a rotational temperature. In a thermal reactor, such as the one that the UTRC method utilizes, rotation is in equilibrium with the other degrees of freedom and accurately represents the bulk gas temperature.

The experimental observation that, in the reactor, under normal operating conditions, the CARS signal due to NH_3 is not detectable above 1000 C, has been reported previously [4]. NH_3 is known to undergo homogeneous thermal decomposition in the gas phase at this temperature and above, but gas kinetic calculations [3] indicate that no significant homogeneous decomposition occurs within the typical residence time (~1 sec) of gas phase species in the reactor. It was therefore expected that significant amounts of unreacted NH_3 would survive the reactor environment, and be present in concentrations amenable to CARS detection. As discussed below, our results indicate that this is not the case; NH_3 is not spectroscopically observable in the reactor. However, a decomposition product, H_2, is detectable.

Heterogenous reactions are thus found to contribute significantly to the global chemistry of the reactor. Two surfaces exist in an unloaded reactor, the bare graphite retort walls, and the Si_3N_4 coating. NH_3 etching of graphite has been demonstrated to occur in the absence of SiF_4. Based on the CARS detection of H_2 at a variety of conditions, the importance of surface (Si_3N_4) catalyzed decomposition of NH_3:

$$2NH_3 \xrightarrow{Si_3N_4} 3H_2 + N_2 \qquad (1)$$

is established as a competing reaction to:

$$4NH_3 + 3SiF_4 \rightarrow Si_3N_4 + 12HF \qquad (2)$$

in the CVD reactor under deposition conditions. Mass spectroscopic measurements, performed on the reactor exhaust, confirm that the primary gas phase species are N_2, H_2 and HF [5]. H_2 is observed spectroscopically both in the presence and absence of SiF_4.

Experimental

The experimental setup has been discussed in detail elsewhere [5]. Collinear narrowband scanning CARS is employed. The frequency doubled output of a Nd:YAG laser (pump beam) is divided, one portion being used to pump a dye laser. The output of the dye laser (probe beam) is combined with the pump beam in a collinear and copropagating manner. The resultant dichroic beam is focused with a 50 cm focal length lens into the CVD reactor shown schematically in Fig. 1. After recollimation at the reactor exit, the signal is separated from the pump and a probe beams by a dichroic mirror. Further noise reduction is accomplished by filters and monochromater, after

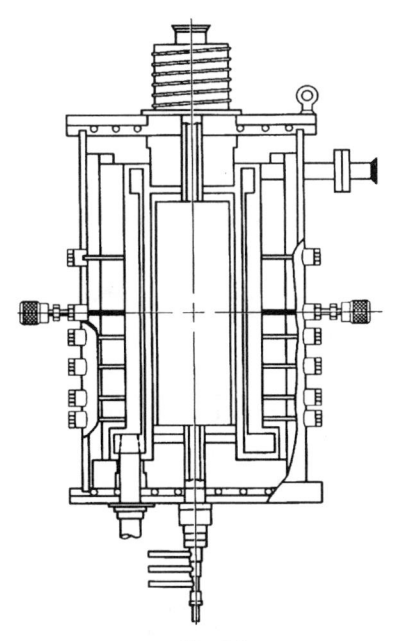

Figure 1
Schematic of diagnostic reactor. All optical access ports are shown plugged with the exception of the second highest. Reactants are injected from the reactor bottom and exhausted at the top.

which the signal impinges on a PMT. The PMT output is processed by a boxcar integrator and stored on a PDP 11/34 microcomputer for display and further processing. The pump and probe beams are further utilized to produce a nonresonant CARS signal by focusing them into a high pressure (50 - 80 psi) cell, normally filled with N_2. This nonresonant signal is recovered, processed and stored in a similar manner to the resonant signal.

In the course of an experiment, the CARS system is aligned at room temperature in such a manner that both resonant and nonresonant signals are maximized by small changes in the position of the beam combiner. During the heating cycle, small changes in the position of the dichroic are required to correct for changes in the optical index of the reactor atmosphere. Thermal expansion of the retort wall can also occur, resulting in a partial eclipsing of the optical access path. This is followed during the experiment by monitoring the magnitude of the nonresonant signal generated in the reference cell. Any fluctuation in laser power, including partial eclipsing of the pump and probe beams, affects the nonresonant signal intensity. The spectra employed in data analysis have all been divided by the nonresonant signal to compensate for any fluctuations in laser intensity that occur.

Results and Discussion

CARS spectra of H_2 were obtained in the reactor at heights of 5.1, 12.7 and 20.3 cm above the injector assembly. Figure 2 shows spectra obtained near reference deposition conditions (1390

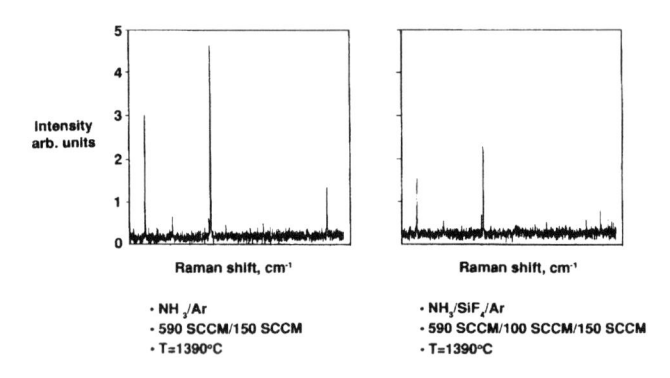

Figure 2
CARS spectra of H₃ produced by the decomposition of NH₃ on a Si₃N₄ surface

C, 2.3 torr, 590 sccm NH₃ and 100 sccm SiF₄) and under similar conditions in the absence of SiF₄. The spectra arise from the Q branch ($\Delta J = 0$) of the $v = 1$ to $v = 0$ vibrational transition near 4395 cm^{-1}. At the elevated temperatures employed for Si₃N₄ deposition, the spectrum is dominated by the odd J rovibrational transitions Q(1), Q(3), and Q(5); with the most intense spectral line being Q(3). The right trace corresponds to deposition conditions, although only wall deposition was occurring during these tests. The retort wall is graphite but has been coated with Si₃N₄ to a level approximately 15 cm above the injectors. The chemistry could be complicated in this instance by possible etching of Si₃N₄ by HF formed as a deposition by product:

$$4NH_3 + 3SiF_4 \rightarrow Si_3N_4 + 12HF \quad (3)$$

$$12HF + Si_3N_4 \rightarrow 3SiF_4 + 6H_2 + 2N_2 \quad (4)$$

the observed H₂ could be attributable to equation 1 and/or equation 4. The left trace corresponds to the chemically simplified conditions in the absence of SiF₄. No deposition occurs, thus the H₂ origin can only be attributable to catalytic decomposition of NH₃ on the Si₃N₄ surface or etching of the graphite surface by NH₃. This is confirmed by two observations. First, no hydrogen CARS signal (from ammonia decomposition) is observable in a "clean" retort. The reactor was first run at reference operating conditions but in the absence of SiF₄, these conditions generate the maximum H₂ in a conditioned (Si₃N₄ covered) retort. No H₂ CARS signal was observable, even at pressures up to 10 torr. SiF₄ was added and the reactor was operated through two normal coating cycles, providing a fresh Si₃N₄ coating on the inner retort wall. The experiment was then repeated (flowing ammonia only) under identical conditions, a large H₂ CARS signal was now observable. This indicates, that in the absence of SiF₄, the hydrogen is not produced by the etching of graphite but by catalytic decomposition on the Si₃N₄ surface. The second collaborating observation is that the NH₃ CARS signal disappears above 1000 C. The absolute sensitivity of our system to ammonia was determined in a pyrex cell at temperatures up to 1200 C, and indicated that NH₃ should be observable in the reactor unless removed by a reaction other than Si₃N₄ formation.

To determine if H₂ production is in direct competition with Si₃N₄ formation, the experiments were repeated at stoichiometric partial flow rates. That is, total flow rate, pressure and temperature

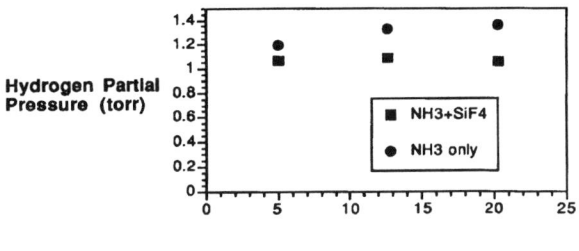

Figure 3
Axial concentration profile of H$_2$ in the reactor

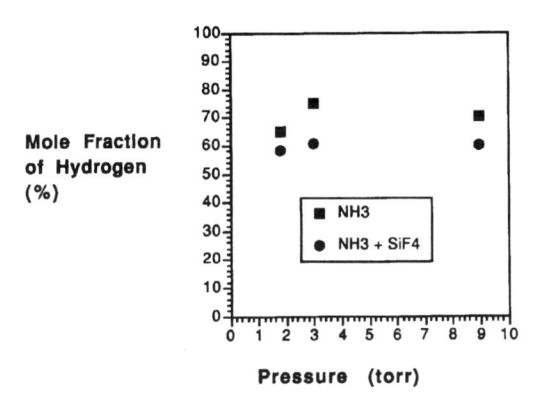

Figure 4
H$_2$ mole fraction (%) as a function of pressure at the 5.1 cm observation port.

were kept constant; but the NH$_3$ and SiF$_4$ flow rates were adjusted to 390 sccm and 290 sccm respectively. Under these conditions H$_2$ is still observed, however the spectrum is near the limit of sensitivity to H$_2$ (0.5 torr, 1400 C), so no quantitative information is available.

Figure 3 depicts the H$_2$ concentration measurements made at reference deposition conditions as a function of axial position downstream of the injector assembly. Each data point represents the average of approximately three separate experimental measurements. As no further decomposition is possible above the 15 cm location, measurements at higher ports were not attempted. The trend apparent from the data in Fig. 3 is that the H$_2$ concentration increases slightly with increasing distance from the injectors when only ammonia is injected but appears to have reached a steady state value by the second observation port. When SiF$_4$ is added however, the H$_2$ concentration has reached this state at the lowest detection port. Since CFD calculations indicate that thermal equilibration occurs within a few cm of the injector [3], the main effect of flowing downstream is an increase in the total number of heterogeneous collisions. One would expect a similar behavior if the collision frequency was raised by increasing the total pressure, while keeping constant the point of observation. Figure 4 shows the effect of increasing total pressure in both cases. As observed in the previous data, the mole fraction of hydrogen, from NH$_3$ only, increases slightly with increasing pressure until reaching a constant value at approximately 0.75 (The stoichiometric limit.) While the H_2 mole fraction (≈ 0.62) produced under deposition conditions (6:1 mole ratio of NH_3 to SiF_4) is independent of pressure under the limited set of conditions investigated.

If the deposition of silicon nitride competes equally with the decomposition of ammonia (and the mole ratio of ammonia to silicon tetrafluoride is 6:1) the global equation can be written as:

$$3SiF_4 + 18NH_3 \xrightarrow[\Delta]{Si_3N_4} Si_3N_4 + 7N_2 + 21H_2 + 12HF \qquad (5)$$

This reaction produces a maximum expected hydrogen mole ratio of 52.5%. The observed ratio is about 59% and appears independent of total pressure. This could indicate that ammonia decomposition occurs faster than silicon nitride deposition or that equation (4) produces H_2 as well as equation (5). The problem with interpreting this data lies in the observation that coating continues to occur up to a distance of approximately 15 cm downstream of the injector assembly. A source of nitrogen must exist at least to this point in the retort. Ammonia decomposition occurs throughout the reactor wherever ammonia encounters the silicon nitride surface. Back diffusion of the hydrogen then occurs to reduce the observed concentration gradient of hydrogen. This effect must be considered in interpreting the H_2 concentration data.

Acknowledgements

This research is being supported by DARPA-DSO through WRDC Contract F33615-89-C-5628. Major Joseph Hager (WRDC) is the contract monitor, and Mr. William Barker (DARPA) is the sponsor. The technician services of Edward Dzwonkowski and Paul Lasewicz contributed greatly to the success of the measurements described herein. In addition, discussions and encouragement from Drs. Alan Eckbreth, Jim Strife, Woo Lee and Meredith Colket III of UTRC, and Drs. Bob Kee, Pauline Ho, Mike Coltrin, Duane Outka, Rich Larson and Greg Evans of Sandia National Laboratory were extremely valuable.

References

1. Strife, J. R., "Development of High Temperature Oxidation Protection for Composites," Annual Report R88-916789-16, Naval Air DC, Warminster, PA, (1988).

2. F. S. Galasso, R. D. Veltri, and W. J. Croft, "Chemically Vapor Deposited Si_3N_4," Powder Met. Int., 11 (7) (1979).

3. Strife, J. R. and Kee, R. J., "Manufacturing Science of Silicon Nitride Chemical Vapor Deposition," Annual Report for WRDC Contract F33615-89-C-5628, (1991).

4. S.O. Hay, R.D. Veltri, W.Y. Lee, and W.C.Roman, Proceedings of the SPIE 1435, paper #51 Los Angeles CA (1991).

5. S.O. Hay, W.C. Roman, and M.B. Colket, J. Mater. Res., 5 (11), 2387-2397 (1990).

LASER INDUCED FLUORESCENCE FOR TEMPERATURE MEASUREMENT IN REACTING FLOWS

R.G. JOKLIK
National Institute of Standards and Technology, Gaithersburg, MD 20899

ABSTRACT

OH vibrational Thermally Assisted Fluorescence (THAF) temperature measurements have been demonstrated in both premixed and diffusion flames. The accuracy of the measurements is generally better than 100 K over a wide range of flame conditions for which the collisional quenching rate varies considerably. Application of this technique for temperature measurement in Chemical Vapor Deposition (CVD) flows, for which the quenching rate is relatively constant, should exhibit greater accuracy. THAF measurements in these flows are limited by signal to noise considerations, and should be possible down to pressures of 10^3-10^4 Pa or less.

INTRODUCTION

The application of laser based optical diagnostic techniques to reacting flows is attractive due to the non-intrusive nature of these techniques and to the excellent spatial and temporal resolution with which measurements can be made. Laser Induced Fluorescence (LIF), with its comparatively large signals, is particularly suitable for making measurements on the many reactive intermediates that are found in such flows. In addition to concentration, LIF measurements can also yield information on internal state distributions, and hence temperature. Several different LIF approaches have been developed for measuring temperature, including excitation scans for measuring ground state temperatures [1-4], saturated and linear two-line fluorescence [5,6], and thermally-assisted fluorescence (THAF) [7-11].

In this work the application of THAF to temperature measurements in flames is described, and its possible application for temperature measurements in reacting flows used for materials synthesis, such as those typical of CVD systems, is discussed. The THAF technique uses the collisional redistribution of population among internal degrees of freedom following laser excitation to measure temperature. This is accomplished by spectrally resolving the fluorescence from the collisionally populated states. The THAF temperature measurements in flames are based on vibrational transfer in $A^2\Sigma^+$ OH following (0-0) band laser excitation from the $X^2\Pi$ ground state. Initial studies employing this approach were carried out by Crosley and coworkers [10,11] in a premixed methane-air flame for a limited number of conditions and demonstrated that temperature measurements could be made successfully, although calibration was required for each flame condition due to the variability of the collisional quenching rate. This early work has been extended through a detailed study of the effects of quenching in a wide variety of methane, ethylene, and acetylene flames [12], and a calibration procedure has been developed that accounts for these effects without requiring knowledge of the local flame conditions. Examples of THAF temperature measurements made in several premixed and diffusion flames are discussed below.

THEORY AND APPARATUS

Vibrational THAF involves the collisional transfer of population from the laser excited level (v'=0 in this case) to other vibrational levels within the laser excited electronic manifold. A two-level model of this process considering only the laser excited level and the adjacent higher level (Fig. 1) can be used to relate the populations of these two levels to temperature. A steady-state population balance for the higher of the two levels (v'=1) is assumed and the principle of detailed balancing invoked to relate the upward and downward rates of vibrational transfer. This yields the following expression for the temperature,

$$T = \frac{-\Delta E_{10}/k}{\ln\left[\left(1+\frac{Q+A}{V_{10}}\right)\frac{N_1}{N_0}\right]}, \quad (1)$$

where ΔE_{10} is the energy difference between the v'= 1 and 0 levels, N_1 and N_0 are their respective populations, k is Boltzmann's constant, Q is the collisional quenching rate, V_{10} is the downward vibrational transfer rate, and A is the spontaneous emission rate. For atmospheric pressure flame measurements Q >> A, and therefore the term multiplying N_1/N_0, which causes a departure from a Boltzmann population distribution, becomes (1+Q/V) and is independent of pressure. Thus temperature may be determined through knowledge of the relative populations of v'=0 and v'=1 and the collisional term Q/V.

The apparatus used for making the OH vibrational THAF measurements employed a standard 90° LIF geometry. The details of the apparatus have been described elsewhere [12,13], and only the main features are described below. The output of a Nd:YAG pumped dye laser system was frequency doubled and tuned to the $Q_1(6)$ rotational transition of the OH A-X (0-0) band at approximately 308.8 nm. The pulse energy delivered to the probe volume was approximately 25 μJ with a spot size of 1-2 mm. The OH (1-0) and (0-1) band fluorescences at 285 nm and 343 nm were simultaneously detected using a dual channel system consisting of a pair of photomultipliers and interference filters.

The measurements in premixed flames were performed in the post flame zones of a series of C_2H_2, C_2H_4, and $CH_4/O_2/N_2$ flames using a water cooled, capillary tube burner. The OH THAF measurements were compared to sodium line reversal measurements [12], and also to equilibrium calculations and OH ground state rotational temperatures determined by absorption. The diffusion flame measurements were performed using an ethylene jet diffusion flame described by Santoro et al. [14] as their flame 2. The OH THAF measurements were compared directly to the thermocouple measurements of Santoro et al.

TEMPERATURE MEASUREMENTS IN FLAMES

In order to make OH vibrational THAF temperature measurements in flames a calibration must first be performed [12]. Briefly, the change in Q/V in a premixed flame as composition is varied is determined by inverting Eqn. 1 and calculating Q/V using the measured OH fluorescences and temperatures obtained independently through sodium line reversal measurements for each flame condition. A linear fit to Q/V vs. temperature is then carried out and used to replace the Q/V term in Eqn. 1. This calibration then allows

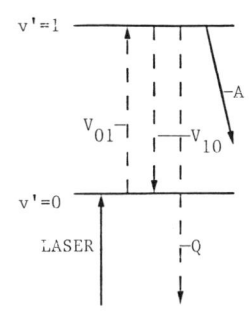

Figure 1. Two level model of vibrational transfer.

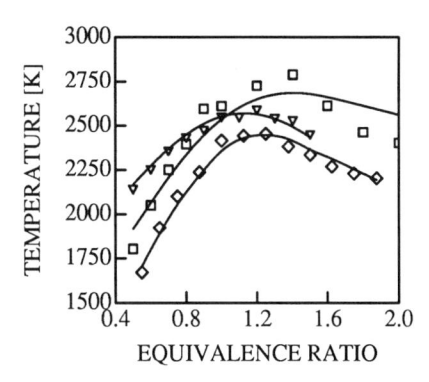

Figure 2. Comparison of OH vibrational THAF temperatures (diamonds-C_2H_2, triangles-CH_4, squares-C_2H_4) to sodium line reversal measurements (solid lines) versus equivalence ratio in premixed flames.

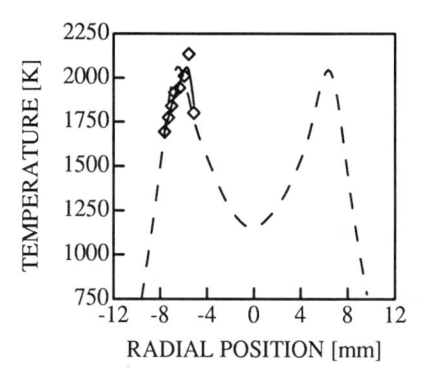

Figure 3. OH vibrational THAF temperatures (diamonds and solid line) and thermocouple measurements (dashed line) in an ethylene jet diffusion flame at a height of 2 cm.

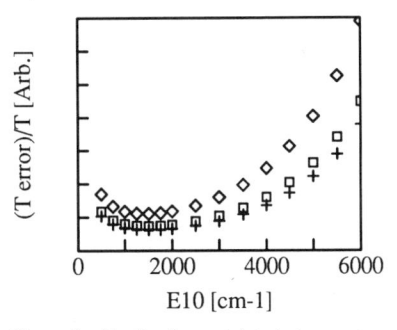

Figure 4. Fractional uncertainty in temperature versus vibrational energy spacing for (Q+A)/V values of 1 (diamonds), 0.3 (squares) and 0.1 (pluses).

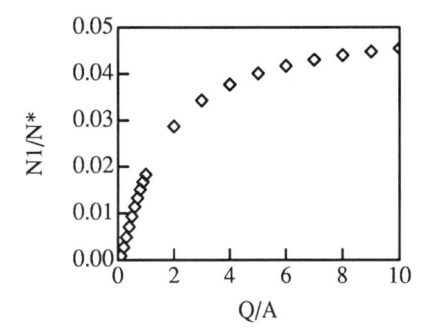

Figure 5. Variation of normalized population of v'=1 versus Q/A, the ratio of quenching rate to spontaneous emission rate. Q/A is proportional to pressure.

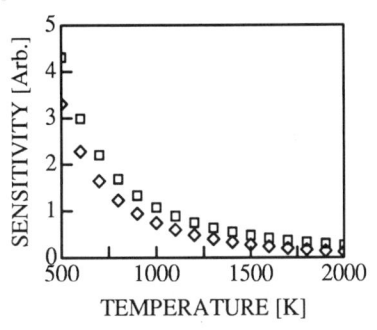

Figure 6. Sensitivity of the OH THAF measurement to changes in temperature as a function of temperature for constant Q/V (squares) and Q/V a linearly increasing function of temperature (diamonds).

calculation of temperature from the OH fluorescences without the need for knowledge of the flame conditions. The extent to which this linear approximation of Q/V with respect to temperature holds determines the accuracy of the THAF measurement. The resultant OH vibrational THAF temperatures in a series of methane, ethylene, and acetylene premixed flames are shown as a function of equivalence ratio in Fig. 2. The solid lines are cubic spline fits to the sodium line reversal measurements. In general, the agreement of the THAF measurements with the sodium line reversal temperature is better than 100 K except in the most fuel rich cases. The 3σ uncertainty of the measurements, which are based on 400 laser shot averages, is approximately 120 K.

OH THAF measurements in the diffusion flame based on a calibration performed in a premixed ethylene flame are shown along a diameter at a height of 2 cm in Fig. 3. This axisymmetric flame consists of an annular combustion zone with fuel from the central jet in the inner core and air on the outside. Traversing outward along a radius one first passes through a sooting zone, which can give rise to scattered radiation and broadband fluorescence, and then through a thin zone where OH is present. The OH THAF measurements shown in Fig. 3 have been corrected for both non-resonant background light and also for self-absorption of the fluorescence, and are denoted by the diamonds and the solid line cubic spline fit. The thermocouple measurements are denoted by the dashed line. The agreement between the two measurements is within 75 K.

TEMPERATURE MEASUREMENTS IN REACTING FLOWS

The results discussed above show that accurate vibrational THAF temperature measurements can be made in high temperature flames characterized by a variable collisional environment. The reacting flows used for materials synthesis, such as those typical of the CVD process, are characterized by lower temperatures, lower pressures, and a less variable collisional environment since the bulk of the flow is composed of a carrier gas such as argon or hydrogen. In order to assess the applicability of the THAF technique to such flows the influence of these factors on signal levels, and hence degree of measurement uncertainty, and on sensitivity to changes in temperature must be considered. The fractional uncertainty in the temperature is given by

$$\frac{S_T}{T} = \frac{kT}{\Delta E}\frac{S_R}{R} = \frac{kT}{\Delta E}\left\{(\frac{S_n}{N_o})^2+(\frac{S_n}{N_1})^2\right\}^{1/2} = \frac{kT\,S_n}{\Delta E\,N_0}\left\{1+\frac{(1+\frac{Q+A}{V})^2}{e^{-2\Delta E/kT}}\right\}^{1/2} \quad (2)$$

where S_T is the resultant uncertainty in the temperature, R is the ratio of the fluorescences from $v'=1$ and 0 and S_R is the uncertainty in that ratio, and S_N is the uncertainty in the populations of $v'=1$ and 0. Constants of proportionality have been omitted in progressing from left to right. Since signal to noise is limited by the amount of population found in N_1, the final form of Eqn. 2 was derived by assuming a constant uncertainty in the measurement of N_1 and N_0, S_N, and substituting an expression relating N_1 to temperature, vibrational energy difference, and population transfer rates. Note that as population loss from quenching or spontaneous emission becomes large compared to the vibrational transfer rate ($Q+A >> V$) the uncertainty in the temperature increases. This can be seen in Fig. 4, in which the fractional temperature uncertainty has been plotted against vibrational energy difference for three values of (Q+A)/V. Since Q and V are both pressure dependent while A is not, the population of v'=1 becomes independent of pressure for Q >> A. This is illustrated directly in Fig. 5, in which the normalized population of v'=1 is plotted against the pressure dependent ratio Q/A for a constant Q/V (which is pressure independent). Significant loss of population starts to occur at a Q/A value of about one. Uncertainty also increases for large $\Delta E/kT$ since not as much population transfer occurs to v'=1 at low temperature or for a large energy difference (Fig. 4). For small $\Delta E/kT$ the decreased sensitivity also causes larger uncertainty. Best performance is obtained for $\Delta E/kT$ around 2. Variation in Q/V with temperature, such as that found for OH in flames, also affects the sensitivity of the technique. This is shown in Fig. 6, from which it can be seen that the variation in Q/V for OH reduces the sensitivity of the technique by about a factor of two at 2000 K. In CVD systems this effect is not expected to occur.

Parameters necessary for assessing the suitability of several candidate species for use in vibrational THAF temperature measurements in silicon processing CVD flows in terms of the considerations discussed above are shown in Table I. The ratio $\Delta E/kT$ is listed for each species for a temperature of 1000 K, and the pressure at which Q/A = 1 based on the quenching cross-sections listed has been calculated. For SiH, SiO, and SiF a quenching cross-section of 1 Å^2 for collisions with Ar is assumed due to the lack of available data. With the exception of SiO, it appears that vibrational THAF should be feasible with these species without a loss of signal strength of more than 50% down to pressures in the range of 10^3 - 10^4 Pa. This comparison is based on differences in the radiative lifetimes (τ_{rad}) and illustrates the effects of pressure. Increases in Q/A due to larger quenching cross-sections will lower the pressure limits shown in Table I but will also result in lower absolute signal levels since the populations in both v'=0 and 1 in the Q >> A limit are inversely proportional to the quenching cross-section. Finally, it appears that the vibrational energy spacing of these species is within acceptable bounds for the temperature range of interest, with SiH probably the most attractive species based on the criteria set forth here.

TABLE I

Radiative lifetimes and quenching rates for some Si containing species.

Species	ΔE (cm^{-1})	$\Delta E/kT$ [a]	$\tau_{rad}=1/A$ (ns)	σ_Q (Å^2)	P(Pa)[b]
OH	2993 [15]	2.1	693 [15]	0.42 (Ar)[16]	4.4 x 10^4
				6.5 (H$_2$)[17]	8.0 x 10^2
SiH	1661 [15]	1.2	534 [18]	†	2.4 x 10^4
SiO	840 [15]	0.59	10-30 [15]	†	7.0 x 10^5
SiF	698 [15]	0.50	230 [15]	†	3.1 x 10^4

† σ_Q = 1 Å^2 assumed.
[a] for T = 1000 K.
[b] for Q/A = 1.

CONCLUSIONS

Vibrational THAF thermometry using OH in flames has been demonstrated over a wide range of conditions in premixed flames. It has also been shown that it is possible to make OH THAF temperature measurements in a diffusion flame based on a calibration performed in a premixed flame. This calibration takes into account changes in the collisional quenching rate as flame conditions change, and is both the limiting factor in the accuracy of OH THAF flame temperature measurements as well as a factor in reducing the sensitivity of the measurements. In considering the application of this technique to CVD flows, it is not expected that such a calibration procedure will be necessary, as collisions are primarily with the carrier gas species. Based on consideration of the factors that limit the sensitivity and uncertainty levels of THAF thermometry and estimation of the relevant parameters for species of interest, it appears that vibrational THAF thermometry should be feasible in CVD flows down to pressures of 10^3 - 10^4 Pa or less.

Acknowledgement: This work was supported by Naval Air Systems Command grant No. N0001991IPB744R.

REFERENCES

1. Anderson, W.R., Decker, L.J., and Kotlar, A.J., Combust. and Flame, **48**, 163 (1982).
2. Crosley, D.R. and Smith, G.P., Combust. and Flame, **44**, 27 (1982).
3. Rea, E.C. Jr., and Hanson, R.K., Appl. Opt., **27**, 4454 (1988).
4. Rensberger, K.J., Jeffries, J.B., Copeland, R.A., Kohse-Hoinghaus, K., Wise, M.L., and Crosley, D.R., Appl. Opt., **28**, 3556 (1989).
5. Lucht, R.P., Laurendeau, N.M., and Sweeney, D.W., Appl. Opt., **21**, 3729 (1982).
6. Cattolica, R., Appl. Opt., **20**, 1156 (1981).
7. Zizak, G., Horvath, J.J., Van Dijk, C.A., and Winefordner, J.D., J. Quant. Spectrosc. Radiat. Transfer, **25**, 525 (1981).
8. Elder, M.L., Zizak, G., Bolton, D., Horvath, J.J., and Winefordner, J.D., Appl. Spec., **38**, 113 (1984).
9. Joklik, R.G., Horvath, J.J., and Semerjian, H.G., Appl. Opt., **30**, 1497 (1991).
10. Crosley, D.R. and Smith, G.P., Appl. Opt., **19**, 517 (1980).
11. Dyer, M.J. and Crosley, D.R., AFWAL-TR-84-2045 (1984).
12. Joklik, R.G., Combust. Sci. and Tech., to appear.
13. Joklik, R.G., to appear in "Temperature, Its Measurement and Control in Science and Industry," Vol. 6, AIP.
14. Santoro, R.J., Yeh, T.T., Horvath, J.J., and Semerjian, H.G., Combust. Sci. and Tech., **53**, 89 (1987).
15. Huber, K.P. and Herzberg, G., "Molecular Spectra and Molecular Structure, Vol. 4. Constants of Diatomic Molecules," (Van Nostrand Reinhold, New York, 1979).
16. Lengel, R.K. and Crosley, D.R., J. Chem. Phys., **68**, 5309 (1978).
17. Copeland, R.A. and Crosley, D.R., Chem. Phys. Lett., **107**, 295 (1984).
18. Bauer, W., Becker, K.H., Duren, R., Hubrich, C., and Meuser, R., Chem. Phys. Lett., **108**, 560 (1984).
19. Nemoto, M., Suzuki, A., Nakamura, H., Shibuya, K., and Obi, K., Chem. Phys. Lett., **162**, 467 (1989).

IN-SITU OPTICAL EMISSION SPECTRA OF Ti, TiN AND TiSi$_2$ PLASMA DURING THIN FILM GROWTH BY PULSED LASER EVAPORATION

S. Pramanick and J. Narayan
Department of Materials Science and Engineering, North Carolina State University, Raleigh NC 27695.

ABSTRACT

The first optical emission spectra of Ti, TiN and TiSi$_2$ plasma during growth of thin films by pulsed laser evaporation has been reported. A KrF excimer laser (248nm) with pulsewidth of 45ns operating on 5Hz repitition rate was used for deposition of refractory metal thin films using varying laser fluence of 4J/cm^2 to 15J/cm^2. Most of the radiative species seen in plasma belongs to atomic neutrals and ionic species such as Ti I, Ti II, Si I, Si II and N II. Emission spectra was mostly dominated by neutral and ionic emission from Titanium (Ti I and Ti II).

INTRODUCTION

Pulsed laser evaporation (PLE) has emerged as the dominant technique for growth of stoichiometric thin films because it has produced high Tc superconducting thin films of better quality than those obtained by all other deposition technique [1,2]. In the last few years, PLE has been used, increasingly, for the deposition of compound thin films such as TiN, TiSi$_2$, CoSi$_2$, BaTiO$_3$, PbTiO$_3$ etc. [3,4]. Lately much of efforts in our group has been devoted to the growth of TiN, TiSi$_2$, CoSi$_2$ thin films. In ULSI processing, silicides are increasingly being used for S/D contact metallization and local interconnects for polysilicon gate with TiN acting as a diffusion barrier. PLE has the advantage of producing silicides [4] and TiN films with a very sharp interface [3], which is essential for low junction leakage of submicron MOSFET.
The deposition mechanism involved in PLE is not very well understood. Most of the experimental and theoretical research by groups working in this field has been devoted to understand the mechanism involved in YBCO high T$_c$ superconducting thin film growth [5,6,7]. In PLE, the physical processes of thin film deposition can be divided into three regimes: 1) interaction of laser beam with bulk target, 2) initial isothermal expansion, and 3) adiabatic expansion of HT-HP plasma leading to the thin film deposition [5]. It has been now known that the YBCO and PrBCO plasmas consists of two components : 1) neutral and ionic atomic beams with a high velocity component, and 2) a low velocity component involving molecular species [6]. This has been independently confirmed by optical emission spectroscopy [6] and quadropole mass spectroscopy [8]. We have studied the optical emission from laser produced plasma of Ti, TiN and TiSi$_2$ during thin film growth to understand the nature of growth by PLE and of the species present in these plasmas. It will be of practical importance to be able to relate the quality of thin film growth with the species present and with the complex transport processes involved in plasma plume.

EXPERIMENTAL

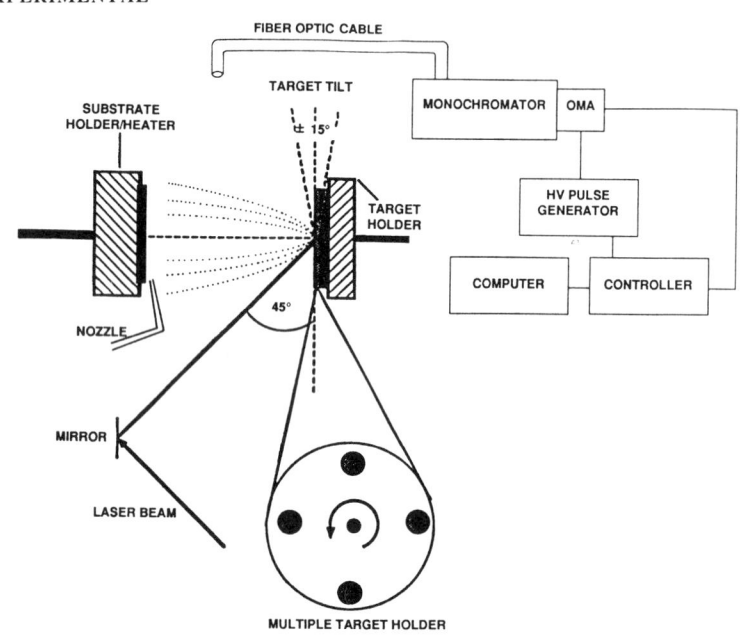

Figure 1. Schematic diagram for pulsed laser deposition and optical emission spectroscopy setup.

A schematic diagram of the experimental setup is shown in fig.1. We have used bulk pellets of Ti, TiN and TiSi$_2$ as targets for pulsed laser evaporation. The optical emission was taken in the range of 200 to 900nm using an intensified diode array based optical multichannel analyzer (OMA). A lambda physik excimer laser operating on 248nm (KrF) wavelength with 45ns pulsewidth was used for generation of plasma plume from the target. Prior to deposition of thin film, the UHV chamber was evacuated to a base pressure of 2×10^{-7} Torr. During the deposition, substrate temperature was kept at 600 °C. We have used laser fluence of 5J/cm^{-2} and a repitition rate of 5Hz during the deposition of thin film. Emission from the plasma plume was collected normal to its forward direction using a UV grade fibre optics cable with aperture of 2.1, connected to the input slit of the monochromator. The OMA and monochromator setup was calibrated using mercury(Hg) and krypton(Kr) lamps.

RESULTS AND DISCUSSION

Optical emission spectra from Ti, TiN and TiSi$_2$ plasma are shown in the range of 200 to 900nm in two segments in Fig.2, Fig.3, and Fig.4, respectively. As shown in Fig.2a, prominent lines in the emission spectra of Ti has been assigned to atomic neutral and ionic species such as Ti I and Ti II, respectively.

Fig. 2.a.

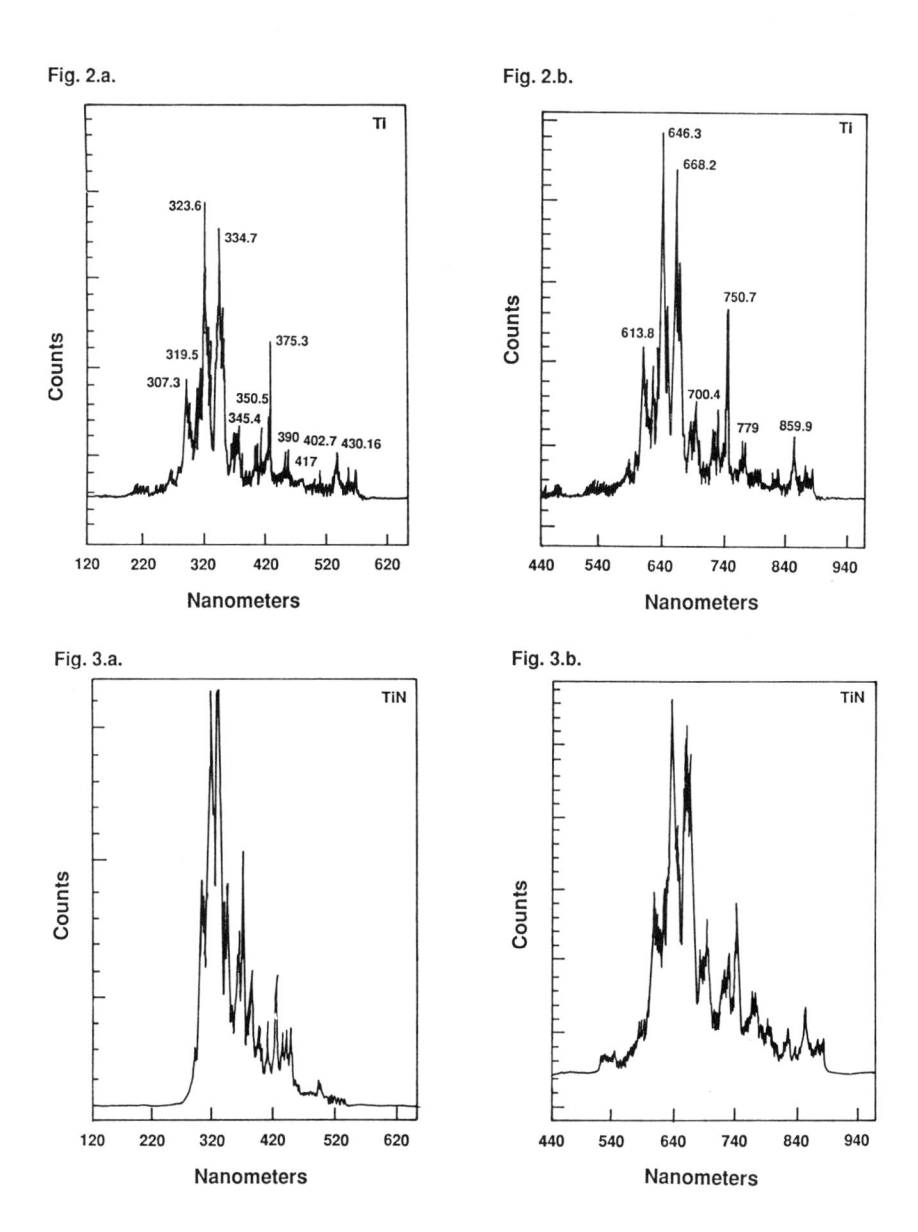

Fig. 2.b.

Fig. 3.a.

Fig. 3.b.

Figure 2. and Figure 3. Emission from laser ablation plasma during thin film growth of Ti and TiN, respectively.

Individual peaks such as, at 307.3nm is assigned to Ti II (306.62nm, 307.29nm, 307.86nm, 308.80 nm), peak at 319.5nm to Ti I (318.64nm, 319.19nm) and Ti II (319.08nm), peak at 323.6nm to Ti II (323.45nm, 323.65nm, 323.90nm), peak at 334.7nm to Ti I (334.18nm) and Ti II (334.90nm,334.94nm), peak at 345.4nm to Ti II (344.43nm, 346.15nm), peak at 350.5nm to Ti I (350.66nm) and Ti II (350.48nm, 351.08nm), peak at 375.3nm to Ti I (375.28nm, 375.36nm) and Ti II (375.93nm, 376.13nm), peak at 390nm to Ti I (390.47nm) and Ti II (390.05nm), peak at 402.7nm to Ti I (402.45nm), peak at 417nm to Ti I (417.1nm) and Ti II (417.19nm), peak at 430.16nm to Ti I (429.57nm,429.86nm,430.05nm.430.59nm), and peak at 454.52nm to Ti I (453.32nm,453.55nm). In Fig.2b, prominent peaks at 613.8nm, 646.3nm, 668.2nm, 700.4nm, 750.7nm, 779nm, and at 859.4nm are due to the second order effect in the monochromator. The above emission lines have been identified using standard spectroscopic tables [9]. In Fig.3, apart from strong Ti emission, possible nitrogen emission peaks has been identified at 399.5nm (N II), 417.6nm (NII) and 44.7nm (NII). Similarly, in Fig.4, apart from abundant Ti emission peaks, emission due atomic silicon neutrals (Si I) and ionic (Si II) species has been assigned to peaks at 334.7nm (Si II), 385nm (Si II), 391.67nm (Si I), 417nm (Si II) and 505nm (Si II).

Emission due to Ti I and Ti II seems to dominate plasma emission from both TiN and TiSi$_2$, which is also energetically favorable. It should be noted

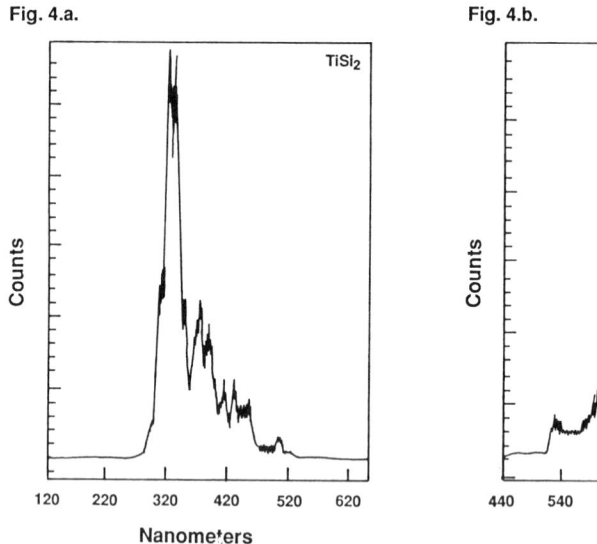

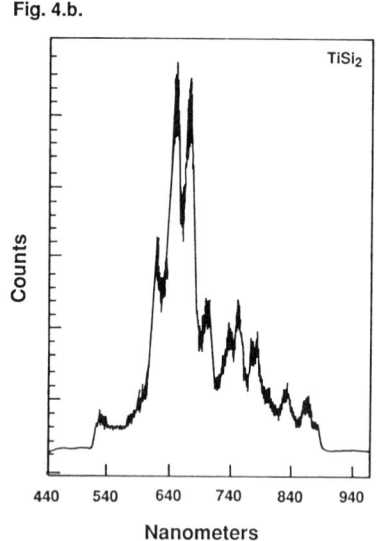

Fig. 4.a. **Fig. 4.b.**

Figure 4. Emission from laser ablation plasma during TiSi$_2$ thin film growth.

here that first ionization energy for Ti (6.82 eV) is much smaller then both N (14.54 eV) and Si (8.15 eV). During TiN deposition, weak and broad molecular fluroscence centred around 345nm has been detected as shown in Fig3a, which is completely absent in Ti emission spectra shown in Fig.1a. A bright visible plasma can be seen during Ti and TiN deposition even at lower laser fluence (less than 10J/cm2), whereas faint visible plasma from TiSi$_2$ can be only seen at laser fluence of 15J/cm^2 and higher. Emission linewidths for Ti and TiN can be seen to be narrower than TiSi$_2$. This is because a 50 μm slitwidth was used for Ti and TiN, whereas a 250 μm slitwidth was used for TiSi$_2$. Also, there is a considerable amount of spectral overlap due to Si emission lines. Although radiative emission due to Ti is dominant in the case of both TiN and TiSi$_2$ plasma, RBS compositional analysis of deposited thin films have shown the ratio of atomic concentrations of nitrogen to titanium is 1 at substrate temperatures above 400 °C and the ratio of atomic concentrations of silicon to titanium is 2 [3,4]. Measured resistivities are on the order of 150 μohm-cm for TiN films and 18 μohm-cm for TiSi$_2$ films [3,4]. In PLE deposition of Ti, TiN and TiSi$_2$ thin films, arrival of high energy atomic beams of Ti, N and Si at the substrate is the dominant contributor to growth. Presence of molecular compounds such as TiN and TiSi$_2$ were not found in the plasmas. We are unable to detect any molecular emissions due to compounds such as TiN and TiSi$_2$. For PrBCO plasma, an abundance of Pr$_2$O$_3$ in the plasma was supported by the presence of strong Pr$_2$O$_3$ molecular fluroscence [6].

The reason for lack of molecules such as TiN and TiSi$_2$ in plasma may be due to the use of much higher laser fluence of 15 J/cm^2. For PrBCO plasma, a much smaller laser fluence of 1.5 J/cm^2 was used. Also, formation of Pr$_2$O$_3$ is much more energetically favorable than formation of TiN and TiSi$_2$ (standard heat of formation of Pr$_2$O$_3$, TiN and TiSi$_2$ are -444.5 Kcal/g-mole, -73 kcal/g-mole and -32 Kcal/g-mole, respectively, @ 300 K [9].). The emission linewidths of neutral and ionic species of Ti and TiN plasma are almost same, which are broadened due to stark broadening seen in high density (~10^{18} cm^{-3}) laser plasma. It can be presumed that the temperature of plasma is about ~ 10,000 Kelvin, which has been measured for similar laser induced plasma using similar energy density. We were unable to mesaure the plasma temperature using local thermal equilibrium approximation [10] as our emission spectra are intregrated spatially between target and substrate.

CONCLUSIONS

We have presented emission spectra of Ti, TiN and TiSi$_2$ plasma during growth of thin film by pulsed laser evaporation. Arrival of high energy elemental atomic beams at the substrate have been found to play the most important role during the growth of TiN and TiSi$_2$ thin films. Emission due to Ti neutral and ionic species seems to dominate emission spectra of Ti, TiN and TiSi$_2$ plasma. RBS compositional analysis showed stoichiometric growth of TiN and TiSi$_2$, even if emission due to nitrogen and silicon ions are very weak.

REFERENCES

[1] D. Dijkkamp, T. Venkatesan, X. D. Wu, S. A. Shaheen, N. Jisrawi, Y. H. Min-Lee, W. L. McLean, and M. Croft, Appl. Phys. Lett. 51, 619 (1987).

[2] R.K. Singh, J. Narayan, A.K. Singh, and J.Krishnaswamy, Appl. Phys. Lett. 54, 2271 (1988).

[3] N. Biunno, J. Narayan, S. K. Hofmeister, A. R. Srivatsa, and R. K. Singh, Appl. Phys. Lett. 54, 1519 (1989).

[4] P. Tiwari, M. Longo, G. Matera, S. Sharan, P. L. Smith, and J. Narayan, TMS meeting proceedings (1991).

[5] R. K. Singh, and J. Narayan, Phys. Rev. B 41, 8843 (1990).

MOLECULAR PRECURSORS TO BORON NITRIDE THIN FILMS: THE REACTIONS OF DIBORANE WITH AMMONIA AND WITH HYDRAZINE ON Ru(0001)

Charles M. Truong, José A. Rodriguez, Ming-Cheng Wu and D. W. Goodman
Texas A&M University, Department of Chemistry,
College Station, Texas 77843-3255

ABSTRACT

The coadsorption and reaction of diborane with ammonia and with hydrazine on Ru(0001) have been studied using X-ray photoelectron spectroscopy (XPS) and thermal desorption mass spectroscopy (TDS). Diborane is found to decompose to atomic boron and hydrogen upon adsorption at T>200K. Multilayers of diborane and ammonia, deposited at 90K on Ru(0001), react when annealed to 600K. The XPS results indicate that boron-nitrogen adlayers can be formed by this reaction. These boron-nitrogen films are boron-rich and decompose at temperatures higher than 1100K. Our TDS studies reveal that hydrazine decomposes extensively to NH_3, N_2, N and H on Ru(0001). Due to its higher reactivity, boron-nitrogen films of B/N stoichiometric ratio near unity are obtained when hydrazine is used rather than ammonia. In our studies, these films were formed by either simultaneously dosing B_2H_6 and N_2H_4 at 450K or by coadsorption of the reactants at 90K and subsequent annealing to 450K. These studies have shown that diborane and hydrazine can be successfully used as molecular precursors in the low temperature deposition of boron nitride thin-films.

INTRODUCTION

Boron nitride thin films have found widespread industrial applications because of its unique combination of chemical and physical properties. This material has low density, good thermal conductivity, excellent chemical inertness and electrical resistivity. Thin films of boron nitride have been employed as dielectrics in the electronic device technology and as hardness coatings for machine tools [1,2]. Boron nitride thin films are commonly produced by pyrolysis or by plasma decomposition of mixtures of boron halides or hydrides with nitrogen or ammonia [3,4].

In this present work, we study the interactions of diborane with ammonia and also with hydrazine on Ru(0001) surface using thermal desorption mass spectroscopy (TDS) and X-ray photoelectron spectroscopy (XPS). Diborane and ammonia are used as volatile chemical precursors to BN thin films prepared in most thermal and plasma-enhanced chemical vapor depositions (CVD and PE-CVD) [5,6]. However, to our knowledge, no CVD or PE-CVD system has been reported that utilize diborane and hydrazine as reactants.

EXPERIMENTAL

The experiments were carried out in an apparatus described in reference [7]. This apparatus consists of a ultra high vacuum chamber (ultimate pressure < 5×10^{-10} torr) equipped with a hemispherical electron analyzer for XPS and Auger spectroscopy, LEED screen and a quadrupole mass-spectrometer for thermal desorption mass spectroscopy. The Ru(0001) sample was spot-welded to a manipulator which was capable of resistive heating to 1500K and electron-beam heating up to 2400K. The sample can also be transferred to an attached high pressure reaction cell. All XPS spectra presented were recorded with Al Kα radiation. All XPS and Auger detections were normal the sample surface. The sample temperature was monitored using a W-5%Re/W-26%Re thermocouple which was spot-welded to the back side of the sample. The Ru(0001) crystal was cleaned using the procedure reported in the literature [8]. C, N and B were removed from the surface by heating in 5×10^{-8} torr of O_2 at 1100K for 5 minutes, followed by electron beam heating to 1650K. This procedure was applied until the impurities were below the detection limit of our XPS and Auger spectrometer.

RESULT AND DISCUSSION

Diborane Adsorption on Ru(0001)

The thermal desorption spectra of figures 1,2 and 3 describe the adsorption and decomposition of diborane on Ru(0001). As can be seen in figure 1 (m/e=27), very low diborane exposures did not lead to any molecular desorption until 1 Langmuir was reached. The features seen in the range 250 to 400K can be assigned to monolayer desorption which saturates at 2 Langmuirs. At larger exposures, a physisorbed state was observed to grow at 150K. Below 2 L, the H_2-TDS spectra of figure 2 display a feature around 380K, similar to that of hydrogen desorption from clean Ru(0001) [9]. This feature likely originates from the associative desorption of adsorbed atomic hydrogen (2 $H_a \rightarrow H_{2,g}$) rather than from B-H bond scission. At larger exposures, these spectra closely follow those of molecular diborane TDS and hence likely result from the cracking of diborane in the mass spectrometer.

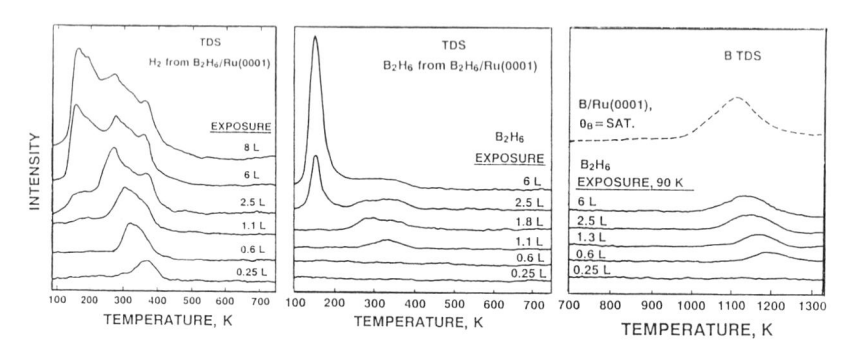

Figures 1, 2 and 3 Thermal desorption spectra of diborane following exposures at 100 K. (H_2-TDS is with m/e=2; B_2H_6 with m/e=27; B with m/e=11)

The decomposition of diborane deposited atomic boron on the Ru(0001). After exposure of diborane at 90K, and subsequent annealing to 450K, atomic boron was the only species on the surface. As shown by the B-TDS spectra of figure 3, no boron desorption was seen for exposures less than 0.5 L. At larger exposures, boron desorbed partially, consistently leaving about 0.2 ML on the surface after heating to 1500K. The top curve for a saturation coverage of boron (1.1 ML) was obtained only by dosing diborane at temperatures higher than 450K.

Diborane and Ammonia Adsorption on Ru(0001)

Shown in figure 4 are thermal desorption spectra taken in one of our diborane and ammonia coadsorption experiments. We have found from the above studies that diborane decomposed very readily on Ru(0001) surface at very low temperatures. To minimize this reaction channel which competes with the surface reaction between diborane and ammonia, we covered the Ru(0001) surface with ammonia at 90K prior to diborane exposures.

In the experiment shown in figure 4, we dosed 2 ML of ammonia to the Ru(0001), followed by 1.5 ML of diborane, at 90K. The NH_3-TDS shows a new feature at 450K which has not been seen for ammonia on clean Ru(0001). The desorption states below 450K are similar to those seen for ammonia desorption from clean Ru(0001): multilayer at 110K, second layer at 135K and monolayer at 285K [10]. In addition, the H_2-TDS shows a new feature near 500K which coincides with a N_2 desorption. This N_2 TDS is similar to the recombinative desorption of atomic nitrogen from clean Ru(0001) [11]. This simultaneous desorption of H_2 and N_2 is strong evidence for the decomposition of a NH_x species on the surface. Since ammonia adsorbs molecularly on Ru(0001), these spectra suggest a direct reaction between ammonia and diborane.

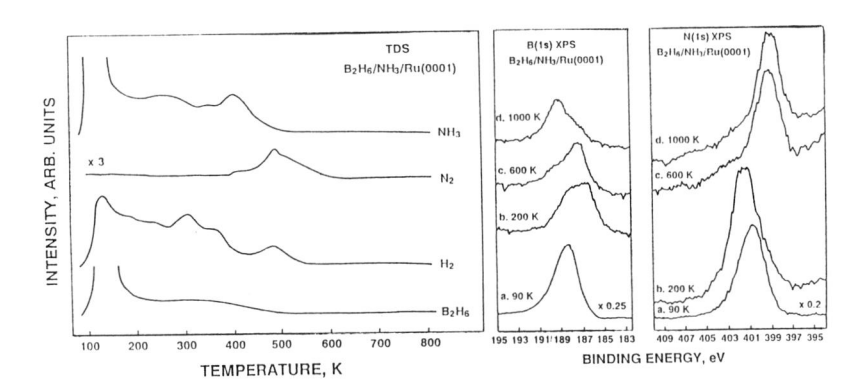

Figure 4. Thermal desorption spectra following diborane coadsorption with ammonia at 100 K. The sample was pre-covered with 2 ML of ammonia, then 1.5 ML of diborane was added.

Figure 5 B(1s) and N(1s) XPS spectra of 1.5 ML diborane with 3 ML of ammonia precoverage.

Repeating this experiment and using XPS, we obtained the spectra of figure 5. The B(1s) and N(1s) spectra (a) indicate no reaction at 90K. Heating to 200K leads to desorption of the physisorbed ammonia and diborane. The B(1s) spectrum shows a new feature (less than 187 eV), lower than chemisorbed diborane. On the other hand, the corresponding N(1s) spectrum shows a feature at higher binding energy (401.5 eV). These shifts suggest the formation of an amino-borane adduct on the surface at this temperature. Further heating lead to an boron-nitrogen adlayer (θ_B=0.38 ML, θ_N=0.22 ML) which is stable at temperatures up to 1100K. The B(1s) and N(1s) binding energies, 189.4 eV and 399.0 eV respectively, are close to those reported for boron-rich boron nitride thin films [1].

Hydrazine Adsorption

Hydrazine adsorption and decomposition on Ru(0001) is shown in the thermal desorption spectra of figure 6. Our results show that hydrazine decomposes into ammonia, hydrogen and nitrogen. Ammonia and hydrazine are the only nitrogen-hydrogen compounds seen desorbing from the substrate. Other nitrogen-hydrogen species were found to result from the cracking of N_2H_4 and NH_3 in the mass spectrometer.

The H_2-TDS of figure 6 show peaks at 185, 320 and 425K. The peak at 185K appears simultaneously with the N_2H_4 desorption and is probably a result of the cracking of hydrazine in the spectrometer. The peaks appearing between 200 and 450K are from the decomposition of chemisorbed hydrazine. The peak at 320K is probably from the associative desorption of hydrogen atoms since it is very similar to H_2 desorption from hydrogen adatoms on Ru(0001) [9]. On the other hand, the peak at 425K is rate-limited by the scission of N-H bonds in a NH_x species.

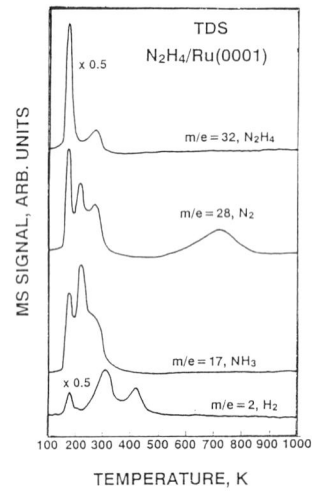

Figure 6 Thermal desorption spectra for multilayers of hydrazine dosed at 90 K.

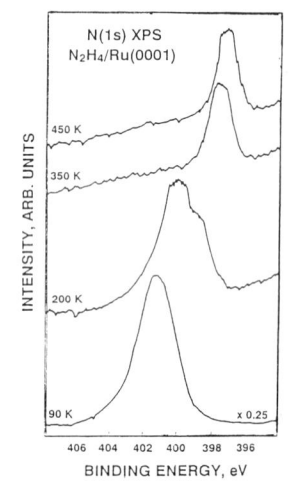

Figure 7 N(1s) XPS for 5 ML of hydrazine dosed at 90 K, and at indicated annealing temperatures.

The NH$_3$ desorption peak at 225K was accompanied by simultaneous desorption of N$_2$. Since ammonia desorption from Ru(0001) does not have a sharp feature at 225K [11], molecularly adsorbed nitrogen desorbs below 130K [12] and adsorbed nitrogen atoms recombine and desorb at temperatures from 500K to 800K (see below), we attribute these N$_2$ and NH$_3$ features at 225K to a nitrogen-hydrogen adspecies (NH$_x$). This NH$_x$ species decomposes at this temperature to produce N$_2$ and NH$_3$ which desorb from the surface. Other NH$_3$ peaks (at 190K and 285K) are consistent with TDS results reported in the literature. In figure 7, we show the N(1s) XPS of 5 ML of hydrazine on Ru(0001) at 90K which is centered at 401.4 eV. At 200K, the hydrazine multilayer desorbs and the chemisorbed species gives a spectrum with two overlapping peaks. The higher binding energy (400.1 eV) component corresponds very closely to our observed spectrum for chemisorbed ammonia on Ru(0001). The lower component, centered at 399 eV, probably belong to partially decomposed hydrazine. At 350K, the N(1s) peak is at 397.9 eV, similar to the value reported for an NH (imide) species on metal surfaces [12]. At 450K, only atomic nitrogen whose N(1s) core-level is at 397.2 eV, remains on the surface.

Diborane and Hydrazine Adsorption on Ru(0001)

In figure 8, we show B(1s) and N(1s) XPS spectra taken after coadsorbing multilayers of B$_2$H$_6$ and N$_2$H$_4$ at 90K. 1.5 ML hydrazine were first dosed on Ru(0001), followed by 1.5 ML of diborane, 1.5 ML of hydrazine and finally another 1.5 ML of diborane. At 90K, the peak positions correspond to values of multilayers of hydrazine and diborane. After the desorption of the multilayers at 200K, N(1s) feature has a component at higher binding energy than seen for physisorbed and chemisorbed N$_2$H$_4$. The B(1s) spectrum also differs from that of chemisorbed diborane. This suggests that there is a reaction between hydrazine and diborane. Further heating to 1100K centers the B(1s) peak at 190.5 eV and the N(1s) peak at 398.6 eV. These peak locations are very near to values reported for boron nitride [13]. The overlayer was found to compose of 0.96 ML of B and 0.92 ML of N.

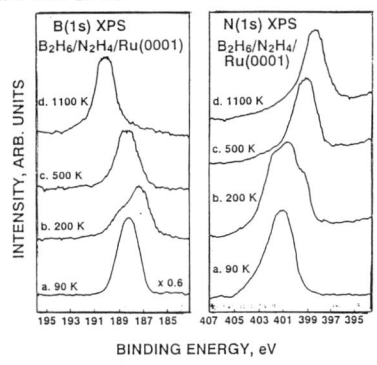

Figure 8 B(1s) and N(1s) XPS spectra for diborane coadsorbed with hydrazine at 90 K.

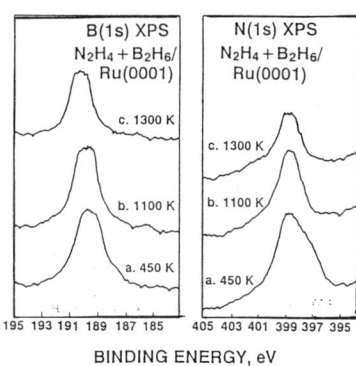

Figure 9 B(1s) and N(1s) XPS spectra for diborane reacting with hydrazine at 450 K.

In a different set of experiments in which diborane and hydrazine were dosed simultaneously at 450K, boron-nitrogen adlayers were produced with B/N ratios varying from 0.6 to 1.8. Figure 9 shows the N(1s) and B(1s) spectra taken in these experiments. At 450K, the film showed a composition of 0.87 ML of B and 1.36 ML of N. At 1100K, the excess in nitrogen is removed and the composition is 0.84 ML of B and 0.81 of N. Further heating to 1300K led to an adlayer with $\theta_B = 0.61$ ML and $\theta_N = 0.59$ ML.

CONCLUSION

We have found that diborane decomposes very readily on Ru(0001). When coadsorbed with ammonia on Ru(0001) surface, diborane reacts to form a boron-nitrogen adlayer which is boron-rich likely due to the stability of ammonia. It was also necessary to pre-cover the Ru(0001) surface with ammonia prior to the diborane exposure, in order to minimize diborane decomposition on the surface.

Near Stoichiometric boron-nitrogen films were formed on Ru(0001) when hydrazine was reacted with diborane likely due to the higher reactivity of hydrazine.

ACKNOWLEDGEMENT

We acknowledge with pleasure the support of this work by the Texas Advanced Research Program.

REFERENCES

1. C. Weissmantel, K. Bewilogua, K. Breuer, D. Dietrich, V. Ebersbach, H. J. Erler, B. Rau, G. Reisse, Thin Solid Films 96, 31 (1988).
2. C. Weissmantel, Thin Films from Free Atoms and Particles, edited by K.J. Klabunde (Academic Press, New York, 1985), Chapter 4.
3. J. Kouvatekis, V. V. Patel, C. W. Miller, P. B. Beach, J. Vac. Sci. Technol. A, 8, 3929 (1990).
4. S. P. S. Arya, A. D'Amico, Thin Solid Films 157, 267 (1988).
5. M. J. Rand, J. Roberts, J. Electrochem. Soc. 115, 423 (1968).
6. M. Hirayama, K. Shohno, J. Electrochem. Soc. 122, 1671 (1975).
7. R. A. Campbell, D. W. Goodman, Rev. Scientific. Instrumentation, submitted.
8. J. E. Houston, C. H. F. Peden, P. J. Feibelman, D. R. Haman, Surf. Sci. 167, 427 (1986).
9. C. H. F. Peden, D. W. Goodman, J. E. Houston, J. T. Yates, Surf. Sci. 194, 92 (1986).
10. C. Benndorf, T. E. Madey, Surf. Sci. 202, 357 (1988).
11. C. Benndorf, T. E. Madey, Chem. Phys. Lett. 101, 59 (1983).
12. J. E. Houston, C. H. F. Peden, D. S. Blair, D. W. Goodman, Surf. Sci. 167, 427 (1986).
13. C. D. Wagner, W. M. Riggs, L. E. Davis, J. F. Moulder and G. E. Muilenberg, Handbook of X-ray Photoelectron Spectroscopy, (Perkin-Elmer Corporation, Eden-Prairie, Minnesota, 1979) pp. 36 and 40.

NEAR-ROOM TEMPERATURE DEPOSITION OF W AND WO₃ THIN FILMS BY HYDROGEN ATOM ASSISTED CHEMICAL VAPOR DEPOSITION

Wei William Lee* and Robert R. Reeves**
* IBM Research Division, Almaden Research Center, San Jose, CA 95120-6099
** Department of Chemistry, Rensselaer Polytechnic Institute, Troy, NY 12180-3590

ABSTRACT

A novel near-room temperature CVD process has been developed using H-atoms reaction with WF_6 to produce tungsten and tungsten oxide films. The chemical, physical and electrical properties of these films were studied. Good adhesion and low resistivity of W films were measured. Conformal WO_3 films were obtained on columnar tungsten using a small amount of molecular oxygen in the gas stream. A reaction mechanism was evaluated on the basis of experimental results. The advantages of the method include deposition of adherent films in a plasma-free environment, near-room temperature, with a low level of impurity.

INTRODUCTION

The development of chemical vapor deposition (CVD) has recently progressed to the point where low-temperature preparation of a wide variety of materials must be considered necessary. As microelectronic technology advances with device features in the submicron range, new metallization materials and processes are needed. For ultra large-scale integration (ULSI) metallization applications, these materials must have low resistivity, low contact resistance, high electromigration resistance, good adherence and step coverage and conformality. In order to satisfy these requirements, the refractory metals (tungsten in particular) as well as their disilicides are of interest [1-5]. Deposition of tungsten and other materials at relatively high temperature could result in undesirable diffusion and redistribution of materials to unwanted areas. Hence, there is a need for low-temperature deposition technology for ULSI applications.

In the present investigation, the deposition of tungsten and tungsten oxide have been studied using hydrogen atom reaction with tungsten hexafluoride. In this process, atomic hydrogen has been generated and reacted with tungsten hexafluoride to deposit W and WO_3 films at near-room temperature. The conventional W deposition process has used molecular hydrogen to react with WF_6 in the temperature range of 300-600 °C. The objective of this research was to investigate the gas phase reaction between H-atom and WF_6 and characterize the deposited films in order to optimize the deposition conditions. Atomic absorption spectroscopy was used for the *in-situ* measurement of W-atoms during deposition and results of these measurements provided a basis for the reaction mechanisms.

EXPERIMENTAL PROCEDURE

Tungsten and tungsten oxide thin films were prepared in a horizontal atom-assisted CVD system. Apparatus for this work was described previously [6]. The pressure of the deposition was typically in the range of 0.01 to 1.0 mm Hg. Hydrogen atoms (10-20%) were produced upstream of the deposition chamber in a U-shaped discharge tube which operated at 400-500 watts and up to 6000 volts A.C. The concentration of hydrogen atoms was indicated by a thermistor placed in the middle of the deposition chamber. Since the resistance of the thermistor decreases with increasing temperature, the heat of recombination of H-atoms on the thermistor could be readily used as a semi-quantitative monitor for H-atom concentrations.

The process gases used were: tungsten hexafluoride (99%); hydrogen (99.99%); and argon (99.99%). Tungsten hexafluoride (2-10%) was diluted in argon and the mixture was fed into the reactor at 0.04-0.4 cm^3/s. Pure hydrogen, or hydrogen mixed with argon was fed into the discharge tube at 0.5-2.5 cm^3/s. For WO$_3$ deposition, oxygen was applied to the mixture using a few percent air and deposition continued on a previously deposited W film. Films of tungsten were deposited onto single-crystal silicon, TiN/Si, SiO$_2$, glass and Teflon substrates. These substrates were in the size of 4-10 cm^2. Substrates were cleaned by the standard "RCA" cleaning procedure [6]. The system was evacuated to a base pressure of 10^{-6} mm Hg and the substrates were then given a few minutes exposure to the hydrogen atoms for further cleaning.

The film thickness was determined by Dektak profilometry, scanning electron microscopy (SEM) cross section pictures and Rutherford backscattering spectroscopy (RBS). Surface morphology was inspected by SEM. Chemical composition of these deposited films was examined using an Auger electron spectrometer (Perkin-Elmer PHI 545C) and RBS. Particle induced X-ray emission (PIXE) was also used to determine the composition of W films by RBS. Tungsten oxide film was examined by electron spectroscopy for chemical analysis (ESCA) using a Surface Science SSX-100 small spot ESCA spectrometer. An X-ray diffractomer (Philips, model 160-143-00) was used for X-ray diffraction (XRD) studies. Sheet resistance measurements were carried out using a four-point probe (Veeco, FPP-100) at room temperature.

Atomic absorption was used to measure tungsten atoms *in-situ* in the deposition system. A monochromator consisting of a photomultiplier and a RCA (1P28A) photomultiplier tube was used for measuring W-atoms at the resonance line at 354.52 nm. A tungsten hollow cathode lamp (Fisher, type 14-386-107) emitted light which passed through the reactor and was focussed on the monochromator.

RESULTS

The average deposition rate for W deposition was 180-240 Å/min. SEM micrographs of these W films showed a uniform and fine-grained surface. The SEM cross-section of these films illustrated that columnar-like W films adhered well to the substrate with no indication of porosity. SEM micrograph of the WO3 films revealed a continuous, conformal blanket-like coverage with no observed crystalline structure.

XRD patterns showed that the W films were crystalline. Tungsten films deposited at low pressure (0.035-0.5 mm Hg with or without argon as a dilute gas) were normally present in the α structure shown in Figure 1. For those tungsten films deposited at higher pressure (ca.. 1.0 mm Hg with argon 50% as a dilutant), XRD patterns indicated mixed α and β phases (see Figure 2).

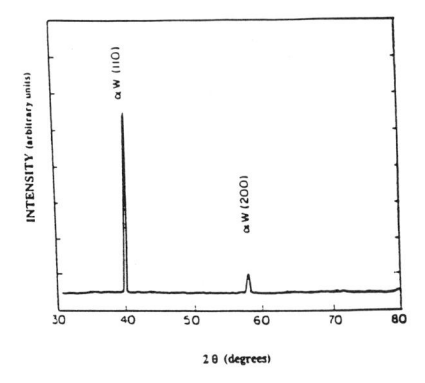

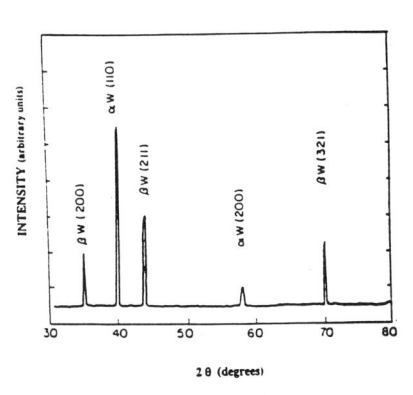

Figure 1. XRD Patterns of W deposited at low pressure (0.035 mm Hg).

Figure 2. XRD Patterns of W deposited at low pressure (1.0 mm Hg).

Auger electron spectroscopy depth profiling of a number of deposited W films showed that these films were almost pure tungsten in bulk form with just traces of carbon and oxygen. A typical AES spectrum is shown in Figure 3. The surface of the films contained an appreciable amount of oxygen indicative of a surface oxide by AES depth profile measurements. The RBS spectra of several samples were studied and results of these samples gave no indication that tungsten diffused into silicon or silicon diffused into the tungsten film. Particle induced X-ray emission (PIXE) was also used for analyses of the deposited W films as part of RBS experiments. Figure 4 shows a typical PIXE spectrum of deposited tungsten film on a silicon wafer. The real and simulation overlay spectra indicated both W and Si characteristic peaks were present in the spectrum. WO3 layer (ca. 2500 Å) on tungsten was examined by ESCA and the analysis corresponded to a stoichiometric atom ratio of tungsten to oxygen of 1:3.

The resistivity obtained from these tungsten films is in the range of 9-33 $\mu\Omega$ cm. For W film thickness greater than 2000 Å, resistivity was approximately 7 $\mu\Omega$ cm, which is close to the bulk resistivity (5.5 $\mu\Omega$ cm).

Free tungsten atom presented in the gas phase reaction region above the substrate were measured by atomic absorption spectroscopy. Based on the measurements, the concentration of tungsten atoms was approximately 10^{10}-10^{11}/cm^3.

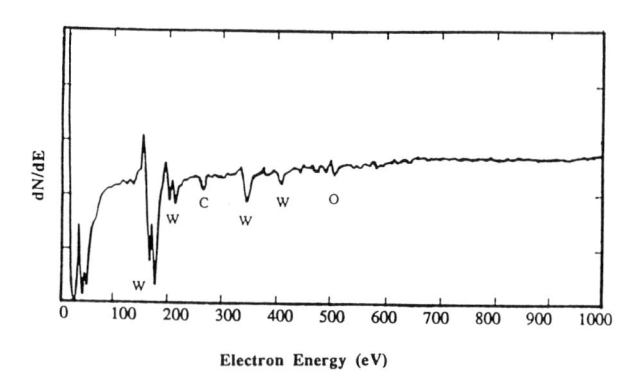

Figure 3. AES spectrum of tungsten film deposited at near-room temperature.

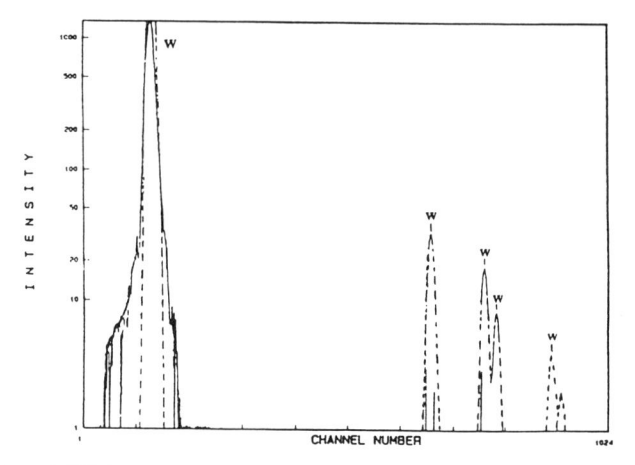

Figure 4. PIXE spectrum of tungsten deposited on silicon.

DISCUSSION AND CONCLUSION

Tungsten and tungsten oxide films have been deposited on various substrates including Teflon at near-room temperature. These reactions were taken place in a plasma-free environment.

The chemistry of the conventional CVD of W films may be summarized by the overall reaction:

$$WF_6 + 3 H_2 \longrightarrow W_{(g)} + 6 HF$$
$$\Delta H = + 225.7 \text{ Kcal/mol.}$$

Using the thermodynamic data [7,8], shown in Table 1, the enthalpy of this reaction was calculated to be endothermic.

Table 1. Thermodynamic Data Related to W Formation

	Enthalpies ΔH_f^0 (kcal/mol)
W-F$_6$(g)	-411.5
H-F(g)	-64.8
W(g)	203
H(g)	52.1

	Bond Strength (kcal/mol)
WF$_5$-F	131±15
W-F(average)	121
H-F	135.3
F-F	37.8
H-H	104

In the process reported here, the tungsten hexafluoride is reduced with H-atoms and the overall reaction can be written as:

$$WF_6 + 6\,H \longrightarrow W(g) + 6\,HF$$
$$\Delta H° = -86.9\ \text{Kcal/mol.}$$

The enthalpies of these reactions indicated that the overall H-atom reaction with WF$_6$ is an exothermic reaction, while that for the molecular hydrogen reaction is endothermic.

The Gibbs free energy for these reactions can also be calculated with the $\Delta G°$ data listed in the JANAF table [9]. For molecular hydrogen reduction:

$$WF_6 + 3\,H_2 \longrightarrow W(g) + 6\,HF$$
at 298K, $\Delta G_r° = +\ 189.63$ Kcal/mol.
at 700K, $\Delta G_r° = +\ 144.19$ Kcal/mol.

However, the atomic hydrogen reaction in the gas phase gives:

$$WF_6 + 6\,H \longrightarrow W(g) + 6\,HF$$
at 298K, $\Delta G_r° = -101.88$ Kcal/mol.

From these calculations, the Gibbs free energy is also favored for the H-atom reaction with WF$_6$ at near-room temperature. Notice here, the individual reaction of six H-atoms occurring should be considered rather than the overall reaction. However, data for each reaction are limited and these individual calculations can not be made at this time.

Since W-atoms (10^{10}-10^{11} /cm^3) were measured during the deposition the mechanism of the reaction could be described as the following:

$$H + WF_6 \longrightarrow WF_5 + HF$$
$$H + WF_n \longrightarrow WF_{n-1} + HF \quad (n=5, 4, 3, 2)$$
$$H + WF \longrightarrow W(g) + HF$$
$$W(g) \longrightarrow W(s)$$

All of these reactions were estimated to be exothermic.

The primary reaction producing the tungsten oxide was assumed to be occurring in the gas phase according to the equation:

$$W + O_2 \longrightarrow WO + O$$

This reaction is exothermic. The WO could then be oxidized in subsequent steps either on the surface or in the gas phase, or both.

The advantages of the method reported here are (1) deposition takes place in a plasma-free environment; (2) film deposits at near-room temperature; and (3) a low level of impurities results in high-quality adherent films. This low-temperature atom-assisted chemical vapor deposition (AACVD) method is suggested as a possible alternative to prepare materials, where molecular hydrogen is normally used as a reducing agent for depositions at relatively high temperatures. This method of producing the oxides may be applicable to other oxides which are of interest for various applications in integrated circuit manufacture.

ACKNOWLEDGEMENTS

The authors gratefully acknowledge the support of the Office of Naval Research and the International Business Machine Co. Thanks go to Dr. H. Bakhru of SUNY at Albany and Dr. Rutten of IBM, Essex Junction for RBS, and ESCA and SEM measurements, respectively. Thanks is also given to Dr. George Tyndall of IBM Almaden Research Center and Dr. Mark Crowder of IBM Storage Systems Products Division for support and encouragement.

REFERENCES

1. R. Blewer, Ed. Tungsten and Other Refractory Metals for VLSI Applications I, (Mat. Res. Soc.; Pittsburgh, PA 1986).
2. E. Broadbent, Ed. Tungsten and Other Refractory Metals for VLSI Applications II, (Mat. Res. Soc.; Pittsburgh, PA 1987).
3. V. Wells, Ed. Tungsten and Other Refractory Metals for VLSI Applications III, (Mat. Res. Soc.; Pittsburgh, PA 1988).
4. R. Blewer and C. McConica, Ed. Tungsten and Other Refractory Metals for VLSI Applications V, (Mat. Res. Soc.; Pittsburgh, PA 1989).
5. S. Wong and S. Furukawa, Ed. Tungsten and Other Refractory Metals for VLSI /ULSI Applications V, (Mat. Res. Soc.; Pittsburgh, PA 1990).
6. W. Lee and R. Reeves, J. Vac. Sci. Tech. $\underline{A9(3)}$, 653, (1991).
7. R. Weast, Ed. CRC Handbook of Chemistry and Physics, 64th ed. (CRC Press, Boca Raton, FL (1985), F180.
8. J. Dean, Ed. Lange's Handbook of Chemistry, 6th ed. (John Wiley & Sons, New York, 1983), P.86.
9. D. Stull and H. Prophet, Eds. JANAF Thermochemical Tables, 2nd ed. (NBS, U.S.CPO. Washington D. C. 1971).

PART III

Microstructure-Process-Property Relationships

EFFECT OF PROCESS CONDITIONS AND CHEMICAL COMPOSITION ON THE MICROSTRUCTURE AND PROPERTIES OF CHEMICALLY VAPOR DEPOSITED SiC, Si, ZnSe, ZnS AND ZnS_xSe_{1-x}

MICHAEL A. PICKERING, RAYMOND L. TAYLOR, JITENDRA S. GOELA AND HEMANT D. DESAI, Morton International, 185 New Boston Street, Woburn, MA 01801

ABSTRACT

Sub-atmospheric pressure chemical vapor deposition (CVD) processes have been developed to produce theoretically dense, highly pure, void-free and large area bulk materials, SiC, Si, ZnSe, ZnS and ZnS_xSe_{1-x}. These materials are used for optical elements, such as mirrors, lenses and windows, over a wide spectral range from the vacuum ultraviolet (VUV) to the infrared (IR).

In this paper we discuss the effect of CVD process conditions on the microstructure and properties of these materials, with emphasis on optical performance. In addition, we discuss the effect of chemical composition on the properties of the composite material $ZnS_x Se_{1-x}$.

We first present a general overview of the bulk CVD process and the relationship between process conditions, such as temperature, pressure, reactant gas concentration and growth rate, and the microstructure, morphology and properties of CVD-grown materials. Then we discuss specific results for CVD-grown SiC, Si, ZnSe, ZnS and ZnS_xSe_{1-x}.

INTRODUCTION

Chemical vapor deposition (CVD) is a well known process to produce theoretically dense crystalline materials for a variety of applications [1-41]. This process has been used to produce metal films (W, Al, Mo, Au, Cu, Pt) for protective coatings [1-5]; ceramic materials (Al_2O_3, TiC, SiC, B_4C, TiB_2, HfC, HfN) used for hard or diffusion barrier coatings [6-11]; semiconductors (GaAs, GaP, InP, PbS, Si) with required doping [12-16]; refractory oxides (ZnO_2) used for thermal barrier [6,17]; films (BN, $MoSi_2$, SiC, B_4C) for protection against corrosion [18-20]; powders (Si_3N_4, SiC) used to fabricate complicated shaped parts by sintering or hot-pressing [1]; materials (Si, GaAs, HgCdTe, CdZnTe) for solid state and energy conversion devices [21-24]; fibers (B, B_4C, SiC) used to fabricate composite materials [24]; transmissive infrared optical materials (ZnSe, ZnS, CdS, CdTe, ZnS_xSe_{1-x}) [25-31]; and monolithic ceramic materials (Si_3N_4, SiC, Si) [32-41].

Five major types of products have been made via CVD technology. These are fibers, thin film coatings, powders, monolithic structures and composites. While there has been much CVD technology development and understanding for the first three types of products, less attention has been directed toward monolithic and composite materials. However, recently there has been a growing interest in developing CVD technologies to produce monolithic optical substrates and structural ceramic composite materials.

The CVD process is capable of producing theoretically dense, high purity and homogeneous materials. This, combined with the fact that the process is readily scalable, makes it attractive for producing high performance optical elements. These elements must be fabricated to optical tolerances, be capable of polishing to high surface finish (≤ 20 Å RMS) and exhibit good optical homogeneity over large areas. Other technologies to fabricate large area crystalline materials such as sintering and hot-pressing produce less homogeneous materials which are not theoretically dense. Although single crystal technology can produce materials with the desired properties, this technique is not readily scalable to large areas, i.e., up to 1-2 m in diameter.

GENERAL DISCUSSION OF CONTROL OF MATERIAL PROPERTIES

Properties of CVD materials can be controlled by adjusting process parameters such as pressure, reactant concentrations, flow rates, deposition geometry, gas temperature, and substrate material and temperature. The most important parameter which controls the properties of the deposited material in a CVD reactor is the deposition temperature. At sufficiently low temperatures, the deposition is kinetically limited and shows a strong dependence on the temperature as shown in Figure 1. The rate determining step in this regime may be either the nucleation or the decomposition of the absorbed reactant species on the surface. Relatively uniform thickness profiles are obtained in the kinetically limited regime, provided the gas and substrate temperature is uniform throughout the deposition zone of the CVD reactor. Many CVD reactors used to produce electronic devices, where deposition thickness uniformity is important, are operated in the kinetically limited regime.

When high growth rates are more important than thickness uniformity, the CVD reactor can be operated in the mass transport limited regime. Here the rate determining step is diffusion or transport of the reactant gases to the substrate surface. This occurs at higher temperatures (see Figure 1), and because the transport phenomenon is relatively independent of temperature the deposition rate does not change much as the substrate temperature is varied. In the mass transport limited regime, the deposition thickness profile depends on the flow pattern of thereactant gases. Therefore, it is important to understand and control the flow in the reactor in order to obtain fairly uniform growth over large areas. Monolithic materials are usually produced in CVD reactors operating in the mass

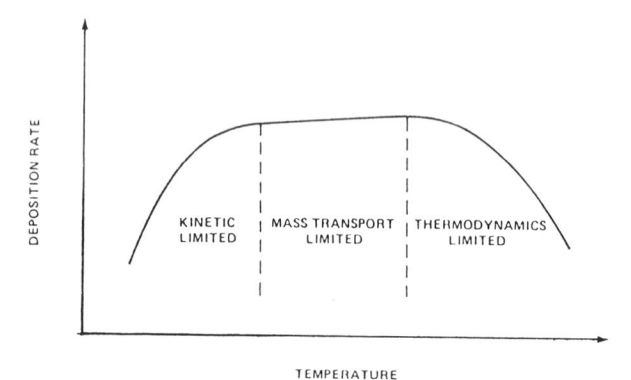

Figure 1.   Different reaction regimes in a CVD process involving an exothermic ($\Delta G < 0$) reaction to produce crystalline materials.

limited regime. The CVD grown materials discussed in this paper, i.e., SiC, Si, ZnSe, ZnS and ZnS_xSe_{1-x}, were all grown in hot-wall CVD reactors operated in the mass transport limited regime.

At high temperatures the net reaction rate decreases as a function of temperature for an exothermic reaction. In this regime, where the reaction rate is thermodynamically limited (see Figure 1), the free energy of the reaction, ΔG, increases with temperature, i.e., becomes less negative, reducing the driving force of the reaction. Epitaxial or single crystal growth is usually performed in this regime, where the reactants and products of the reaction are essentially at or near equilibrium ($\Delta G=0$) with each other. For endothermic reactions (ΔG is positive), the net reaction rate increases with temperature in the thermodynamically limited regime, because ΔG decreases with temperature. If ΔG becomes very negative, a gas phase reaction becomes more likely, and if it occurs, produces a powder deposit instead of a crystalline material. At very high temperatures, ΔG can become positive for an exothermic reaction and instead of deposition, etching can occur.

The structure of the deposited material can be controlled by varying the substrate temperature and reactant gas concentration [1]. In general, for an exothermic reaction (ΔG is negative), as the temperature increases, the structure of the deposited material changes from a gas phase nucleation powder, to fine grained polycrystals, to dendrites, to epitaxial growth. The dependence on reactant gas concentration is the opposite, i.e., as the gas concentration increases, the structure changes from epitaxial growth to gas phase nucleation powder at very high reactant gas concentrations. Therefore, the two most important parameters which control the microstructure, morphology and grain size of the deposited material is the deposition temperature and reactant gas concentration. For bulk CVD materials, a fine-grained polycrystalline structure is generally preferred because good material uniformity and high strength is obtained with this structure at a relatively high deposition rate. High strength and material property homogeneity are especially advantageous for optical applications, along with good polishability which is obtained on theoretically dense, fine-grained materials.

The CVD polycrystalline structure consists of columnar grains possessing some degree of preferred orientation. This preferred orientation, which is more prevalent at high deposition temperature and low reactant gas concentrations, develops even though the substrate surface is randomly oriented. Preferential growth of certain crystallographic planes at the expense of less favorably oriented grains occurs during deposition [1]. Generally speaking, the grains become more oriented as the growth thickness increases. Therefore, in most CVD materials the growth starts randomly and becomes more oriented with time (and thickness). For optical applications it is desirable to have the grains randomly oriented, since this produces a more homogeneous and isotropic material at a macroscopic scale. Although completely random grain orientation is not easily achieved in CVD processes, one can choose conditions which are less favorable to oriented growth.

Some control of mechanical properties can be achieved by adjusting the CVD process parameters. The fracture toughness and strength of a polycrystalline material is inversely related to the square root of the grain size [10,42]. Therefore, to produce a high strength material one must adjust the CVD conditions to produce a fine-grained material. However, there is some minimum grain size where the fracture toughness and strength reach a maximum. This maximum point varies from material to material. However, a good rule-of-thumb is that for brittle materials this maximum occurs at an average grain size of 1-10 µm. For example, this maximum is reported to occur in CVD-SiC at an average grain size of ≈ 3 µm [10]. In general, as the deposition temperature decreases or the growth rate increases, the grain size decreases. More specifically, fine-grained CVD materials are produced by maximizing the parameter, J/DN_o, where J is the incoming reactant gas flux (atoms cm^{-2} s^{-1}), D is the surface diffusion coefficient of the adsorbed atoms (cm^{-2} s^{-1}) and N_o is the number of surface sites per cm^2 [32,43].

Most crystalline materials have more than one stable crystal structure. For example, SiC, ZnSe and ZnS all have cubic (β) and hexagonal (α) crystal structures which are stable over a wide temperature range. In general, either or both of these crystal structures can be present in the CVD grown material. For most optical applications, it is desirable to produce a phase pure material to minimize stress, reduce property inhomogenieties and provide a single phase material for polishing. Since most of the thermal and optical properties, such as thermal conductivity, thermal expansion, thermal stability, refractive index and extinction coefficient are isotropic in cubic (β) crystals, this structure is preferred over the hexagonal (α) structure in optical applications. By appropriate adjustment of the CVD process parameter, one can obtain a relatively phase pure (≥ 95%) material.

Properties of CVD materials can also be controlled by incorporating appropriate dopants or a second chemical phase in the material. Dopants or a second chemical phase are usually incorporated by adding an appropriate amount of a volatile compound to the carrier gas stream along with other reagents. Since the deposition occurs on a molecular scale, layer by layer, the dopants and/or second phase are dispersed uniformly through the material as it grows. For example, manganese doped layers of ZnS and ZnSe have been produced for applications as optoelectronic devices [44]. Also, doping of ZnS and ZnSe with excess zinc, indium or aluminum has been proposed as a technique to obtain conducting layers of ZnS or ZnSe [45,46]. When a second dispersed phase is incorporated in CVD materials, the resulting composite may or may not possess properties which fall in between the properties of the two phases. Examples of the former include alloys ZnSSe [25,47] and CdZnTe [22,48] and those of the latter include composites $SiC\text{-}TiSi_2$ [49] and $Si_3N_4\text{-}TiN$ [50,51].

DISCUSSION OF SPECIFIC PROCESS-PROPERTY RELATIONSHIPS

In this section we discuss specific process-property relationships for bulk, monolithic, CVD grown SiC, Si, ZnSe, ZnS and ZnS_xSe_{1-x}. All of these materials have been grown using a low pressure CVD process in hot-walled CVD reactors at Morton International Inc./CVD Incorporated (MI/CVD). These CVD production reactors are capable of producing large area, i.e., up to ≈ 5 m^2, bulk material up to ≈ 8 cm thick in a single deposition. These materials were first developed in small research CVD reactors and subsequently scaled to large production reactors. The primary use of these materials commercially has been as optical components for applications ranging from the VUV to the IR regions of the electromagnetic spectrum.

CVD Silicon Carbide

CVD-SiC has been identified as a leading mirror substrate material for a wide variety of optical applications because of its properties, which include high thermal conductivity, high stiffness, low CTE and excellent polishability. For many applications, it is important to know the temperature dependence of the thermal conductivity over a wide range and to establish the relationship between grain size and thermal conductivity.

The thermal conductivity of electrical insulators and semiconductors is determined by two main factors, the harmonic and anharmonic phonon-phonon interaction and the scattering of the phonons by the crystal boundaries. The phonon-phonon interactions dominate at high temperature and the scattering of the phonons by crystal boundaries dominates at low temperature. For polycrystalline materials, such as CVD-SiC, the thermal conductivity at low temperatures should be proportional to the grain size of the individual crystallites [52]. Recently, we have experimentally established the relationship between the grain size and thermal conductivity of CVD-SiC [53]. This was done by first depositing SiC at different temperatures in a CVD process involving the thermal pyrolysis of methyltrichlorosilane in excess hydrogen and argon. By changing the deposition temperature over a narrow range (1300-1350 C) and adjusting the CVD growth parameters, we were able to produce material with varying grain size without altering the crystal structure (cubic) or the chemical composition of the deposited material.

The average grain size, thermal diffusivity and heat capacity of CVD-SiC produced at three different temperatures, i.e., 1300, 1325 and 1350 C, were measured. The grain size measurements were made from micrographs of etched surfaces of the material. The thermal diffusivity was measured at the University of Dayton Research Institute (UDRI) by the flash lamp technique. Heat capacity data were obtained by differential scanning calorimetry (DSC) using a sapphire sample as the reference.

The thermal diffusivity as a function of temperature is shown in Figure 2, along with the average grain size, d, for CVD-SiC deposited at 1300, 1325 and 1350 C. Notice that as the deposition temperature increases the grain size increases and the thermal diffusivity also increases. Using the data from Figure 2 and the measured heat capacity of the material we can calculate the thermal conductivity, which is the product of the density, thermal diffusivity and heat capacity. Since CVD-SiC has a low CTE, we assume that the density does not change significantly over the temperature range of the measurements, i.e., we assume the density is constant as a function of temperature. The calculated thermal conductivity is given in Figure 3. Two regimes of temperature dependence are observed with a maximum at ≈ 200 K. From the position and value of the conductivity maximum, we can obtain a rough estimate of the grain size [52]. We get a calculated grain size of 5.1, 6.0 and 11.3 for deposition temperatures, respectively, of 1300, 1325 and 1350 C. These values agree quite well with the measured grain size of 6.8, 8.4 and 17.2 determined from optical micrographs.

Transparent CVD-SiC

Silicon carbide is a good candidate material for transmissive optics applications in the 0.5-5.5 µm wavelength region due to its lightweight, high values of hardness, flexural strength, elastic modulus, and thermal conductivity, low value of thermal expansion coefficient, high oxidation resistance and inertness to many chemicals such as common acids and bases. The α-phase single crystal SiC shows good transmission (> 60%) in the wavelength range 0.5-5.0 µm [54].

However, attempts to fabricate β-phase (cubic) SiC with good transmission in the vis-IR region have met with limited success. We report here preliminary results of our attempts to fabricate polycrystalline β-SiC of improved optical properties by the CVD process.

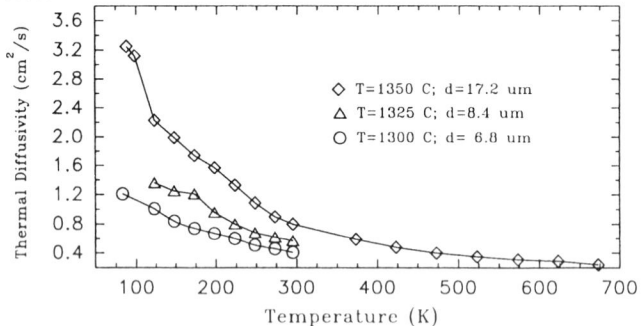

Figure 2. Thermal diffusivity vs. temperature of CVD-SiC material produced at different deposition temperatures, T = 1300, 1325 and 1350 C; and corresponding average grain size, d = 6.8, 8.4 and 17.2 μm.

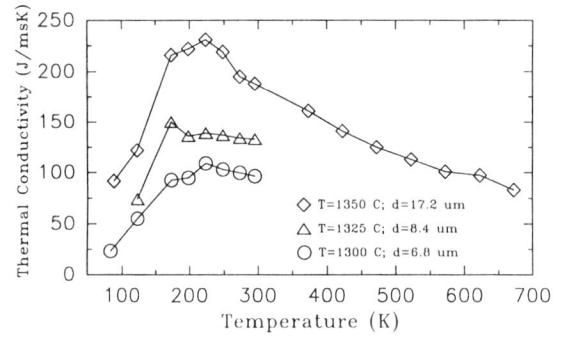

Figure 3. Thermal conductivity of CVD-SiC vs. temperature for material produced at different deposition temperatures, T = 1300, 1325 and 1350 C; and corresponding average grain size, d = 6.8, 8.4 and 17.2 μm. Thermal conductivity was calculated from data in Figure 2, heat-capacity data and density of material (see text.)

A mixture of methyltrichlorosilane (MTS), H_2 and Ar was used to fabricate β-SiC. Argon was used as a carrier gas to transport the MTS vapors to the reaction zone. The deposition was performed on graphite substrates which were coated with a coating of amorphous carbon. The CVD process parameters were varied as follows: MTS/H_2 molar ratio: 4-30, substrate temperature: 1175-1475 C, furnace pressure: 10-200 torr, Ar/H_2 molar ratio: 10-0.2. Our experiments have shown that the most important parameter that governs the fabrication of transmissive β-SiC is the substrate temperature. When the substrate temperature is ≥ 1375 C, β-SiC with improved transmission is obtained. At lower substrate temperatures (≤ 1375 C), β-SiC, Si rich SiC or Si is obtained.

Figure 4 shows a comparison of vis-IR transmission of CVD-SiC prepared at two different substrate temperatures (a) 1450 C and (b) 1300 C. From this figure, it can be seen that there is a dramatic improvement in transmission as the substrate temperature increases from 1300 C to 1425 C. The fall off in IR transmission as the wavelength increases, is due to the presence of free carriers in the material.

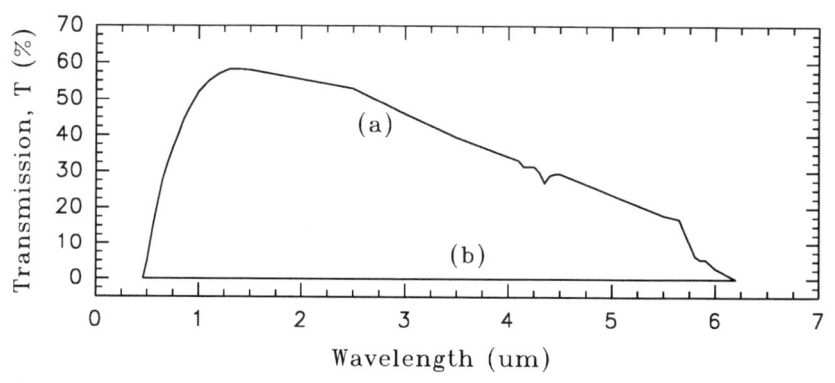

Figure 4. Comparison of transmission of CVD-SiC fabricated at two different substrate temperaturers (a) 1450 C and (b) 1300 C. Sample thickness = 0.18 mm.

CVD Polycrystalline Silicon

Silicon is a good candidate material for transmissive optics applications such as lenses, windows, domes, prisms, etc. in the 3-5 µm wavelength range and reflective optics applications such as LIDAR mirrors, solar collectors and concentrators, and astronomical telescopes due to its attractive properties such as low coefficient of thermal expansion, high thermal conductivity, low density, high value of elastic modulus, high and constant transmission in the 3-5 µm wavelength range and excellent polishability [39,41,54-65]. For many of these applications, Si blanks in the 1-m-dia range are required which are beyond the current state-of-the-art of the single crystal and casting processes. The sintered and hot pressed forms of Si are usually porous and do not produce a high quality optical surface.

Recently, at MI/CVD, a scalable and subatmospheric pressure CVD process based upon pyrolysis of $SiHCl_3$ in excess H_2 was developed to fabricate polycrystalline Si for large scale optics applications [62-65]. This material exhibited good physical, thermal and mechanical properties which were comparable to that of single crystal Si but also showed a large scattering loss in the near infrared (1-7 µm) wavelength region which made its use unsuitable for transmissive optics applications. The CVD procecss parameters were varied over a wide temperature, pressure and molar concentration regimes but significant improvement in transmission in the near infrared could not be obtained. A post deposition treatment for the CVD-Si was then developed which significantly improved its infrared transmission and microstructure as shown below.

The post deposition treatment involved annealing the CVD-Si samples at 1573 K for five hours [65]. The improvement in transmission as a consequence of this annealing was independent of size, thickness and CVD process conditions used to fabricate samples, furnace heating and cooling rates (50-300 C/hr), and the flow rate of inert gas (Ar) during annealing. Figure 5 shows a comparison of infrared transmission of annealed and unannealed samples of thickness 3.43 mm and 3.8 mm, respectively. From this figure we see a dramatic improvement in the transmission of annealed samples thorughout the infrared region. The transmission of the annealed sample is very close to that of semiconductor grade, single crystal Si.

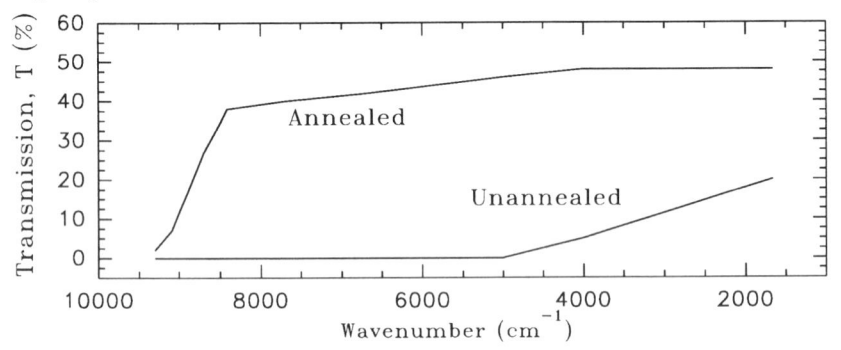

Figure 5. A comparison of infrared transmission of annealed and unannealed samples of CVD-Si. Thickness, annealed sample = 3.43 mm, unannealed sample = 3.81 mm.

The annealing treatment also changed the microstructure of CVD-Si as shown in Figures 6 and 7. Figure 6 shows a comparison of microstructure taken parallel to the growth direction of (a) unannealed and (b) annealed samples. In Figure 6(a), the columnar growth characteristic of a CVD material is clearly visible. These growth columns totally disappear after annealing and distinct grain boundaries appear (Figure 6(b)). These grain boundaries indicate that a rearrangement of the crystallites (or recrystallization) has occurred thereby producing material with more uniform grain structure.

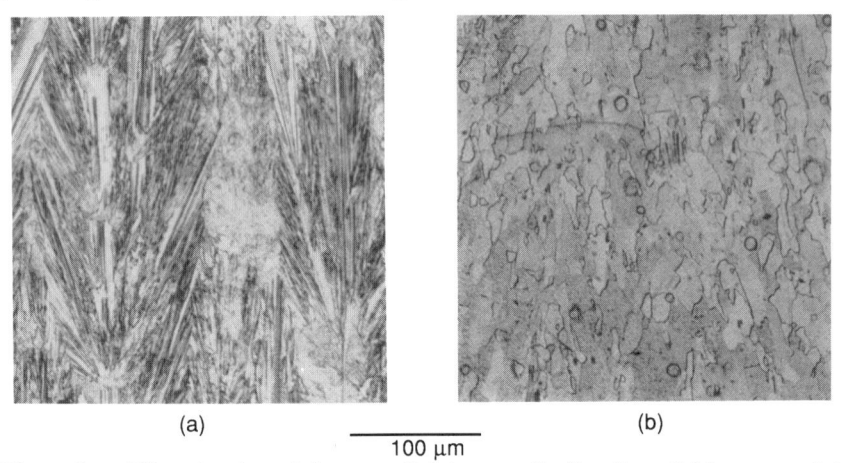

Figure 6. Microstructure taken parallel to growth direction of (a) unannealed sample and (b) annealed sample of CVD-Si.

Figure 7 shows a comparison of microstructure taken perpendicular to the growth direction of unannealed and annealed CVD-Si samples. One can clearly see large grains in the form of "rosettes" in a matrix of small grains in the microstructure of the unannealed sample. This rosette pattern exists throughout the material and essentially shows that the material is microstructurally non-uniform with two quite different grain size distributions. After annealing, however, the rosette pattern disappears and the grain structure becomes more uniform.

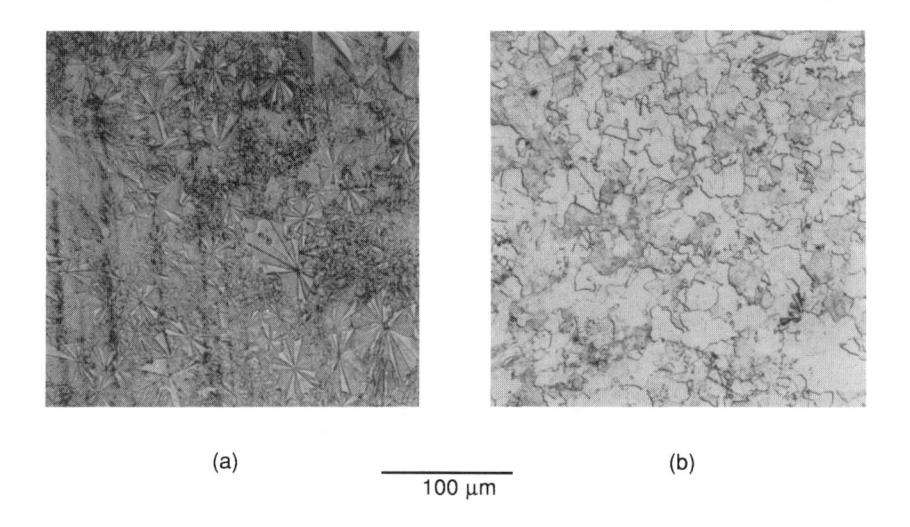

(a) 100 μm (b)

Figure 7. Microstructure taken perpendicular to growth direction of (a) unannealed sample and (b) annealed sample of CVD-Si.

CVD Zinc Selenide

CVD ZnSe is widely used as transmissive optical elements for commercial and military IR systems and as output windows in commercial carbon dioxide (CO_2) lasers. CVD ZnSe is used as a window material in many IR systems in which the windows experience pressure and/or temperature gradients. Because of this, a high strength material is desirable. One factor that effects the strength and fracture toughness of brittle materials is the grain size. Therefore, early in the development of this IR material, the relationship between the deposition temperature, grain size and strength was established [26]. This data is shown in Figures 8 and 9. In Figure 8 the average grain size vs. deposition temperature is shown for CVD ZnSe. The grain size was determined by the intercept method on optical micrographs of CVD ZnSe samples which were polished, and subsequently etched in hydrochloric acid, Notice, there is a strong grain size dependence on temperature, i.e., the grain size increases more than an order of magnitude as deposition temperature changes from 650 to 825 C. The relationship between the flexural strength and grain size is shown in Figure 9. Here the flexural strength is plotted versus the reciprocal of the square root of the grain size ($d^{-1/2}$). Notice, in Figure 9 that the points appear to lie on a straight line, which indicates that polycrystalline CVD ZnSe follows the well known Petch relationship over this range, where the grain size changes from 25 to 400 µm.

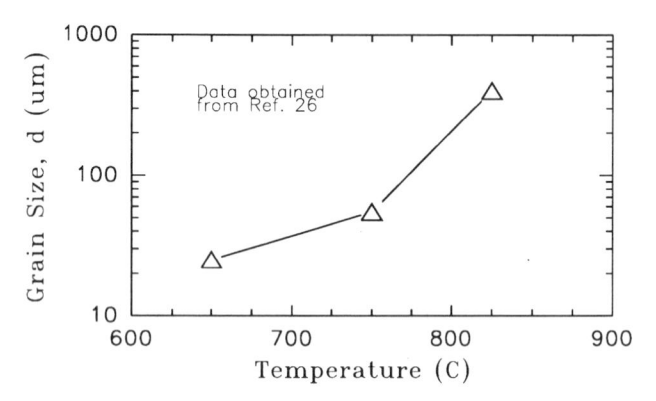

Figure 8. Average grain size, d, of CVD ZnSe vs. deposition temperature. Data obtained from Ref. 26.

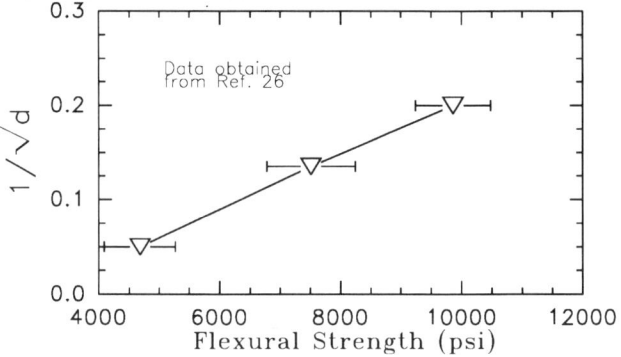

Figure 9. The reciprocal square root of grain size, $d^{-1/2}$ ($\mu m^{-1/2}$) vs. the flexural strength for CVD ZnSe. Data obtained from Ref. 26.

CVD Zinc Sulfide

CVD ZnS is used as a high durability window material in military systems operating in the 3-5 and 8-12 µm spectral band. In many applications such as imaging and targeting, optical scatter and absorption are of great importance. In the development of this IR material, the relationship between the processing conditions and optical transmission were established. The IR transmission of four ZnS samples (thickness = 0.9 cm) is shown in Figure 10. These measurements were carried out on a Perkin-Elmer spectrophotometer (Model no. 1430) operating in the double-beam mode. The deposition temperature for the four samples was varied from 670 to 735 C and the molar ratio of the reactants H_2S/Zn was varied from 1.0 to 0.6. The most prominent feature in the transmission spectrum is the broad absorption band centered at 1600 cm^{-1}, which has been tentatively attributed to vibrational modes associated with zinc hydride radicals. [66]

Although the deposition temperature has an influence on both the scatter and the 6.2 μm absorption band, it is the H$_2$S/Zn molar ratio which exerts a greater control on the optical properties of CVD ZnS. An effort to understand the cause of scatter in CVD ZnS is currently underway.

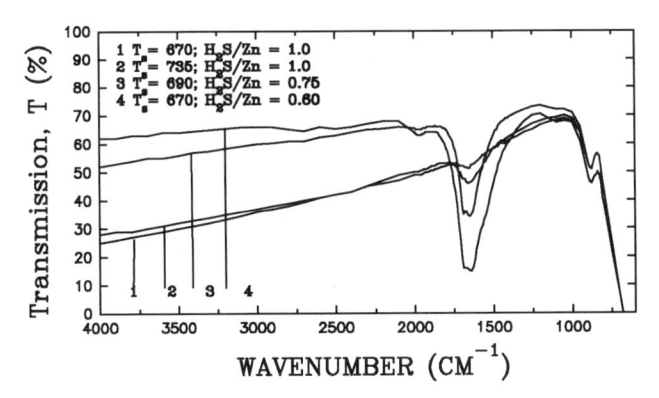

Figure 10. IR transmission of CVD ZnS produced at different temperatures and H$_2$S/Zn molar ratios.

CVD Zinc Sulfo-Selenide

In a sophisticated lens system, the use of optical elements with a variable index of refraction can reduce the complexity of the components, increase reliability and reduce cost. Gradient-index optical materials have long been available for the visible portion of the electromagnetic spectrum, but not for IR portions. Recently, MI/CVD developed an IR gradient index material from mixtures of ZnS and ZnSe via a CVD process [25]. These latter two materials are widely used transmissive optical materials in IR systems. Over their useful transmissive range, the refractive index of ZnS and ZnSe differ by ≈ 0.2 units. In addition, ZnS and ZnSe are completely miscible in the crystalline state, i.e., the alloy ZnS$_x$Se$_{1-x}$ exists for $0 \leq x \leq 1$. By combining the latter two properties, we were able to produce an alloy with a controlled and variable x.

In order to produce a gradient index material, numerous experimental CVD depositions were performed to establish important functional relationship between the composition of the reactant gases and solid composition. Zinc sulfide and zinc selenide were produced by the following CVD reactions:

$$Zn_{(g)} + H_2S_{(g)} \longrightarrow ZnS_{(s)} + H_{2(g)} \qquad (1)$$

$$Zn_{(g)} + H_2Se_{(g)} \longrightarrow ZnSe_{(s)} + H_{2(g)} \qquad (2)$$

where the subscript (g) and (s) refer to gas and solid species respectively. By varying the ratio of the flow rates of the gaseous reagents, Q(H$_2$S)/Q(H$_2$Se), while maintaining a constant Zn$_{(g)}$ flow rate in the CVD reactor we were able to produce alloys, ZnS$_x$Se$_{1-x}$, with x varying from 0 to 1. The composition and properties of these alloys were measured to establish the required relationships to fabricate a gradient index material. The most important relationships required to produce a controlled and quasi-continuous index gradient are: alloy composition and deposition rate as functions of gas phase composition; and the refractive index as a function of alloy composition.

The gas phase composition, expressed in terms of the flow rate of H_2S gas, $Q(H_2S)$, over the total flow rate of reactant gases, $Q(H_2S) + Q(H_2Se) = Q_T$, is plotted vs. the mole fraction (x) of ZnS in the alloy in Figure 11. The chemical composition of the solid alloy, ZnS_xSe_{1-x} was determined by SEM/EDAX analysis. Because the reactants, H_2S and H_2Se were premixed before entering the deposition zone in the CVD reactor, the gas phase composition should be proportional to the flow rates. The solid circles in Figure 11 were obtained under identical run conditions, i.e., substrate temperature of 710 C, total gas pressure of 30 torr, $Q(Zn) = 0.31$ slpm and $Q(H_2S) + Q(H_2Se) = 0.3$ slpm. It is clear from Figure 11 that ZnSe is preferentially deposited, i.e., when the flow rates of H_2S and H_2Se are equal, the mole fraction of ZnS in the alloy is 0.30 and not 0.50. Similar behavior was observed by Stutius for a different CVD reaction using metalorganic compounds as reactants and under different CVD conditions [67]. The solid triangles in Figure 11 represent compositional data obtained at two different deposition temperatures, 750 and 670 C. Since these latter data points appear to give similar results to the points (solid circles) obtained at a deposition temperature of 710 C, we conclude there is little temperature dependence on the composition of the alloy over the temperature range 670-750 C. This result indicates that fluctuations in temperature and small-temperature gradients in the CVD reactor should have little effect on the solid composition of the gradient material (alloy).

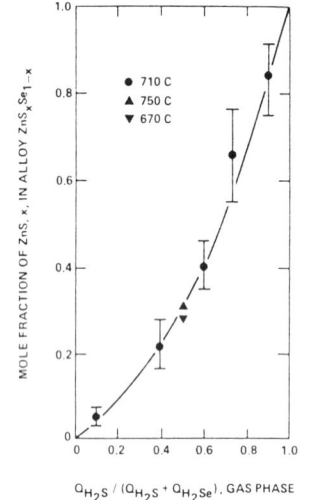

Figure 11. Mole fraction of ZnS in alloy ZnS_xSe_{1-x} vs. gas phase com-position of reactant gases H_2S and H_2Se. Solid line is a least-squares fit to data.

ZnSe and ZnS do not deposit at the same rate under identical CVD conditions. Therefore the deposition rate of the alloy depends on the gas phase composition of the reactant gases. Figure 12 is a plot of the deposition rate, R_D, vs. the gas phase composition. To a good approximation, the deposition rate is a linear function of the gas phase composition. The solid line in Figure 12 is a linear least-squares fit to the data.

The refractive index of the alloy, $ZnS_{2x}Se_{1-x}$, should be a linear function of the solid composition at wavelengths removed from optically allowed transitions. This has been confirmed experimentally, and is shown in Figure 13. Here the refractive index (left-hand ordinate) and molar percent of ZnSe in the alloy (right-hand ordinate) are plotted vs. the thickness of a ZnS$_x$Se$_{1-x}$ gradient index materials deposited over a thickness of ≈ 4 mm. The index measurements were made at a wavelength of 0.647 μm (Kr laser) using a specially designed interferometer [68] and the composition profile data was measured using SEM/EDAX. From these data, we conclude that the refractive index of the alloy is given by:

$$\eta = \eta_{ZnSe}(1-x) + \eta_{ZnS}(x) \quad (3)$$

where η_{ZnSe} and η_{ZnS} are the refractive indices of ZnSe and ZnS, respectively, and x is the mole fraction of ZnS in the alloy, ZnS$_x$Se$_{1-x}$.

Finally, the grain size of several alloys was measured and this data is shown in Figure 14. Notice, that the grain size is not simply proportional to the composition. There is an abrupt change in grain size at a mole fraction of 10-15% ZnS in the alloy and the grain size remains relatively constant as the mole fraction of ZnS changes from 20 to 100%. We do not understand the fundamental reason for this observed behavior. However, it appears that a small amount of ZnS produces a grain size in the alloy similar to pure ZnS.

ACKNOWLEDGEMENT

This work was supported by Morton International Inc. internal research funding; U.S. Air Force Contracts F33615-87-C-5227, F33615-85-C-5004 and F33615-85-C-5010; NASA Contracts NAS1-18222 andNAS1-18476; and U.S. Army Contracts DAAB07-87-C-F108 and DAAH01-84-C-0084. The authors are grateful to the government sponsoring agencies for support of this research.

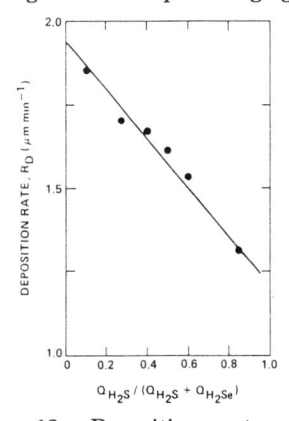

Figure 12. Deposition rate, R$_D$, of alloy ZnS$_x$Se$_{1-x}$ vs. gas phase composition of reactant gases H$_2$S and H$_2$Se. Solid line is a least-squares fit to data.

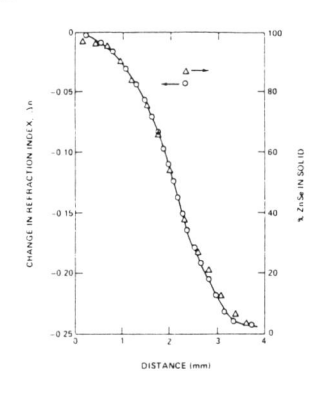

Figure 13. Change in refractive index, Δn (left-hand ordinate) and mole% ZnSe in alloy ZnS$_x$Se$_{1-x}$ (right-hand ordinate) vs. distance from substrate, d, along deposition axis for gradient index material grown by CVD.

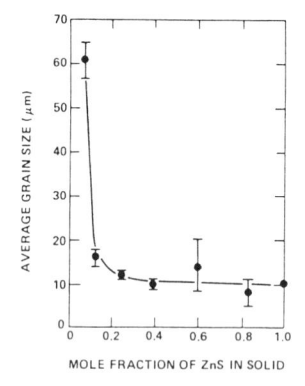

Figure 14. Average grain size vs. mole fraction ZnS in alloy ZnS_xSe_{1-x}.

REFERENCES

1. W.A. Bryant, J. Mater. Sci. 12 (1977), 1285.
2. J.Y. Tsao and D.J. Ehrlich, Appl. Phys. Lett. 45 (1984), 617.
3. T.F. Deutsch and D.D. Rathman, Appl. Phys. Lett. 45 (1984), 623.
4. T.H. Baum and C.R. Jones, Appl. Phys. Lett. 47 (1985), 538.
5. F.A. Houle, C.R. Jones, T. Baum, C. Pico and C.A. Kovac, Appl. Phys. Lett. 46 (1985), 538.
6. W.J. Lachey, D.P. Stintor, G.A. Cerny, A.C. Schaffhauser and L.L. Fehrenbacher, Adv. Ceram. Mater. 2 (1987), 24.
7. D.P. Stinton, W.J. Lackey, R.J. Lauf and T.M. Besmann, Ceram. Eng. Sci. Proc. 5 (1984), 668.
8. J. Saraie, J. Kwon and Y. Yodogawa, J. Electrochem. Soc.: Solid-State Sci. Tech. 132 (1985), 890.
9. A.A. Cochran, J.B. Stephenson and J.G. Donaldson, J. Metals 22 (1970), 37.
10. K. Nichara, Ceram. Bull. 63 (1984), 1160.
11. M.J. Hakim, in "Proceedings of the 5th International Conference on CVD," edited by J.M. Blocher, Jr., H.E. Hinterman and L. Hall (The Electrochemical Society, Princeton, NJ, 1975).
12. J. Bloem and L.J. Giling, in "Current Topics in Material Science," Vol. 1, edited by E. Kaldis (North Holland, Amsterdam, 1978), p. 147.
13. G.W. Cullen and J.F. Corboy, J. Crystallogr. Growth 70 (1984), 230.
14. G.R. Srinivasan, J. Crystallogr. Growth 70, 201.
15. M.J. Ludowise, J. Appl. Phys. 58 (1985), R31.
16. V.M. Donnelly, D. Brasen, A. Appelbaum and M. Geva, J. Appl. Phys. 58 (1985), 2022.
17. M. Balog and M. Schieber, Thin Solid Films 47 (1977), 109.
18. J. Schlichting, Powder Metall. Int. 12 (1980), 14.
19. T. Matsuda, N. Uno, H. Nakae and T. Hirai, J. Mater. Sci. 21 (1986), 649.
20. N.J. Archer, in "High Temperatuare Chemistry of Inorganic and Ceramic Materials," edited by F.P. Glassu and P.E. Potter, Special Publ. No. 30 (Chemical Society, London, 1976), p. 167.

21. T. Arizumi, in "Current Topics in Material Science," Vol. 1, edited by E. Kaldis (North Holland, Amsterdam, 1978), p. 343.
22. J.S. Goela and R.L. Taylor, SPIE Proc. 659 (1986), 161.
23. P.W. Kruse, Semiconductors and Semimetals 18 (1981), 1.
24. H.E. Debolt, in "Hardbook of Composites," edited by G. Lubin (Van Nostrand Rheinhold, NY, 1982), p. 171.
25. M.A. Pickering, R.L. Taylor and D.T. Moore, Applied Optics 25(19) (1986), 3364.
26. R. Donadio, A. Swanson and J. Pappis, in "Proceedings of the 4th Conference on Infrared Laser Window Materials," edited by C.R. Andrews and C.L. Strecher (Air Force Materials Laboratoary, Wright-Patterson AFB, OH, 1975), p. 494.
27. Y.M. Yim and E.J. Stofko, J. Electrochem. Soc. 119 (1972), 381.
28. C.A. Klein, B. diBenedetto and J. Pappis, Opt. Eng. 25(4) (1986), 519.
29. J.S. Goela, R.L. Taylor, M.J. Lefebvre, P.E. Price, Jr. and M.J. Smith, in "Laser Induced Damage in Optical Material: 1983," NBS Special Publ. 688, edited by H.E. Bennett, A.H. Guenther, D. Milarn, B.E. Newman (National Bureau of Standards, Boulder, CO, 1983), p. 106.
30. M.A. Pickering and R.L. Taylor, SPIE Proc. 576 (1985), 16.
31. M.A. Pickering, R.L. Taylor and A.L. Armirotto, SPIE Proc. 618 (1986), 110.
32. J.S. Goela and R.L. Taylor, in "Proceedings of the ASME/JSME Thermal Engineering Joint Conference," Honolulu, March 1987, edited by P.J. Marto and I. Tanasawa (American Society of Mechanical Engineers, NY, 1987), p. 623.
33. R.A. Tanzilli and J.J. Gebhardt, SPIE Proc. 297 (1981), 59.
34. R.E. Engdahl, SPIE Proc. 315 (1981), 123.
35. M.A. Pickering and R.L. Taylor, in "Proceedings of the Topical Meeting on High Power Laser Optical Components," 24-25 Oct. 1988, NWC TP-7017 Part 1, Unclassified Papers (Naval Weapons Center, China Lake, CA, 1989), p. 259.
36. M.A. Pickering, R.L. Taylor, J. Keeley and G. Graves, Nucl. Instrum. Methods in Phys. Res. A291 (1990), 95.
37. A. Collins, J. Keeley, M.A. Pickering and R.L. Taylor, Mat. Res. Soc. Symp. Proc. 168 (1990), 193.
38. M.A. Pickering, R.L. Taylor, J. T. Keeley and G. Graves, SPIE Proc. 1118 (1989) 2.
39. J.S. Goela, M.A. Pickering, R.L. Taylor, B.W. Murray and A. Lompado, SPIE Proc. 1330 (1990), 25.
40. J.S. Goela, M.A. Pickering, R.L. Taylor, B.W. Murray and A. Lompado, Applied Optics 30 (22) (1991), 3166.
41. J.S. Goela and R.L. Taylor, SPIE Proc. 1118 (1989), 14.
42. R.W. Rice, S.W. Freiman and P.F. Becker, J. Am. Ceram. Soc. 64(6), 1981, 345.
43. A. Emmanuel and H.M. Pollock, J. Electrochem. Soc.: Solid State Sci. Tech. 12 (1973), 1586.
44. P.J. Wright, B. Cockayne, A.F. Cattell, P.J. Dean and A.D. Pitt, J. Crystallogr. Growth 59 (1982), 155.
45. L.C. Olsen, R.C. Bohara and D.L. Barton, Appl. Phys. Lett. 34 (1979), 528.
46. P. Besomi and B.W. Wessels, Appl. Phys. Lett. 37 (1980), 955.
47. P.J. Wright and B. Cockayne, J. Crystallogr. Growth 59 (1982), 148.
48. J.S. Goela and R.L. Taylor, Appl. Phys. Lett. 51 (1987), 928.
49. D.P. Stinton, W.J. Lockey, R.J. Lauf and T.M. Besmann, Ceram. Engng. Sci. Proc. 5 (1984), 668.

50. S. Hayaashi, T. Hirai, K. Hiraga and M. Hira-Bayashi, J. Mater. Sci. 17 (1982), 3336.
51. T. Hirai and S. Hayashi, J. Mater. Sci. 17 (1982), 1320.
52. R. Berman, Proc. Phys. Soc. LXV, 12-A (1982), 1029.
53. A.K. Collins, M.A. Pickering and R.L. Taylor, J. Appl. Phys. 68(12), (1990), 6510.
54. S. Singh, J.R. Potopowicz, L.G. Van Witert and S.H. Wemple, Appl. Phys. Lett. 19, 53 (1973).
55. J. Bloem and L.J. Giling, "Mechanism of the Chemical Vapor Deposition of Silicon," in Current Topics in Material Science, Vol. 1, ed. E. Kaldis (North Holland, 1978, p. 147.
56. F.M. Anthony and A.K. Hopkins, in SPIE Proc., 297 (1981), 196.
57. E.A. Maguire, N.T. Dionesotes and R.L. Gentilman, Fabrication of Large Mirror Substrates by Chemical Vapor Deposition (Air Force Wright Aeronautical Laboratories, Rept. No. AFWAL-TR-86-4128, Raytheon Research Division, Rept. No. RAY/RD/M-4410, December 1986).
58. J.S. Goela and R.L. Taylor, J. Am. Cer. Soc. 72(9), (1989) 1747.
59. J.S. Goela and R.L. Taylor, Appl. Phys. Lett., 54(25), (1989), 2512.
60. J.S. Goela and R.L. Taylor, SPIE Proc., 1062 (1989), 37.
61. J.S. Goela and R.L. Taylor, SPIE Proc., 1047 (1989), 198.
62. J.S. Goela and R.L. Taylor, Polycrystalline Silicon Improved Materials Property Data Base for Cooled Laser Mirrors (Air Force Weight Aeronautical Laboratories, Rept. No. AFWAL-TR-86-4131, CVD Incorporated, Rept. No. TR-031, March 1987).
63. J.S. Goela and R.L. Taylor, Fabrication of Lightweight LIDAR Mirrors (NASA SBIR Phase I Final Report, CVD Incorporated, Technical Rept. No. 9069-1, March 1987).
64. J.S. Goela and R.L. Taylor, J. Mat. Sci. 23 (1988), 4331.
65. J.S. Goela and R.L. Taylor, "Post Deposition Process for Improving Optical Properties of CVD-Si," J. Am. Cer. Soc. (to appear).
66. H.G. Lipson, Appl. Opt. 16 (1977), 2902.
67. W. Stutius, J. Electron. Mater. 10(1) (1981), 95.
68. J.J. Miceli, Gradient Index Optics: Materials, Fabrication and Testing (Ph.D. Thesis, U. Rochester, 1983).

CVD OF SILICON NITRIDE PLATE FROM $HSiCl_3-NH_3-H_2$ MIXTURES

J. W. LENNARTZ AND M. B. DOWELL
Union Carbide Coatings Service Corporation, 12900 Snow Road, Parma, OH 44130

ABSTRACT

Preferred conditions for deposition of thick α-Si_3N_4 plate from $HSiCl_3-NH_3-H_2$ on the vertical surfaces of a low-pressure, hot-wall CVD reactor were identified by means of a designed experiment. The design included the range of temperatures 1300°C-1500°C, pressures 0.5-2.0 Torr, and residence times 0.01-1.0 sec. The vertical deposition surfaces received a viscous, laminar flow of well mixed, thermally equilibrated reactants. Plates 0.05-0.5 mm thick were produced on multiple vertical substrates 350 cm^2 in area at deposition rates 5-70 µm/hr. Plates 0.5-4.0 mm thick were produced on horizontal substrates at deposition rates of 60-120 µm/hr. When NH_3 flows in stoichiometric excess, deposition rates on vertical surfaces increase approximately linearly with the flow rate of $HSiCl_3$ but depend little on temperature, as would be expected if the reaction proceeds under mass transport control with product depletion. Multiple correlation analyses show that thickness variations in the deposit are reduced by increasing the temperature and decreasing the gas residence time. CVD silicon nitride plate produced under the optimized conditions exhibits theoretical density and is free of pores and cracks. It exhibits a columnar morphology in which the <222> and <101> crystallographic directions are oriented preferentially normal to a surface, which consists of well-defined trigonal facets 10-50 µm across. Crystallite sizes determined by X-ray line broadening range from 0.06-1.0 µm. This CVD plate is gray and contains approximately 0.5 w/o C and 0.5 w/o O as principal impurities.

INTRODUCTION

Silicon nitride is well recognized as an important engineering ceramic. Its high hardness, chemical inertness, resistance to creep and thermal shock, and strength at high temperatures recommend its use in advanced turbines. Its high electrical resistivity and dielectric strength find use in electrical insulators, including some in integrated circuits. Silicon nitride sputtering targets and crucibles for growth of low-oxygen single crystals are two uses which require relatively thick sections free from metallic impurities. The chemical vapor deposition process is of interest for producing such articles.

CVD silicon nitride has been described previously[1-4] as have some of its uses.[5] A statistical optimization approach to production of thick CVD silicon nitride from $HSiCl_3$, NH_3 and N_2 has been reported.[6] We now report preferred conditions for deposition of thick CVD Si_3N_4 plate from $HSiCl_3$-NH_3-H_2 gas mixtures on vertical, non-impingement surfaces, by use of a similar method.

EXPERIMENTAL

Experiments were carried out in a vertical, hot-wall, low-pressure CVD reactor as depicted in Figure 1. Gases are introduced into the reaction chamber through a cooled coaxial nozzle, the annulus of which carries $HSiCl_3/H_2$ mixtures. The overall reaction is

$$3\ HSiCl_3(g) + 4\ NH_3(g) = Si_3N_4(s) + 9\ HCl(g) + 3\ H_2(g) \quad (1)$$

The reaction chamber was a graphite cylinder 38 cm diameter by 33 cm long, shown in Figure 2. Gases introduced through the nozzle centered in the base impinged on horizontal deposition plate A (15.2 cm square) and exited around the periphery of the base after passing over rectangular deposition plates B and C, (15.2 cm x 22.8 cm) oriented vertically as shown in Figure 2. Deposition plates were machined from UCAR graphite, Grade ATJ. The average thermal expansion coefficients of this graphite and of CVD Si_3N_4 are 3.5-4.0 x 10^{-6} $°C^{-1}$ and 3.0-3.5 x 10^{-6} $°C^{-1}$ respectively.

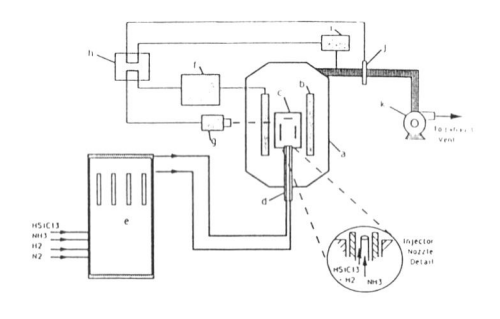

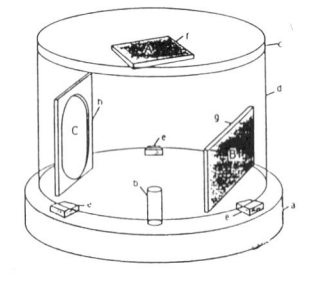

Figure 1

Diagram of CVD System.

a) vacuum chamber, b) graphite furnace element, c) graphite assembly, d) gas injector, e) flow tubes and gas control panel, f) furnace power supply, g) pyrometer, h) DCP controller, i) pressure transducer, j) pneumatic Fisher valve, k) vacuum blower, l) water-sealed ring pump.

Figure 2

Graphite Assembly Used in Deposition Trials.

a) base, b) 3 cm diameter injector hole, c) lid, d) 38 cm diameter by 33 cm long cylinder, e) 1.2 cm spacer block, f) 15.2 cm square substrate (Plate A), g and h) 15.2 cm by 22.8 cm substrates (Plates B and C).

A partial factorial experimental design was generated in which wall temperature, total pressure, and incoming gas flow rates of $HSiCl_3$, NH_3, and H_2 were varied, and total weight of deposit was at least 8 Kg (Si_3N_4 basis) in each run. The design was restricted to ranges of each variable given in Table I, based on prior work. The statistics program EDO[7] generated a one-sixth factorial experiment consisting of 18 trials in order to evaluate the main effects of each of the five parameters and the interactions between them. To these eighteen trials were added five replications and one other variation. After this set of 24 trials, a set of preferred conditions was selected and demonstrated.

TABLE I

Parameter Levels for Experimental Design

Parameter	Units	Lower	Intermediate	Upper
H_2 Flow Rate	slpm*	3.0	6.0	15.0
NH_3 Flow Rate	slpm	0.8	-	2.0
$HSiCl_3$ Flow Rate	slpm	0.4	0.7	1.0
Total Pressure	mm Hg	0.5	-	2.0
Wall Temperature	°C	1300	1400	1500

* standard liter per minute

RESULTS

Analysis of the variable ranges in Table I, and of our reactor geometry, reveals that gases flow over the vertical deposition plates in thermal equilibrium with the reactor wall in viscous and laminar manner. Table II compares the power requirement to attain equilibrium product distribution at wall temperature for actual gas ratios, if power supplied by conductive heat transfer from a tube whose area is equivalent to that of the reactor wall and from the impinged deposition plate A. Convective heat transfer from the reactor wall amply satisfies the power requirement, thus establishing that vertical plates B and C are at the same temperature as the reactor wall. Since convective heating by plate A is insufficient to satisfy the requirement, the temperature of this impinged plate may be as much as 20° less than that of the wall. Power requirements for isothermal expansion of gases and power supplied by radiative heating, may be shown to be small.

TABLE II

Energy Demand for Complete Reaction and High Product T* (in W)

Overall Reaction	ΔH Reaction (1 or 2)	ΔH Decomp. of NH_3	Additional heat to raise products to T (°K) = 600 800 1400 1773**
(1) (High k_f mixture)	8.9	8.5	108 181 409 560
(2) (Low k_f mixture)	22.4	21.3	59 100 231 318

* Assuming reactants enter at 298 °K.
** Highest wall temperature in this study

Convective Heat Transfer to Gases in 'Equivalent' Tube (in W)

Overall Reaction T (°K) =	500	1500
(1)	4040	1050
(2)	1420	310

Convective Heat Transfer to Gases at Plate A (in W)

Overall Reaction T (°K) =	500	1500
(1)	670	150
(2)	150	36

$15 H_2 + 0.8 NH_3 + 0.4 HSiCl_3$
$\rightarrow 0.133 N_2 + 15.8 H_2 + 1.2 HCl \ (+ 0.133 Si_3N_4)$ \hfill (1)

$3 H_2 + 2.0 NH_3 + 1.0 HSiCl_3$
$\rightarrow 0.333 N_2 + 5.0 H_2 + 3.0 HCl \ (+ 0.333 Si_3N_4)$ \hfill (2)

Flow characteristics of the jet and of the gases passing the vertical walls are given in Table III. Flow is characterized as viscous when the mean free path for molecular collisions is small compared to reactor dimensions, i.e., the dimensionless Knudsen number Kn <0.01. These conditions apply in the present work, where 0.5<P<2 Torr. Gas flows can be shown to be incompressible except near the nozzle by evaluating the dimensionless Mach number Ma, the ratio of flow to sonic velocity. Although Ma >0.5 at the nozzle, indicating that flow is compressible there, Ma <0.4 and is incompressible when the jet has expanded to 6 cm dia. Jets impinging on horizontal plate A have greater diameters and their cross-sectional areas were always <0.2 that of the reactor lid, thus meeting the criteria of freely expanding jets which entrain gas and lose momentum.[7] Reynolds numbers 220 <Re <1200 imply that gas flow in the jets are laminar. Other considerations indicate that flows remain laminar and that actual pressures on the vertical plates are in close agreement with measured pressures.

Table III
Dimensionless Ratios

Knudsen

$$\lambda / L \quad <0.01$$

Mach

$$v / (\gamma RT/M)^{1/2} \quad \begin{matrix} = 0.4\text{-}5 \text{ nozzle} \\ \leq 0.4 \text{ jet} \end{matrix}$$

Reynolds

$$\frac{vL\rho}{\mu} = 220\text{-}1200$$

CVD silicon nitride plates 0.5-4 mm thick were produced by impingement on horizontal substrates at deposition rates of 60-120 μm/hr. Expressions for deposition rates were derived by multiple correlation techniques and are shown in Table IV, where the Student t-statistic for significance of each term is given in parentheses. Rates of thickness increase r_t on horizontal plate A can be expressed as a function of temperature and of flows and pressures of $HSiCl_3$, and can be transformed into a kinetic equation in which all power terms are statistically significant. In this kinetic equation, deposition rates increase with fractional powers of $HSiCl_3$ concentration and decrease with powers of NH_3 and H_2 concentrations. Temperature dependence yields an apparent activation energy 36 kcal/mol, which may be compared with 53 kcal/mol when $SiCl_4$-NH_3-H_2 impinge a graphite substrate.[4] We interpret these results as indicating near-kinetic control of the thickness increase at the center of the impingement plate. Kinetic expressions describe the smaller rates of thickness increase near the edge of the plate less well than they do that at the center, however, suggesting that kinetics control deposition rate only where the jet impinges.

CVD plates 0.05-0.5 mm thick were produced on vertical substrates at deposition rates 5-70 μm/hr. Rates of thickness increase depend on pressure, flow and concentration of HSiCl$_3$ as shown in Table IV. This empirical rate expression cannot be transformed into a kinetic expression in which the terms retain statistical significance, however, and no temperature dependence is detected by multiple correlation analysis. These results indicate that deposition on vertical, non-impingement substrates proceeds under mass transport control, and that HSiCl$_3$ is the depleted species.

Although deposition rates are lower on vertical than on horizontal surfaces, vertical deposits can be made uniformly thick more easily than horizontal ones. Correlation analysis reveals that uniformity on vertical surfaces is enhanced by decreasing the gas residence time and by increasing the temperature.

TABLE IV
Deposition Rate Expressions Derived from Multiple Correlation Techniques

Empirical Basis:

		n_t	Ω
Plate A r$_t$ =	20500 + 24000 Tw (2.8) - 0.298 Ptw2 (3.1) - 8040 Tw2 (2.7) + 142 PSx2 (4.9) + 5420 PSx2 (15)	50	0.96
Plate B n =	10.2 + 55.4 FSx (7.3) + 2121 XSx (2.8) - 44.3 Si/N (9.4) - 175 PSx2 (3.8)	58	0.91

Kinetic Theory Basis:

$$r = k_0 \, e^{(-E/RT)} \, P^n \, X_{Si}^a \, X_N^b \, X_H^c \quad (1)$$

	ln k$_0$	E/R	n	a	b	c	n_t	Ω
Plate A ln n =	14.4	-18.4 (8.7)	-0.296 (4.0)	-0.64 (5.0)	1.12 (5.0)	-3.83 (4.2)	0.24	0.93
Plate B ln n =	3.30	-1.23 (0.4)	-0.315 (3.0)	0.16 (0.9)	-0.47 (1.4)	-0.06 (0.1)	0.14	0.96

Average Kinetic Data Derived from Various Deposition Rates
Based on Eqn. (1) rate expression.

Data Set	E (kcal/mol)	n	a	b	c
r$_t$. Plate A	29	0.3	0.7	-0.6	*
(with exclusions)	36	0.3	0.6	-1.1	-3.8
r$_t$. Plate B	*	*	*	*	*
(with exclusions)	*	-0.3	*	*	*
r$_m$. Plate A	17	0.3	0.4	*	*
(with exclusions)	14	0.3	0.7	*	*
r$_m$. Plate B	*	*	*	*	*
(with exclusions)	*	*	*	*	*

* = Insufficient correlation to data set or, in effect, = 0

Preferred conditions for deposition of thick CVD silicon nitride on vertical faces were determined to be 1500°C at 0.5 Torr total pressure, under 67% stoichiometric excess NH$_3$. Deposits up to 0.9 cm thick were grown at 130 μm/hr. on each vertical plate. These crystalline deposits were free of pores and cracks and exhibited densities 3.187 g/cm^3. Columnar morphology of the preferred condition is similar to Figure 3 in which the (222) crystallographic direction is oriented normal to a surface which consists of well-defined trigonal facets 10 μm to 800 μm across as shown in Figure 4. Crystallite size as determined by X-ray line broadening is approximately 1000 A. This CVD plate is gray and translucent, and contains 0.4 w/o C and 0.2 w/o 0 as principal impurities.

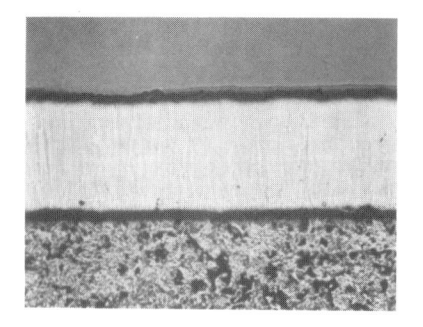

Figure 3

Typical Polycrystalline CVD Si$_3$N$_4$ Having Columnar Growth Pattern (Cross-Section in Bright Field)

Figure 4

As-Deposited Trigonal Surface of CVD Si$_3$N$_4$ from Production Demonstration

REFERENCES

1. H. Fischer, Z. Phys. Chem. (Leipzig) 246, 357 (1971).

2. F. Galasso, U. Kuntz and W. J. Croft, J. Am. Ceram. Soc. 55, 431 (1972).

3. F. Galasso, R. D. Veltri, and W. J. Croft, Am. Ceram. Soc. Bull. 57, 453 (1978).

4. K. Niihara and T. Hirai, J. Mats. Sci. 11, 593, 604 (1976); 12, 12233, 1243 (1977).

5. M. Watanabe et al., in Semiconductor Silicon, (Electrochem. Soc., Pennington, NJ, 1981) p. 126.

6. J. S. Shinko and J. W. Lennartz, Proc. 10th Int. Conf. Chem. Vapor Deposition, (Electrochem. Soc., Pennington, NJ, 1987, p. 1106.

7. EDO, an Experimental Design Optimization Program (Statistical Studies Inc., Cleveland, OH 1988)

8. T.T. Elrod, Heating, Piping Air Cond. 26, 149 (1954).

CHEMICAL VAPOR DEPOSITION OF MULTIPHASE BORON-CARBON-SILICON CERAMICS

E. Michael Golda and B. Gallois
Department of Materials Science and Engineering
Stevens Institute of Technology, Hoboken, NJ, 07030.

ABSTRACT

Specific compositions of boron-carbon-silicon ceramics exhibit improved abrasive wear and good thermal shock resistance, but require bulk sintering at temperatures in excess of 2100K. The formation of such phases by chemical vapor deposition was investigated in the temperature range of 1073K-1573K. Methyltrichlorosilane (CH_3SiCl_3), boron trichloride, and methane were chosen as reactant gases, with hydrogen as a carrier gas and diluent. The coatings were deposited in a computer-controlled, hot-wall reactor at a pressure of 33 MPa (200 Torr). Below 1473K the coatings were amorphous. At higher temperatures non-equilibrium reactions controlled the deposition process. The most common coating consisted of a silicon carbide matrix and a silicon boride, SiB_6, dispersed phase. Multiphase coatings of $B+B_4C+SiB_6$ and $SiC+SiB_6+SiB_{14}$ were also deposited by controlling the partial pressure of methane and boron trichloride. Non-equilibrium thermodynamic analysis qualitatively predicted the experimentally deposited multiphase coatings.

INTRODUCTION

Boron-carbon-silicon ceramics have traditionally been fabricated either by sintering or by electric arc melting. Both processes require temperatures in excess of 2100K. Dispersion strengthened ceramics can be formed in this system which exhibit superior properties compared to the single phase material. Proper selection of the matrix and the dispersed phase can be used to control specific materials properties. A boron carbide matrix with a dispersed silicon carbide phase exhibits a significant improvement in abrasive resistance compared to pure boron carbide.[1] A silicon carbide matrix with a boron carbide dispersed phase exhibits superior thermal cycling stability compared not only to pure boron carbide but to other refractory oxides as well. Boron-carbon-silicon ceramics also exhibit good oxidation resistance up to 1473K.[2]

Surface modification, in which the ceramic is applied as a coating, is an alternative processing approach. Chemical vapor deposition (CVD) is one of the most versatile techniques for applying ceramic coatings. Dispersion strengthening can be accomplished by the simultaneous deposition, or codeposition, of two or more ceramics. In spite of the potential for improved coating performance, there is a dearth of published information on CVD of boron-carbon-silicon ceramics. Niemyski et al. [3] used boron trichloride, silicon tetrachloride and carbon tetrachloride as precursors. The deposition pressure was 100 MPa (760 Torr) at temperatures greater than 1873K. X-ray diffraction identified the deposit as a bulk substitution compound, $B_{12}(C,Si,B)_3$. In the present work we report on the formation of boron-carbon-silicon ceramics using different precursors at lower deposition temperatures and at a lower pressure.

EXPERIMENTAL PROCEDURES

The depositions were conducted in a rectangular channel, hot-wall, horizontal quartz reactor at 33 MPa (200 Torr) over a temperature range of 1073K to 1573K. The graphite rectangular channel acted as a susceptor, which was heated by a high-frequency induction generator coupled to a temperature control system. Substrates were isostatic molded graphite plates 1 cm by 1 cm in size. The deposition process was automated; the throttle valve, the flowmeters, and the valves were all interfaced to a microcomputer.

Methyltrichlorosilane (CH_3SiCl_3 or MTS), boron trichloride, and methane were used as precursors, with hydrogen as a reducing agent and the carrier gas for methyltrichlorosilane. The precursor mixture was characterized by two ratios: the boron ratio and the hydrogen ratio. The boron ratio, $BCl_3/(BCl_3+MTS)$, expressed the percentage of boron in the precursor gas mixture at the reactor inlet. The hydrogen ratio, H_2/BCl_3+MTS, compared the amount of hydrogen to the total amount of precursors at the reactor inlet. The total flow was fixed at 500 sccm.

The coatings were examined by scanning electron microscopy. The phases were identified by X-ray diffractometry. Microprobe analysis of the constituents was performed with a windowless, multi-crystal system.

RESULTS AND DISCUSSION

Figure 1 is a map of the experimental depositions at different deposition temperatures as a function of the boron ratio at a hydrogen ratio of 10. The structure of the coating was determined by the deposition temperature. Above 1473K the coatings were polycrystalline. The coatings were silicon-rich, consisting of a matrix of silicon carbide (SiC) and elemental silicon, with a uniformly dispersed silicon hexaboride (SiB_6) second phase.

Amorphous coatings were deposited between 1173K and 1473K. Amorphous coatings annealed in argon at 1523K crystallized when the annealing time was greater than 10 minutes. X-ray diffraction analyses of the annealed films showed both SiC and SiB_6 peaks. The 1473K transition temperature between amorphous and polycrystalline coatings was approximately 250K higher than the transition temperature reported for a poly-crystalline silicon carbide coating.[4] The addition of boron clearly stabilized the amorphous phase.

The amount of elemental silicon in the polycrystalline coatings was inversely proportional to the deposition temperature. In addition, the silicon content was directly

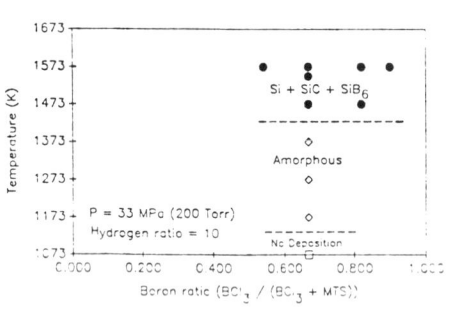

Fig. 1. Map of experimental deposits using BCl_3 and MTS.

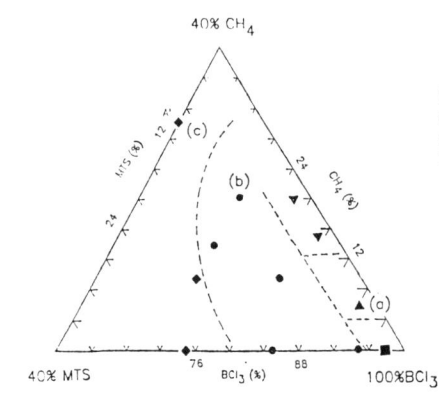

Fig. 2. Map of experiemtnal deposits using BCl$_3$, MTS and CH$_4$. Hydrogen ratio = 5. P = 33 MPa. T = 1548K.
(■) B+B$_{25}$C+SiB$_6$,
(▲) B+B$_4$C+B$_{25}$C+SiB$_6$,
(▼) SiB$_6$+SiB$_{14}$+B$_4$C,
(●) SiC+SiB$_6$+SiB$_{14}$,
(◆) SiC+SiB$_6$. Phase field boundaries are approximate.

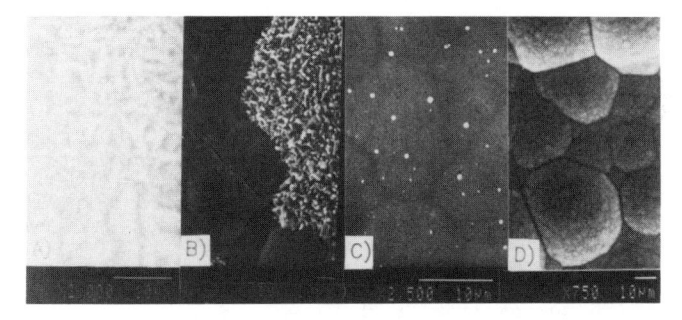

Fig. 3. Phase field morphologies. Precursors, depostion conditions and inlet gas mixtures are shown on Fig. 2. (a) (■), (b) (●), (c) (◆), (d) SiC+SiB$_6$, 20%BCl$_3$:20%MTS:60%CH$_4$.

proportional to the hydrogen ratio. No elemental silicon was deposited at hydrogen ratios less than 10.

In order to eliminate elemental silicon in the coatings, the hydrogen ratio was reduced to 5 and methane was added to the gas stream as an additional source of carbon. Figure 2 is a map of the phases observed at 1423K at 33MPa (200 Torr) as a function of the composition at the reactor inlet. Five distinct phase fields were identified by X-ray diffraction: B+B$_{25}$C+SiB$_6$, B+B$_4$C+B$_{25}$C+SiB$_6$, SiB$_6$+SiB$_{14}$+B$_4$C, SiC+SiB$_6$+SiB$_{14}$, and SiC+SiB$_6$. The boron-rich deposits, B+B$_{25}$C+SiB$_6$, B+B$_4$C+B$_{25}$C+SiB$_6$, and SiB$_6$+SiB$_{14}$+B$_4$C, had extensive transgranular cracking and good substrate adhesion as qualitatively measured by the scotch tape pull test. The SiC+SiB$_6$+SiB$_{14}$ coating had extensive transgranular cracking and branching with very poor substrate adhesion. The largest phase field was the SiC+SiB$_6$ coatings. These coatings had very little transgranular cracking and good substrate adhesion. The presence of the four phase field B+B$_4$C+B$_{25}$C+SiB$_6$, which violates the phase rule, may be due to fluctuations in the gas composition with time or to nonequilibrium phenomena. There was no evidence of the formation of the bulk substitutional compound $B_{12}(C,Si,B)_3$ by X-ray diffraction reported by Niemyski et al.[3]

Figure 3 shows representative morphologies of the various phase fields. Figure 4(a) illustrates the acicular morphology common for all of the boron-rich coatings with no apparent second phase. Figure 4(b) is a typical specimen in the $SiC+SiB_6+SiB_{14}$ phase field. The coating is composed of a matrix of SiC nodules with diameters greater than 75 m and vertical reliefs as great as 50 m. The voids between nodules are filled with boron-rich whiskers which have a length/diameter ratio of approximately 10/1 and taper towards the whisker tip. Figures 4(c) and 4(d) show the morphology of the $SiC+SiB_6$ phase field, a SiC matrix and a distinct, evenly dispersed SiB_6 second phase. The density of the dispersed phase is directly proportional to the boron ratio.

THERMODYNAMIC EQUILIBRIUM ANALYSIS

Thermodynamic modelling can be used to predict the optimum conditions for obtaining a deposit with a specific composition. Thermodynamic equilibrium analyses performed in this report used Solgasmix-PV.[5] The program ran on a personal computer and determined the equilibrium gas composition and condensed phases for a specific condition by minimizing the free energies of all the possible system constituents which could form. The JANAF Tables were the reference for the heats of formation and entropies of the 83 possible gaseous species and the following condensed phases: B, C, Si, SiC, B_4C.[6] Thermodynamic data for the silicon boride phases: SiB_3, SiB_6, and SiB_{14}, were optimized and rationalized by Dirkx [7] and were used in the calculation.

Figure 4 is an analysis of the effect of the hydrogen ratio on the composition of the deposited phases as a function of the boron ratio at the reactor inlet at a deposition temperature of 1548K and a pressure of 33MPa (200 Torr). The figure is a CVD diagram, which graphically displays the predicted condensed phase(s) deposited at equilibrium as a function of experimental parameters. The CVD diagram shows that it is thermodynamically possible to codeposit multiphase boron-carbon-silicon ceramics and that the composition is a function of the boron ratio.

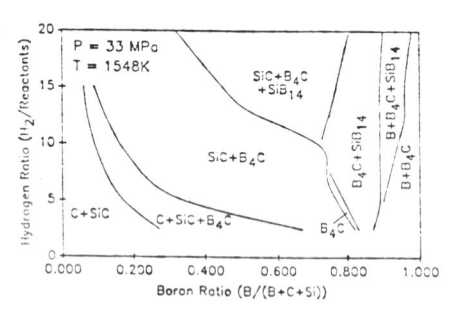

Fig. 4. CVD diagram of the boron-carbon-silicon system using BCl_3 and MTS.

At a hydrogen ratio of 10 and a boron ratio of 0.67 the codeposition of $SiC+B_4C$ is predicted. Under the same conditions, the experimental coatings consist of $Si+SiC+SiB_6$.(Fig. 1) Thermodynamically it is possible to vary the coating composition by changing the boron ratio. Experimentally, however, the com-position of the $Si+SiC+SiB_6$ coating was not a function of the boron ratio. The difference between the two sets of data indicates that the deposition process occurs under non-equilibrium conditions.

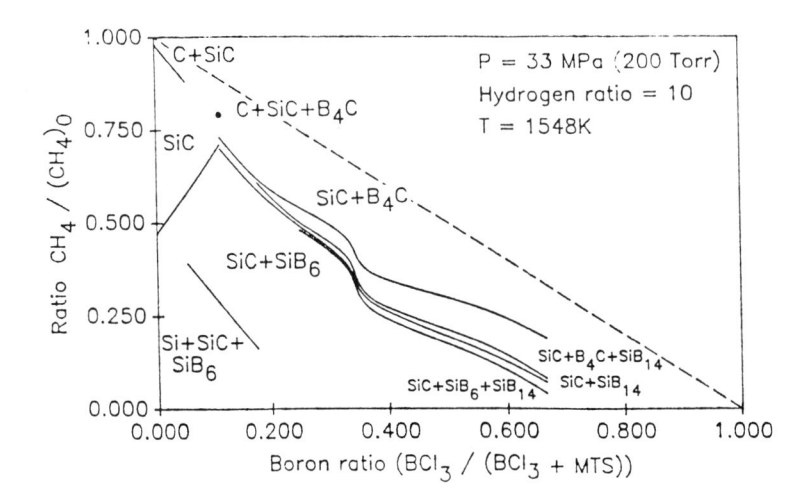

Fig. 5. CVD diagram of non-equilibrium deposition using BCl$_3$ and MTS.

NON-EQUILIBRIUM THERMODYNAMIC ANALYSIS

Thermodynamic equilibrium calculations use the precursor composition at the reactor inlet as input parameters and determine the equilibrium phases in the homogeneous (gas phase) and heterogeneous (gas phase and condensed phase) systems. However, kinetic and mass transport effects generally alter the gas phase composition at the gas/solid interface, the heterogeneous system. These non-equilibrium conditions can be modeled using Solgasmix-PV by assuming that a "local" thermodynamic equilibrium exists at the interface.[8]

One kinetic effect which could alter the composition of the heterogeneous reaction is the homogeneous (gas phase) decomposition of MTS. At temperatures above 1173K, MTS decomposes to form a methane and a chloride intermediates. The methane molecule is very stable at temperatures below 1473K. The formation of a stable carbon intermediate with a slow reaction rate could lower the carbon concentration in the coating.[9,10,11]

Figure 5 is a CVD diagram of a non-equilibrium analysis. The dashed line represents the amount of carbon available as methane in the equilibrium heterogeneous reaction as a function of the boron ratio. Non-equilibrium heterogeneous reaction conditions are modeled by holding the boron ratio constant and artificially decreasing the carbon content from the equilibrium value. The abscissa is the ratio of the amount of methane avail-able for the non-equilibrium heterogeneous reaction, CH_4, to the amount of methane available after the homogeneous decomposition of an inlet gas composition consisting of 100% methyltri-chlorosilane, $(CH_4)_0$. As the methane content of the non-equilibrium reaction decreases the predicted phase fields become increasingly silicon-rich. This non-equilibrium analysis qualitatively predicts the deposition of three of the phase fields observed experimentally: $SiC+SiB_6$, $SiC+SiB_6+SiB_{14}$, and $Si+SiC+SiB_6$.

SUMMARY

Boron-carbon-silicon coatings were deposited from mixtures of methyltrichlorosilane, boron trichloride and methane. The structure of the coating underwent a transition from an amorphous to a polycrystalline coating at 1473K. The most commonly observed coating was a silicon carbide matrix with a uniformly dispersed silicon hexaboride second phase. Other multiphase coatings were deposited by controlling the partial pressures of methane and boron trichloride. The significant difference between the phase fields predicted by the thermodynamic equilibrium analysis and the experimental data demonstrated that deposition occurred under non-equilibrium conditions. Non-equilibrium thermodynamic analysis qualitatively predicted the silicon-rich experimental coatings.

ACKNOWLEDGEMENTS

The authors gratefully acknowledge the support of the Army Research Office, Division of Materials Science, under contract DAAG29-85-K-0124. E. Michael Golda was partially supported by an Army Research Office fellowship, DAAL03-86G-0070.

REFERENCES

1. H. Holleck, J. Vac. Sci. Technol., A4(6), 2661-2669 (1986)
2. G.V. Samasov, et al., Boron, Its Compounds and Alloys, AEC-tr-5032 (Book 1), The Academy of Sciences of the Ukranian S.S.R. (1960)
3. T. Niemyski, S. Appenheimer, J. Panczyk and A. Badzian, J. Crystal Growth, 5, 401-404 (1969)
4. S. Motojima, H. Yagi and N. Iwamori, J. Materials Science Letters, 5, 13-15 (1986)
5. T.M. Besmann, Solgasmix-PV, A Computer Program to Calculate Equilibrium Relationships in Complex Chemical Systems, ORNL/TM-5775 (Oak Ridge National Laboratory, Oak Ridge, Tennessee, 1977)
6. M.W. Chase, Jr., C.A. Davis, J.R. Downey, Jr., et al., JANAF Thermochemical Tables, 3rd Edition, J. of Phys. and Chem. Ref. Data, 14, (1985)
7. R.R. Dirkx and K.E. Spear, Calphad, 1987, 167-175
8. K.E. Spear, Pure and Appl. Chem., 54(7), 1297-1311 (1982)
9. G. Fischman and W.T. Petuskey, J. Am. Ceram. Soc., 68(4), 185-190 (1985)
10. M.H. Back and R.A. Back, in Thermal Decomposition and Reactions of Methane in Pyrolysis:Industrial Theory and Practice, edited by L.F. Albright, B.L. Crynes and W.H. Corcoran (Academic Press, New York, 1983)
11. D.E. Cagliostro and S.R. Riccitiello, J. Am. Ceram. Soc., 73(3), 607-614 (1990)

MODIFICATION OF OPTICAL SURFACES EMPLOYING CVD BORON CARBIDE COATINGS

Richard A. Lowden, Laura Riester, and M. Alfred Akerman
Oak Ridge National Laboratory, P. O. Box 2008, Oak Ridge, Tennessee 37831-6063

ABSTRACT

Non-reflective or high emissivity optical surfaces require materials with given roughness or surface characteristics wherein interaction with incident radiation results in the absorption and dissipation of a specific spectrum of radiation. Coatings have been used to alter optical properties, however, extreme service environments, such as experienced by satellite systems and other spacecraft, necessitate the use of materials with unique combinations of physical, chemical, and mechanical properties. Thus, ceramics such as boron carbide are leading candidates for these applications. Boron carbide was examined as a coating for optical baffle surfaces. Boron carbide coatings were deposited on graphite substrates from BCl_3, CH_4, and H_2 gases employing chemical vapor deposition (CVD) techniques. Parameters including temperature, reactant gas compositions and flows, and pressure were explored. The structures of the coatings were characterized using electron microscopy and compositions were determined using x-ray diffraction. The optical properties of the boron carbide coatings were measured, and relationships between processing conditions, deposit morphology, and optical properties were determined.

INTRODUCTION

Space telescopes and other high-performance optical instruments must have baffle and vane surfaces designed to minimize the effect of stray light on system performance.(1-3) Optical baffles are the non-reflective structures and surfaces within an optical device that act to narrow the sighting area by limiting the source from which light is detected, and enhance performance by minimizing off-axis and stray light that might enter the device and reach the detector(s). Optical baffles are essential in the management of light that enters the apparatus and thus play a significant role in determining the performance of an optical system.

An optical baffle is typically non-reflective or "black." A variety of "black" coatings have been developed for optical baffle surfaces.(1,4,5) These include paints, anodized surfaces, and etched metal coatings. However, many of the current materials and coatings cannot withstand the extreme environments and severe mechanical and thermal loadings associated with applications such as infrared sighting systems for missiles and projectiles, or telescopes on specialized satellites. High gravitational forces and vibration during launch, and thermal or mechanical loads during service, cause the current materials to fail. The coatings flake and spall diminishing the properties of the baffle surface as well as producing particles that can eventually cover lenses, mirrors, and detectors, rendering the system inoperative. In addition, many of the coatings are fragile and difficult to handle. Simply touching the surface smears or removes the coating, decreasing its effectiveness. Thus, there exists a need for improved diffuse-absorptive coatings for high performance applications, coatings which are stronger, more adherent and damage resistant, and therefore able to survive extreme environments.

Boron carbide fulfills many of the requirements for improved baffles and thus was selected for this study.(6) Boron carbide is a low density material with exceptional mechanical properties including high hardness, strength, and modulus. It exhibits poor reflectance from visible through infrared and functioned well in earlier baffle and solar absorption applications.(5,6) Boron carbide has a high melting temperature, good thermal conductivity, and high specific heat resulting in excellent survivability against directed energy threats. A broad range of compositions exist, ie. B_4C, $B_{13}C_2$, B_8C, $B_{25}C$, etc., and these are readily deposited using chemical vapor deposition techniques.(7-11)

Chemical vapor deposition was chosen as the processing method for it is an extremely versatile process. A variety of parameters can be controlled to produce boron carbide coatings with a broad range of compositions and morphologies.(9-13) Gas composition, temperature, and pressure can be changed to modify the texture and morphology of the final coating with nodular, faceted, and whisker-like surface features possible.(10,11) Correlations between deposition conditions, coating morphology, and optical performance at 10.6 μm were investigated.

EXPERIMENTAL PROCEDURES

Deposition

Boron carbide coatings were deposited from gas mixtures containing boron trichloride (BCl_3), methane (CH_4), and hydrogen (H_2) at a pressure of 3.3 kPa and temperatures of 1273 to 1673 K. The effects of deposition parameters including temperature, gas composition, and total reactant flow on the structure and composition of the deposits were examined. The ratio of boron to carbon (BCl_3 to CH_4) in the input gas mixture was varied from 0.25 to 8.0; with the sum of these reactant species held constant at 75 sccm. The hydrogen to chlorine ratio in the input gas was held constant at 30; the total H_2 flow was altered to control this ratio. Deposition times were 45 min for all samples, and details of the experiments are given in Figure 1.

Upon completion of these scoping deposition experiments, additional coating runs were conducted. It was found that rough or textured coatings were produced at temperatures of 1473 and 1573 K and BCl_3:CH_4 ratios of 0.25 to 1.00. Subsequent experiments focused on this processing window. As before, deposition time, pressure, and H:Cl were held constant as temperature and input gas composition were varied. Specific details of the additional deposition experiments are also shown in Figure 1.

A horizontal cold-walled system was used for the deposition of the coatings. The furnace consisted of a fused silica tube (54 mm OD, 2 mm wall, 75 cm long) with custom-fabricated stainless steel end caps. Graphite substrates 6 mm thick and 25 mm in diameter, were cut from round stock, washed in water and acetone, and then dried in an atmospheric oven. A single substrate was supported within the reactor using a graphite rod inserted into a hole drilled into the rear of the substrate and was heated inductively using an RF generator. Substrate temperatures were measured using an optical pyrometer and a black body hole drilled in the substrate. Corrections were applied for absorption by the glass prism and quartz window.

Reactant gases were introduced to the system through an end of the furnace and flow rates were controlled using mass flow controllers. Hydrogen was purified by passage through a titanium sponge getter, while the other gases were metered in their supplied forms. Boron trichloride flow was difficult to control due to its low vapor pressure at room temperature, thus the BCl_3 bottle and lines were heated to obtain reproducible flows. Exhaust gases were scrubbed by passage through a soda-lime bed. Pressure control was accomplished using a gas ballast technique. A capacitance manometer monitored the furnace pressure which was held constant by injecting argon gas into the pump inlet to regulate effective pumping speed.

Characterization

Optical and scanning electron microscopy were used to characterize coating surface morphologies, features, and structures. X-ray diffraction was used to examine the composition of the deposits. Reflectance was measured over the wavelength range of 2 to 55 μm employing a Perkin-Elmer 983G IR spectrophotometer. The instrument is a computer-controlled, dual-beam, ratio-recording spectrophotometer for measuring reflectance or transmittance of flat optical components with a resolution of 0.03 μm and an accuracy of $\pm$ 0.1 μm. Bidirectional Reflectance Distribution Function (BRDF) measurements at 10.6 μm and angles from 0° to $\pm$ 30° were conducted using an advanced optical scatterometer. The instrument measures optical scatter as a function of angle and position of the specimen with an illuminated spot size of 5 mm.

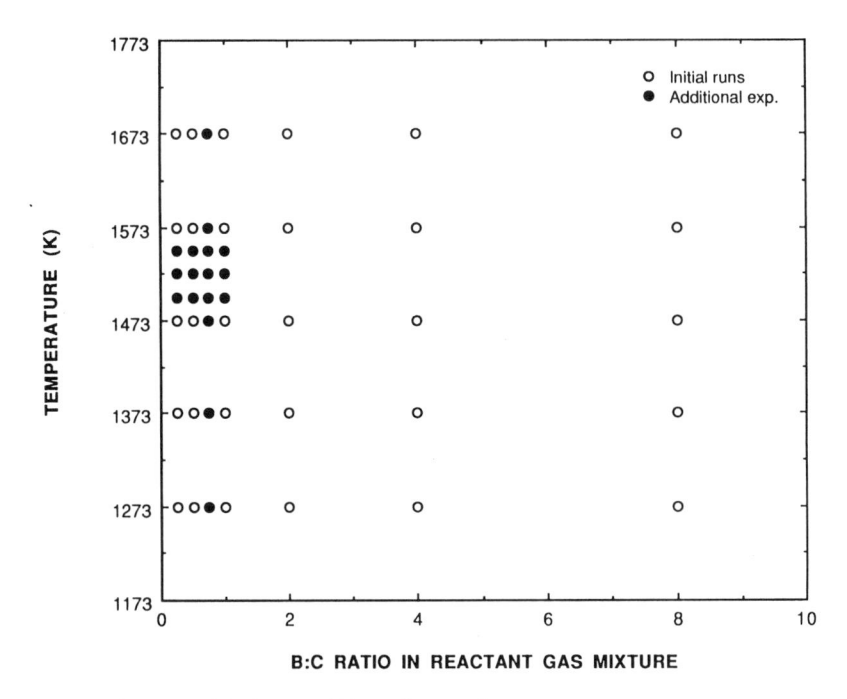

Figure 1. Summary of the boron carbide coating experiments.

RESULTS

Coatings ranging in thickness from 15 - 60 μm were deposited on the graphite coupons. Deposit finishes ranged in color from shiny silver-gray to flat black. Upon closer examination it was found that the scoping experiments produced coatings with a variety of morphologies as shown in the electron micrographs in Figure 2. The majority of the coatings were characterized by smooth nodular or slightly faceted surfaces, however, cylindrical, pyramidal, and cubic structures with sizes ranging from < 1 to 20 μm were observed on certain specimens. The highly textured coatings were produced at processing conditions of 1473 to 1573 K, BCl_3:CH_4 of 0.25 to 1.00, H:Cl of 30, and a pressure of 3.3 kPa.

The goal of the project was to optimize the properties of the coating for a specific infrared wavelength, ie. 10.6 μm, thus the optical properties of coatings with highly textured or rough surfaces were measured. In general, less than 1 % reflectance from 2 to 55 μm was observed. Deposits with fine, whisker-like structures were not examined optically for these coatings typically were fragile and not adherent. The off-angle properties of selected specimens were then characterized. The results of the BRDF measurements at 10.6 μm from 0 to $\pm$ 30° for select coatings are shown in Figure 3. The optimum coating exhibits a low BRDF response over a broad range of incident angles, and is a straight line, characteristic of a Lambertian surface when cosine corrected.

X-ray diffraction identified only rhombohedral boron carbide in the coatings with the rough morphologies. The exact composition of the rhombohedral phase, B_4C or $B_{13}C_2$, has been the subject of numerous studies, and continues to be a topic of debate. Although it is difficult to distinguish between the two phases, it appears that the composition of the deposits, the fraction of B_4C or $B_{13}C_2$ in the coating, was influenced by reactant gas composition and temperature. The B_4C phase seemed to be favored at higher methane concentrations and temperatures.

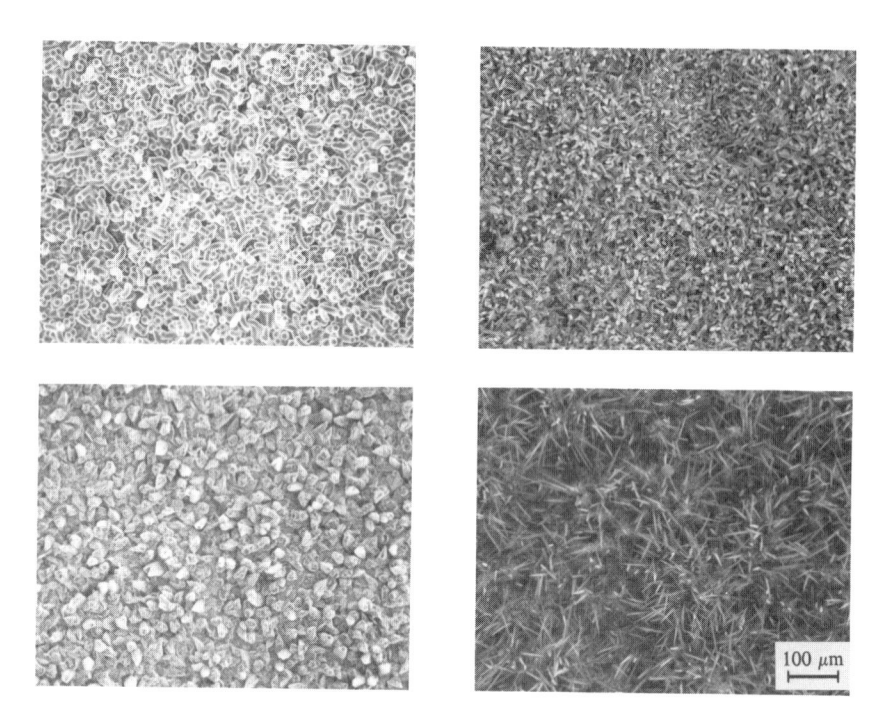

Figure 2. A variety of textured surfaces were observed.

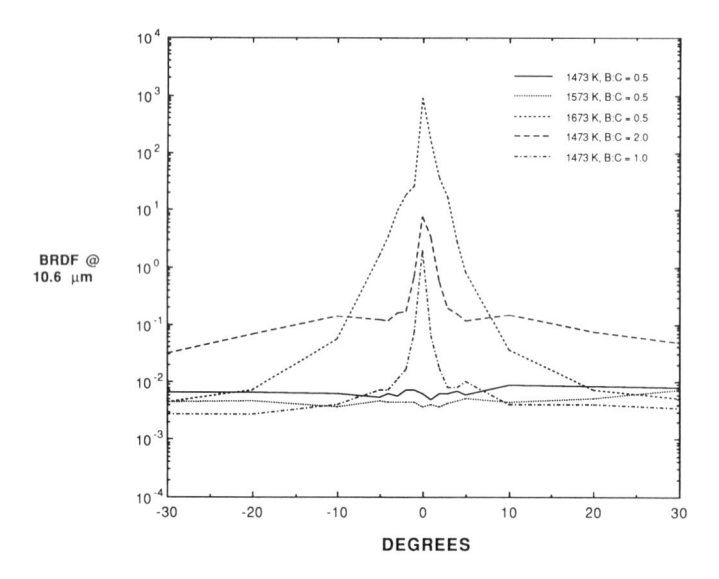

Figure 3. BRDF of various baffle coatings at 10.6 µm.

DISCUSSION

The wavelength of interest in this study was 10.6 μm, thus surface features were tailored for the far-infrared spectrum. Scoping deposition experiments produced coatings with surface features and textures that exhibited low reflectance as indicated by the specular and BRDF measurements. Deposition conditions were then adjusted to optimize the properties of the coating for the given wavelength. Close control of the microstructure was achieved through variations in processing. Highly-sloped and faceted structures with sizes on the same order as the wavelength exhibited the best optical baffle properties.

The optical properties of select boron carbide coatings are compared to other materials in Table 1. The use of BRDF at 0° to compare samples is appropriate since departures from Lambertian perfromance are readily apparent. The polished metal sample is a case in point. Martin Black is a standard to which optical baffle coatings are often compared.(1,4) Martin Black is an anodized aluminum surface that is microrough. It exhibits a matte black finish and contains a proprietary aniline dye. It was developed for the Skylab program and has been used in a variety of space instruments.(4) The coating was originally developed for ultraviolet and visible applications, but has seen extensive use in infrared systems. The Martin Black surface contains a variety of surface feature sizes and exhibits good broadband properties. The properties of boron carbide coatings approach those of the standard. However, it is expected that due to the high strength, stiffness, and hardness of boron carbide, these materials will exhibit higher durability and survivability in the given applications. The coatings are easily handled and appear to be extremely rugged. The ability of the boron carbide materials to withstand man-made threats is currently being assessed.

Table 1. Optical performance of various materials at 10.6 μm.(1,4)

Material	BRDF @ 0° (sr^{-1})	Specular Reflectance (%)
Polished metal	1.7×10^5	98.4
B_4C on metal	3.7×10^2	8.1
CVD Diamond	2.6×10^2	9.9
Plasma-Sprayed Be	4.8×10^{-2}	1.6
Plasma-Sprayed B	1.5×10^{-2}	0.7
Martin Black	$\approx 1 \times 10^{-3}$	0.7
B_4C on graphite	1.9×10^{-2}	---
B_4C on graphite	6.1×10^{-3}	---
B_4C on graphite	6.8×10^{-3}	0.7
B_4C on graphite	4.3×10^{-3}	0.5
B_4C on graphite	3.7×10^{-3}	0.5

CONCLUSIONS

The morphology and size of surface features for boron carbide coatings were modified by altering deposition conditions. Rough or textured deposits exhibited low broadband reflectance and a flat BRDF response in the far-infrared (10.6 μm). The coatings with the optimum optical baffle properties possessed highly-sloped and faceted surface features and were produced at temperatures between 1473 and 1573 K and BCl_3:CH_4 ratios of 0.25 to 1.00. The size and shape of the surface structures was influenced by deposition conditions, and thus the performance of the baffle coating could be optimized through changes in processing.

ACKNOWLEDGEMENTS

Special appreciation is due the U. S. Army Strategic Defense Command in Huntsville, Alabama for their support of this effort through PMA A1SO4. Appreciation is also extended to the Optics Characterization Laboratory at Oak Ridge National Laboratory for the optical property measurementa as well as to the High Temperature Materials Laboratory, also at ORNL, for use of the SEM and X-ray facilities.

REFERENCES

1. R. D. Seals, C. M. Egert, and D. D. Allred, "Advanced Infrared Optically Black Baffle Materials," SPIE Vol. 1330, *Optical Surfaces Resistant to Severe Environments*, 164-177 (1990).

2. W. J. Smith, *Modern Optical Engineering*, 2nd Edition, McGraw-Hill, Inc., New York, 139-142 (1990).

3. M. Kuhl, K. Gindele, and M. Mast, "Determination of the Characterizing Parameters of Rough Surfaces for Solar Energy Conversion," SPIE Vol. 653, *Optical materials Technology for Energy Efficiency and Solar Energy Conversion*, 228-235 (1986).

4. S. M. Pompea, D. F. Shepard, and S. Anderson, "BRDF Measurements at 6328 Angstroms and 10.6 Micrometers of Optical Black Surfaces for Space Telescopes," SPIE Vol. 967, *Stray Light and Contamination in Space*, 236-247 (1988).

5. T. M. Besmann and A. Ismail Abdel-Latif, " Modification of Optical Properties with Ceramic Coatings," *Thin Solid Films* 202, 51-59 (1991).

6. G. V. Samsonov and I. M Vinitski, Handbook of Refractory Compounds, IFI-Plenum, New York, 117-185 and 273-277, (1980)

7. L. G. Vandenbulcke, "Theoretical and Experimental Studiers on the Chemical Vapor Deposition of Boron carbide," Ind. Eng. Chem. Prod. Res. Dev. 24, 568-575 (1985).

8. D. P. Stinton, T. M. Besmann, and R. A. Lowden, "Advanced Ceramics by Chemical Vapor Deposition Techniques," *Ceramic Bulletin* 67 [2], 350-355 (1988).

9. U. Jansson, J.-O. Carlsson, B. Stridh, S. Soderberg, and M. Olsson, "Chemical Vapor Deposition of Boron Carbides I: Phase and Chemical Composition," *Thin Solid Films* 124, 101-107 (1985).

10. D. N. Kevill and T. J. Rissmann, "Preparation of Boron-Carbon Compounds, Including Crystalline B_2C Material, By Chemical Vapor Deposition," *J. Less-Common Metals* 117, 421-425 (1986).

11. T. M Besmann, "Chemical Vapor Deposition in the Boron-Carbon-Nitrogen System," *J. Am. Ceram. Soc.* 73 [8], 2498-2501 (1990).

SiC THIN FILMS BY CHEMICAL CONVERSION OF SINGLE CRYSTAL Si

Chien C. Chiu, Chi Kong Kwok, and Seshu B. Desu
Department of Materials Science and Engineering
Virginia Polytechnic Institute and State University
Blacksburg, VA 24061

ABSTRACT:

The reaction of (100)Si with C_2H_2 in a hot wall CVD reactor has been studied using a X–ray photoelectron spectroscopy, and a scanning electron microscopy. The growth of the SiC films was observed through the behavior of Si_{2p} peaks and their plasmons. Smooth surface morphology with a monolayer of SiC was obtained at 950ºC for 7 minutes and defects were observed for longer reaction times at this temperature. For higher reaction temperatures (*e.g.* 1000ºC), defects were observed for reaction times as short as 10 seconds. The formation of defects was correlated to the out–diffusion of Si in the carborization process.

INTRODUCTION:

Growth of single crystal β–silicon carbide (SiC) on a (100)Si substrates by Chemical Vapor Deposition (CVD) is generally carried out at high temperatures. In growing SiC single crystals by a one step process, some problems were encountered which are related to the large mismatch (20%) in lattice constants, and the differences in thermal expansion coefficients between the deposited SiC film and the underlying Si substrate (8%) [1]. Therefore, the resulting SiC films often exhibit poor morphology. In addition, peeling of SiC film from the Si substrate is ofetn observed after the deposition. Furthermore, in depositing SiC films etching problems were reported for Cl–based precursor systems (*e.g.* $CH_2Cl–SiH_4– H_2$ and CH_3SiCl_3) because of the formation of Cl and CH_3 radicals [2].

Growth of a buffer layer has been proved to be a necessary step in obtaining good quality SiC films on Si substrates [3]. This buffer layer is generally grown at relatively low temperatures, after which the substrate is raised to higher temperatures for carrying out the bulk growth of SiC films. The buffer layer is grown by either reacting Si substrate with a hydrocarbon gas or by sputtering SiC onto the Si substrate [4]. Table I shows various hydrocarbon gases and the reaction parameters that are used for the conversion of Si surfaces to SiC layers. Note that the reactions had been carried out in high vacuum and cold–wall systems which are not compatible with the low pressure chemical vapor deposition (LPCVD) processes which are often carried out in hot wall reactors to increase the throughput. Therefore, in this work the reaction between acetylene (C_2H_2) and (100)Si substrate was studied to obtain the optimum parameters to grow buffer layer in a conventional horizontal hot wall CVD reactor in a low pressure. Emphasis will be given for the formation of surface defects during the reaction of C_2H_2 with Si substrates.

EXPERIMENTAL PROCEDURE:

Single–crystal (100)Si substrates were used in this study. Before introducing the substrates into the reaction chamber, the organic contamination on the surface of Si substrate was first washed out by using acetone and then by dipping into methanol. Following these procedures, the residual surface oxides were etched by 10 wt% HF for 10 seconds. Finally the substrates were cleaned by distilled water. Prior to transferring these Si substrates into the reactor, the surfaces were dried by dry nitrogen.

The reaction chamber was first evacuated to the based pressure of 10^{-3} torr. And the wafers were heated to the reaction temperatures, range from 950 to 1000ºC, in a flowing hydrogen gas at a pressure of 1.8 torrs. After the temperature equilibriation the hydrogen flow was turned off and acetylene gas was introduced at a pressure of

0.1–0.12 torr. After the reaction was completed, the substrate was furnace cooled in hydrogen at a pressure of 1.8 torrs. The surface characterization was carried out by using a Kratos x-ray photoelectron spectroscopy (XPS) with a MgKα x-ray source. Scanning electron microscopy (SEM) was used to obverse the surface morphology of the reacted Si substrates.

RESULTS AND DISCUSSION:

I. Reaction products

The course of the reaction between Si and C_2H_2 was monitored by following the changes in Si_{2p} and C_{1s} XPS peaks and the plasmon loss features of Si_{2p}. Figure 1 depicts the Si_{2p} spectral region for the (100)Si surface before and after the reaction with C_2H_2 at the reaction temperatures of 950 and 1000°C for various reaction times. The Si_{2p} spectrum for a clean (100)Si substrate is shown in Figure 1(A). In Figure 1(A), The major peak is the Si_{2p} peak and the less intense feature which has the energy loss of 17 eV was the bulk plasmon loss feature from Si substrate. The small shoulder in Figure 1(B), which was about 3.5 eV higher than the major Si_{2p} peak, was believed to be from SiO_2 on the surface of the substrate due to the contamination from the atmosphere. As the reaction temperature increased from 950°C to 1000°C, the intensity of this Si plasmon feature decreased, while that of a second plasmon loss feature at about 22.5 eV increased, as shown in Figure 1(B)–1(D). This latter feature corresponds to the bulk SiC plasmon as discussed by Bozso and co-workers [16].

The Si_{2p} plasmon loss feature at about 22.5 eV provides the evidence for the formation of SiC. Additional information can be obtained by following the deconvolution of Si_{2p} peaks of Figure 1, as shown in Figure 2. The Si_{2p}, with the binding energy of 99.1 eV, from clean Si substrate is shown in Figure 2(A) for comparison with that of reacted Si substrates. As depicted in Figure 2, the relative intensities of Si_{2p} from the Si substrate decreased as the reaction temperature increased from 950°C to 1000°C (Figure 2(B)–(D)). Furthermore, for samples which were reacted at 1000°C, the Si_{2p} peak from Si substrate disappeared if the reaction times were longer than 40 seconds. For reaction times laonger than 40 seconds at 1000°C, the spectra of the samples were essentially similar to the ones depicted in Figure 2(D). It was also observed that the relative intensities of Si_{2p} peaks from SiC were smaller for the reaction parameters of 950°C for 5–11 minutes than those prepared at 1000°C for less than 5 minutes. This indicated the SiC films obtained at 1000°C were thicker than those prepared at 950°C. By deconvoluting the Si_{2p} spectra of Figure 2, and assuming that the mean free paths of the photoelectrons in Si (λ_{Si} = 2.25nm) and SiC are the same, the surface coverage of the reaction product can be estimated by using the following equation [10]:

$$R = (\rho_{Si}/\rho_{Sic}) \times \{\exp(-d/\lambda_{Si}\cos\theta)/[1-\exp(-d/\lambda_{Si}\cos\theta)]\} \tag{1}$$

where R is the intensity ratio of Si_{2p} from Si substrate to that from SiC layer; d is the thickness of the SiC layer; and ρ_{Si} and ρ_{Sic} are the densities of Si atoms in Si substrate and in the SiC layers, respectively; θ is the angle between the normal direction to the sample surface and the photoelectron detector. Using the equation (1), it was found that the surface coverage at 950°C (for 5 to 11 min) is limited to about 0.5 to 1 monolayer of SiC, while at 1000°C (for 10 to 20 sec) the SiC thickness is around 30Å, which corresponds to several monolayers. Our inability to observe the Si_{2p} from Si substrate, for reaction times longer than 40 seconds at 1000°C, can be attributed to thicker SiC product layers in comparison to the nominal XPS sampling depth.

Table II shows the binding energies of Si_{2p} peaks for specimens that were reacted at different conditions. For specimens arected at 950°C, the binding energies and the energy differences among these peaks agreed very well with the reported values. However, for the samples prepared at 1000°C, the Si_{2p} peaks of SiC and SiO_2 were slightly shifted to higher binding energies (0.4 to 0.5 eV). The shift to higher binding energies is also confirmed by careful measurement of the plasmon loss features in Figure 2(C). This shift to the higher binding energies can be attributed to the differential charging of SiC and SiO_2 in comparison with Si.

The stoichiometry of the SiC product layers was determined from the

Table I: The conversion of Si to SiC

C–source	Temp (°C)	Pressure (torr)	Chamber type	References
C_2H_2	900–1100	10^{-6}–10^{-4}	cold wall	5
C_2H_2	800–1100	10^{-7}–5×10^{-4}	cold wall	6
C_2H_2	1130–1370	10^{-5}–10^{-2}	cold wall	7
C_2H_2	1225–1380	5×10^{-6}–3×10^{-3}	cold wall	8
C_2H_2	900–1200	2×10^{-6}–3×10^{-6}	cold wall	9
C_2H_2	900	7.6×10^{-6}	cold wall	10
C_2H_2	950	10^{-6}	—	11
*C_2H_2	837–1037	7.5×10^{-10}	cold wall	12
C_2H_4	1280–1330	7.5×10^{-7}	cold wall	13
C_2H_4	1360	7.5×10^{-7}	cold wall	14
C_2H_4	1327	—	—	15
*C_2H_4	667	4×10^{-5}	cold wall	16
*C_2H_4	697	6×10^{-10}	cold wall	17
C_3H_8	1000–1170	—	—	1
C_3H_8	1360	—	cold wall	18

* : molecular beam

Table II: Binding energy of Si_{2p}

T(°C)	Time	Si_{2p} (eV) Si	SiC	SiO_2	ΔBE (SiC–Si)
950	5–11 min	99.1	100.5	102.6	1.4 eV
1000	10–20 sec	99.1	101	103	1.9 eV
1000	40 sec ~ 5min	—	101	103	—

Table III: Times for C and Si atoms to diffuse through the SiC barrier

T(°C)	C in SiC τ(5Å) (sec)	Si in SiC τ(5Å) (sec)
900	1×10^{11}	4×10^{13}
950	8×10^{9}	2×10^{12}
1000	5×10^{8}	1×10^{11}

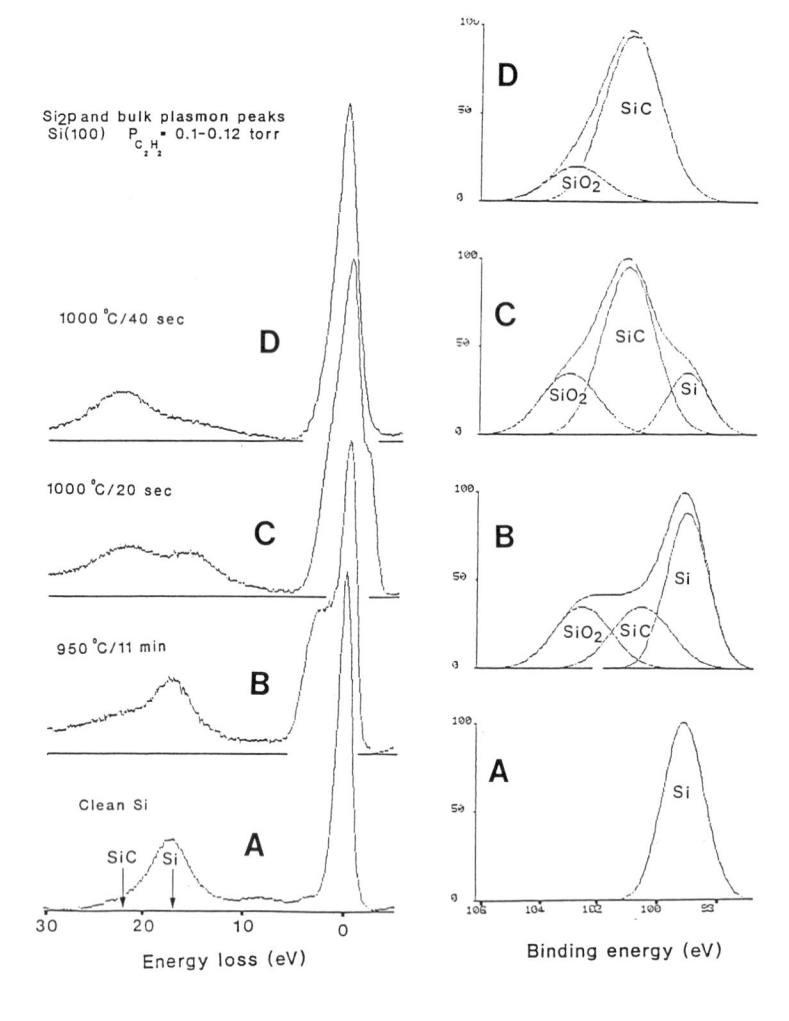

Figure 1: Si$_{2p}$ XPS peaks and associated plasmon loss features for (A) clean Si and Si reacted with C$_2$H$_2$ at (B) 950°C for 11 min (C) 1000°C for 20 sec (D) 1000°C for 40 sec.

Figure 2: Si$_{2p}$ XPS peaks deconvolution for (A) clean Si and Si reacted with C$_2$H$_2$ at (B) 950°C for 11 min (C) 1000°C for 20 sec (D) 1000°C for 40 sec.

deconvoluted Si_{2p} and C_{1s} peaks. In general, the converted SiC layers obtained in this study were slightly Si–rich, which can be attributed to the rapid supply of Si atoms when comapred to carbon atoms [12]. The Si–rich SiC layers were also observed for the reaction between Si and other hydrocarbon gases [10].

II. SEM morphology

The morphology of the (100)Si surfaces reacted at 950°C are shown in Figure 3. As Shown in Figure 3(A), no defects were observed for the samples reacted for 7 minutes. The presence of defects was initially observed only when the reaction time was increased to 9 minutes at 950°C (Figure 3(B)). Figure 4 shows the morphology of the reacted (100) Si surfaces at 1000°C. In constrast to the samples reacted at 950°C, the presence of defects can be observed for reaction times as short as 10 seconds at 1000°C. Initially the defects were nondescriptive, as shown in Figures 4(A). With increasing reaction time, the defects increased in size and depth (Figures 4(B)). It was also observed (Figure 4) that the defects began to acquire regular edges with increasing time.

The high magnification micrograph of defects, for the reaction times of 10 seconds and 3 minutes at 1000°C, were shown in Figure 5. As can be seen from Figure 5, the edges of defects were more well–defined for reaction times of 3 min. than those reacted for just 10 seconds. According to Newman et al. [19], these edges of the defects were believed to be lying parallel to silicon <110> directions. The formation of these defects is realted to the etching of the substrate which exposes the (111) planes of Si due to the slowest out–diffusion rate of Si from these planes. The resulting pyramidal shape defects can be seen (Figure 5(B)) in accordance with this model.

III. Defect formation mechanism

It is believed that the defect formation is closely related to the out–diffusion of Si during the reaction. Strinespring et al. [20] argues that the Si atoms must be supplied by out–diffusion if the product SiC layer exceeded a monolayer in thickness becuase of the nonreactivity of the hydrocarbon gases with SiC. Since the diffusivities of either Si or C through the SiC layer are extremely small, it is thought that the presence of defects would provide a bypass route for the out–diffusion of Si atoms. Therefore, it is expected that the presence of defects can be observed if the reaction times are greater than 9 min. at 950°C, becuase under these conditions the SiC thickness will be about one monolayer(Equation (1)).

The need for the formation of defects may also be illustrated by considering the diffusion times (τ) for a carbon or a silicon atom. The following approximate equation may be used to estimate these valves [21]:

$$\tau = \frac{l^2}{D} \quad (2)$$

where D is the bulk diffusivity and l the thickness of the SiC product layer. Table III shows the values of τ, assuming $l = 5$Å (i.e. about one monolayer of SiC) at different temperatures used in this study. The values of measured diffusivities of C in SiC [22] and Si in SiC [23] were used for estimating τ values. Looking at the τ values listed in Table III, one can easily conclude that monolayer of SiC is a very effective diffusion barrier. Therefore, the occurence of defects is a necessary condition for the out–diffusion of Si and for continued formation of SiC with thicknesses equal to several monolayers. This phenomenon is consistent with the report that 5Å of SiC layer could act as a barrier layer to block the growth of metal silicides on Si surfaces [24].

One of the possible mechanisms for the formation of defects on (100) Si substrates can be described as follows. A Si–rich, monolayer thick SiC layer is first obtained without the formation of defects. Since this SiC monolayer is an effective diffusion barrier, defects are formed to provide the bypass for Si out–diffusion. The sizes of the defects increases with increasing reaction times and temperatures and they finally grow into the Si substrate [6]. Meanwhile the thickness of the SiC layer is increased. Since these newly formed SiC layers again act as a diffusion barriers, defect sizes will be increased to provide paths for the out–diffusion of Si. In other words, the size and the depth of the defects increases with increasing time at a given temperature.

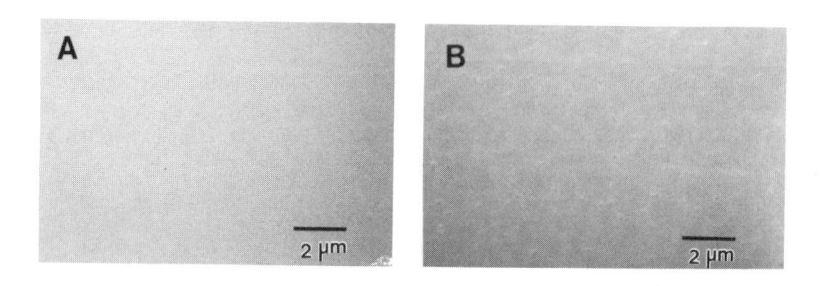

Figure 3: Surface morphology of Si substrates reacted at 950°C for (A) 7 min (B) 9 min.

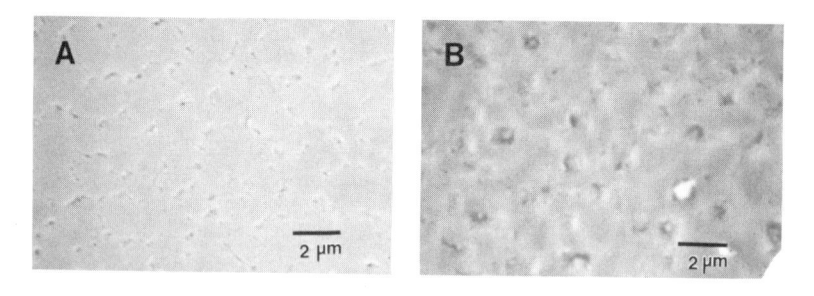

Figure 4: Surface morphology of Si substrates reacted at 1000°C for (A) 10 sec (B) 5 min.

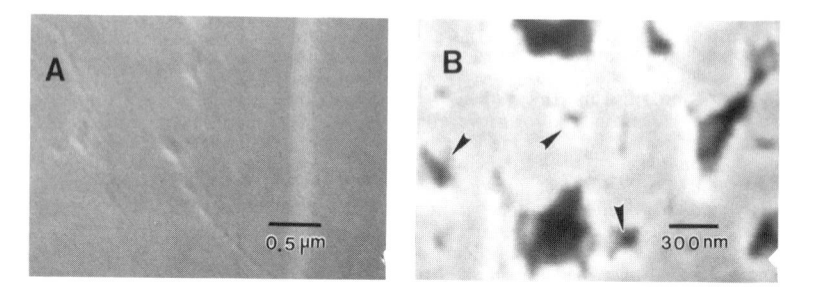

5: Surface morphology with high magnification of Si substrates reacted at 1000°C for (A) 10 sec (B) 3 min.

SUMMARY:

In a horizontal, hot—wall CVD reactor, the chemical conversion of (100)Si surfaces into SiC was achieved at temperatures between 950°C and 1000°C and at low pressures using C_2H_2 as the carbon source. The growth of the SiC films was observed through the behaviour of Si_{2p} peaks and their plasmons. Because the Si atoms were supplied faster than the C atoms, the converted SiC layers were Si—rich. An excellent quality SiC layer was obtained with smooth morphology for a 7 minutes reaction at 950°C. Defects were observed when the reaction times were longer than 9 minutes at 950°C. However, for the specimens reacted at 1000°C, the defects were formed for reaction times as short as 10 seconds. A monolayer thick SiC is formed for 950°C reaction whereas, several monolayer thick SiC is formed at 1000°C. It is argued that the presence of defects provide effective paths for the Si out—diffusion in the carborization process. A possible mechanism for the formation of defects was proposed.

REFERENCES:

1. T.T. Cheng, P. Pirouz, and T.A. Powell in Chemistry and Defecvts in Semiconductor heterostructures, edited by M. Kawabe, p 229
2. K. Ikoma, M. Yamanaka, H. Yamaguchi, and Y. Schichi, J. Electrochem. Soc. 138, 3208 (1991)
3. A. Addamiano and J.A. Sprague, Appl. Phys. Lett. 44, 525 (1984)
4. S. Nishino, J.A. Powell, and H.A. Will, Appl. Phys. Lett. 42, 460 (1983)
5. A.J. Learn and I.H. Khan, Thin Solid Films. 5, 145 (1970)
6. C.J. Mogab and H.J. Leamy, J. Appl. Phys. 45, 1075 (1974)
7. F.W. Smith, Surf. Sci. 80, 388 (1979)
8. F.W Smith and B. Meyerson, Thin Solis Films, 60, 227 (1979)
9. I.H. khan and A.J. Learn, Appl. Phys. Lett. 15, 410 (1969)
10. T. Sugii, T Aoyama, and T. Ito, J. Electrochem. Soc. 137, 989 (1990)
11. K.E. Haq and I.H. Khan, J. Vac. Sci. and Tech. 7, 490 (1970)
12. I. Kusunoki, M. Hiroi, T. Sato, Y. Igari, and S. Tomoda, Appl. Surf. Sci. 45, 171 (1990)
13. H.J. Kim, R.F. Davis, X.B. Cox, and R.W. Linton, J. Electrochem. Soc. 134, 2269 (1987)
14. H.J. Kim, H—S Kong, J.A. Edmond, J.T. Glass, and R.F. Davis,in Ceramic Transactions Vol 2, Silicon Carbide '87 edited by J.D. Cawley and C.E. Samler.
15. P.Liaw and R.F,Davis, J. Electrochem. Soc. 132, 642 (1985)
16. F. Bozso and J.T. Yates, Jr., J. Appl. Phys. 57, 2771 (1985)
17. P.A. Taylor, M. Bozack, W.J. Choyke, and J.T. Taylor, Jr., J. Appl. Phys. 65, 1099 (1989)
18. M. Iwami, M. Hirai, M. Kusaka, Y. Yakota, and H. Matsunami, Jpn. J. Appl. Phys. 28, L293 (1989)
19. R.C. Newman and J. Wakefield, in Solid State Physics in Electronics and Telecommunication, edited by M. Desirant and J.L. Michels, p. 318
20. C.D. Stinespring and J.C. Wormhoudt, J. Appl. Phys. 65, 1733 (1989)
21. C.D. Stinespring and W.F. Lawson, Surf. Sci. 150, 209 (1985)
22. J.D. Hong and R.F. Davis, J. Am. Ceram. Soc. 63, 546 (1980)
23. J.D. Homg, R.F. Davis, and D.E. Newbury, J. Mater. Sci. 16, 2485 (1981)
24. M.A. Taubenblatt and C.R. Helms, J. Appl. Phys. 59, 1992 (1986)

EFFECT OF HIGH TEMPERATURE ANNEALING ON THE MICROSTRUCTURE OF SCS-6 SiC FIBERS

X. J. Ning, P. Pirouz and R. T. Bhatt*
Department of Materials Science and Engineering, Case Western Reserve University Cleveland, OH 44106
*NASA Lewis Research Center, Cleveland, OH 44135

ABSTRACT

The effect of annealing the SCS-6* SiC fiber for one hour at $2000°C$ in an argon atmosphere is reported. The SiC grains in the fiber coarsen appreciably and the intergranular carbon films segregate to the grain junctions. It would appear that grain growth in the outer part of the fiber is primarily responsible for the loss in fiber strength and improvement in fiber creep resistance.

INTRODUCTION

SiC fibers are potential candidates as reinforcements for high-temperature ceramic-matrix composites. Thus, an understanding of the fiber microstructure and its relation to the mechanical properties of the fiber is essential. In particular, it is important to know the changes that take place in the microstructure of the fibers that have undergone high temperature exposure. Such changes may have a significant effect on the high temperature mechanical properties of the fiber. Thus, it has been reported that the tensile strength of SCS-6 fiber decreases significantly after high temperature annealing [1,2]. Also, DiCarlo found that the creep resistance of thermally annealed fibers at temperatures greater than $1400°C$ was enhanced [1]. In this paper, a microstructural characterization of SCS-6 fibers after annealing at $2000°C$ in argon will be presented. The microstructure of the annealed fiber is compared to that of an as-received, un-annealed, fiber.

The microstructure of the SCS-6 fiber has been already investigated in detail [3]. Figure 1 shows a schematic diagram of the microstructure of an SCS-6 fiber cross-section.

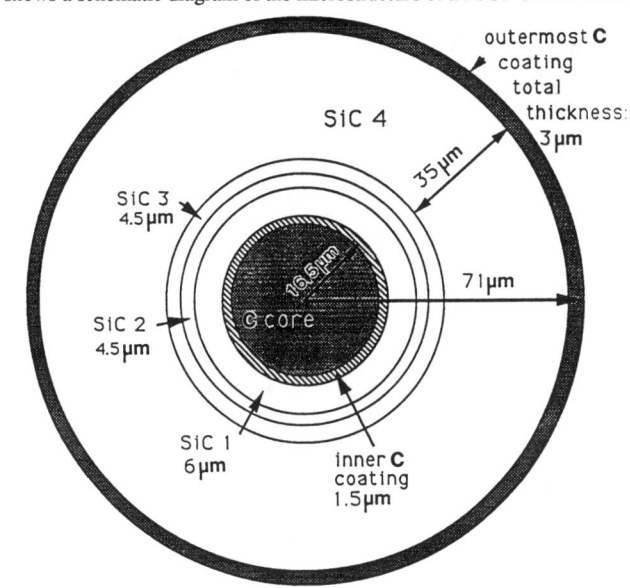

FIG. 1. Schematic diagram of the cross-sectional microstructure of an as-received SCS-6 fiber (from [3]).

---

* Textron Specialty Materials, Lowell, MA

EXPERIMENTAL PROCEDURE

SCS-6 fibers were annealed for 1 hour at 2000°C in a graphite element furnace. The fibers were enveloped in a graphite sheet in order to avoid contamination. The furnace temperature was measured with an optical pyrometer with an accuracy of ±10°C. Annealing was carried out in an argon atmosphere (0.1 MPa) of commercial purity. Annealed fibers were mounted in an epoxy mold, mechanically polished and plasma etched for optical microscopy. For preparing TEM specimens, a technique described previously [3] was used.

RESULTS

The tensile strength of the fiber decreased from ~4 *GPa* to 0.7 *GPa* after annealing. Fig. 2 is a cross-sectional optical micrograph of the annealed SCS-6 fiber. The coarsening of SiC grains in the outer parts of the fiber, which corresponds to the SiC-4 region in Fig. 1, can be clearly seen.

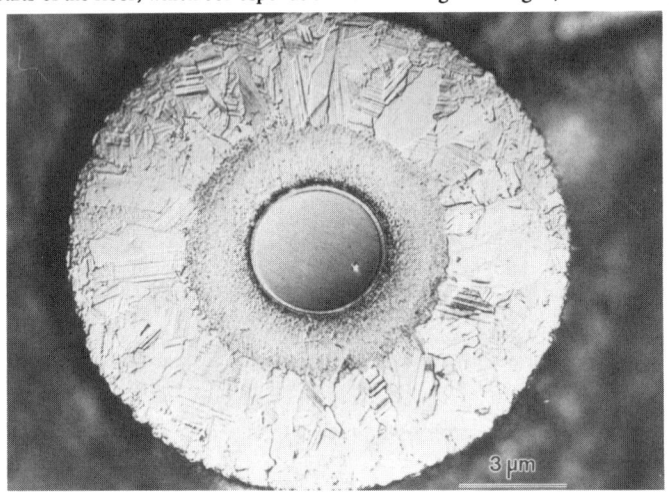

FIG. 2. Optical micrograph of the annealed SCS-6 fiber cross-section. The fiber was plasma-etched after mechanical polishing.

According to [3], the core of the SCS-6 fiber - consisting of a carbon monofilament - and also the inner pyrolitic coating of this core, consist of turbostratic carbon blocks [4,5]. The TEM investigation of the annealed fiber showed that there are no noticeable changes in the microstructure of the core carbon monofilament and the inner pyrolytic carbon coating of the fiber. This is both in terms of the size and the distribution of the turbostratic carbon blocks.

The SiC parts of the fiber, consisting of heavily faulted cubic, β-phase, can be divided into four regions based on differences in grain size and the C/Si ratio [3]. Within the resolution of the SAM (Scanning Auger Microscopy) technique, the C/Si ratio of the SiC layers decreases on moving away from the center to the outer regions until the SiC becomes stoichiometric in the SiC-4 region. After annealing, the β–SiC grains in the SiC-1 layer immediately adjacent to the pyrolytic carbon coating, have transformed into equiaxed grains with sizes in the range 50-150 *nm*. Fig. 3 shows a TEM micrograph of this region. This should be contrasted with the small rod-shaped particles with lengths of 5 to 15 *nm* and an aspect ratio of ~10 in the as-received fiber. In Fig. 3, each SiC grain is heavily faulted on that set of {111} planes which is more or less normal to the fiber radial directions.

In going away from the center, (~1 *μm* from the interface with the inner carbon coating) the grain size increases in the SiC-1 region of the as-received fiber: they become longer (50-150 *nm*) with the same aspect ratio of ~10. In fact, the columnar grains become progressively longer in moving toward the interface with the SiC-2 region. In the annealed fiber, however, the size of the equiaxed round-shaped grains is roughly uniform throughout SiC-1 (Fig. 3).

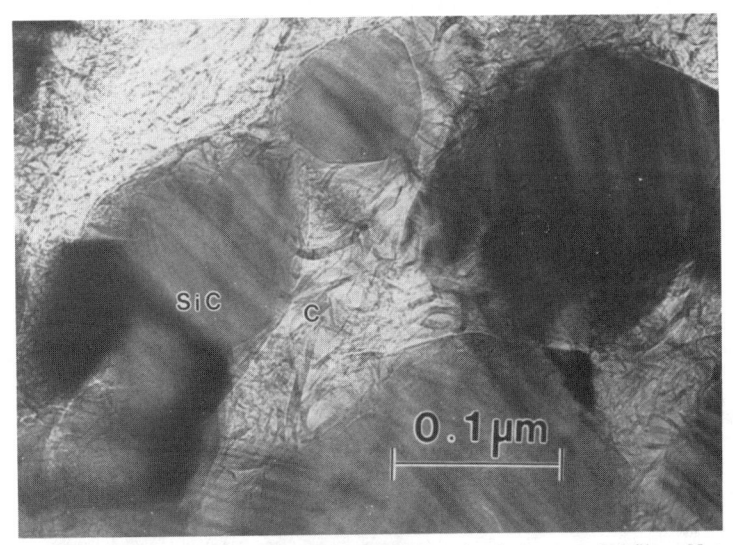

Fig. 3. Cross-sectional TEM micrograph of the SiC-1 region in the annealed fiber. Note the coarsened SiC grains and the segregated carbon.

In [3], it was suggested that the SiC grains are always stoichiometric and the non-stoichiometry in SiC-1, SiC-2, and SiC-3 regions arises from the presence of carbonaceous films at the SiC grain boundaries. The present investigation supports this suggestion. Fig. 3 shows that in these regions of the annealed fiber, the carbon films have clearly migrated from the β-SiC grains and have segregated to the SiC grain junctions. At these junctions, the carbon forms a distinctly separate phase with a ribbon morphology (Fig. 3). Each ribbon is a block of turbostratic carbon with an interplanar spacing of 0.34 nm. The basal planes of the ribbon are preferentially oriented parallel to the C/SiC interface. This can be clearly seen in Fig. 4 which is a HREM micrograph of the interface between a carbon ribbon and a SiC grain in the SiC-1 region.

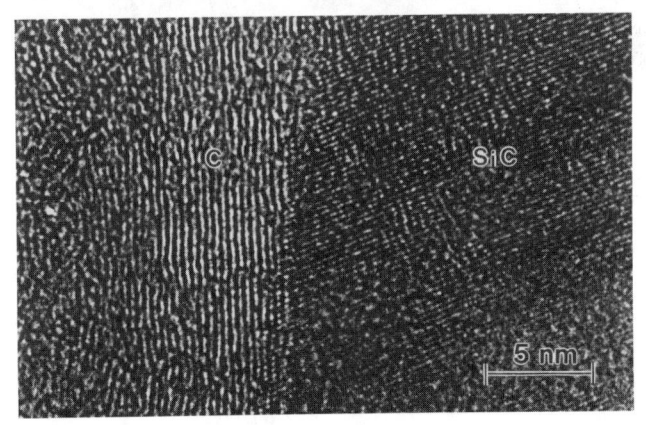

Fig. 4. HREM micrograph of the interface between a segregated carbon ribbon and a coarsened SiC grain.

From an analysis of the TEM micrographs, the atomic concentration of carbon in this region is estimated to be 57.7% which is consistent with Auger results obtained from the as-received fiber [3,5]. This implies that the carbon concentration does not change with annealing.

Unlike the SiC-1 layer, the microstructures of SiC-2 and SiC-3 layers are essentially the same as that of the as-received fiber: they consist of columnar grains with an aspect ratio of 5-10. The one noticeable difference, however, is that the TEM images of the SiC grains are much sharper in the annealed fiber. We think that the reason for this lies in the fact that the carbon, instead of forming a thin film covering the SiC grains, has segregated at the junctions of the SiC grains during annealing. The estimates of carbon concentration in these two regions are consistent with those of the as-received fiber as determined by Auger spectroscopy [3]. In the latter case, the carbon could not be actually imaged, because it was distributed diffusely around the SiC grains. The morphology and the structure of carbon segregants in these two regions are similar to those in the SiC-1 region, except that the carbon concentration is lower (55.3 atom % and 53.8 atom %, respectively).

After annealing, the microstructure of the inner part of SiC-4 immediately adjacent to SiC-3 remained unchanged within a thickness of ~3 μm. This can be clearly seen in Fig. 5.

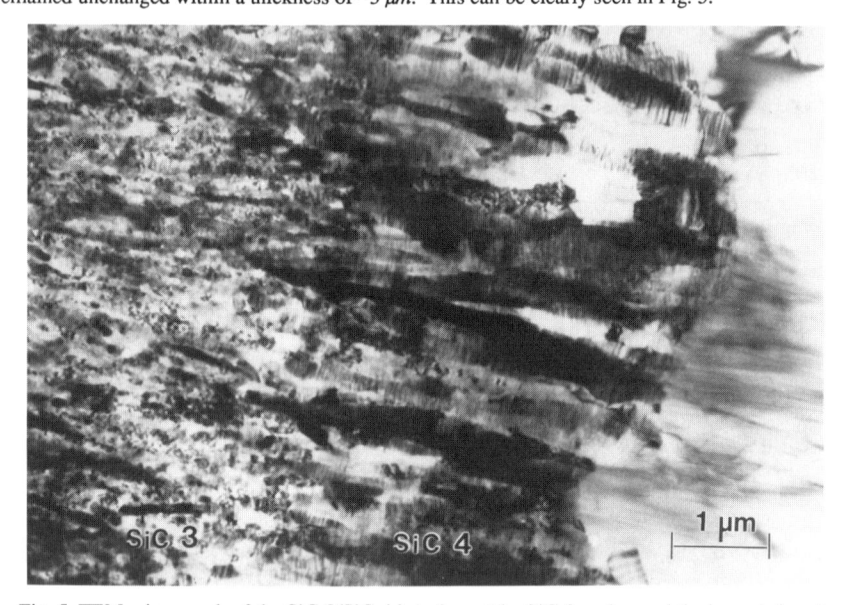

Fig. 5. TEM micrograph of the SiC-3/SiC-4 interface. The SiC-3 region and the inner (~3 μm) part of the SiC-4 region are very similar to the same regions in the as-received fiber, except for the carbon segregation to the junctions of the SiC grains in the SiC-3 layer. The grains in the outer parts of SiC-4 are greatly coarsened.

The contrast of β-SiC grains in the SiC-3 region of the annealed fiber is much sharper than it was in the as-received fiber presumably due to carbon segregation and the consequent removal of the carbon films enveloping the grains during the anneal.

The grains in the rest of the SiC-4 layer are greatly coarsened by a factor of 5-20. Unlike the as-received fiber, where the shape of the grains is rod-like, the grains change to an equiaxed morphology with a diameter of a few microns (on average ~3 μm). Fig. 6 is an image of the coarsened SiC grains in the SiC-4 region. The degree of preferred orientation in this region is lower as compared to the same region in the as-received fiber. No carbonaceous region has been found in the SiC-4 region both before and after annealing. The absence of intergranular carbon films in this region is consistent with the suggestion that the SiC-4 region is stoichiometric in the as-received fiber [3]. The driving force for the particle coarsening during annealing is partially from thermoactivation, and partially from the relaxation of residual stresses which arise during the CVD deposition [1].

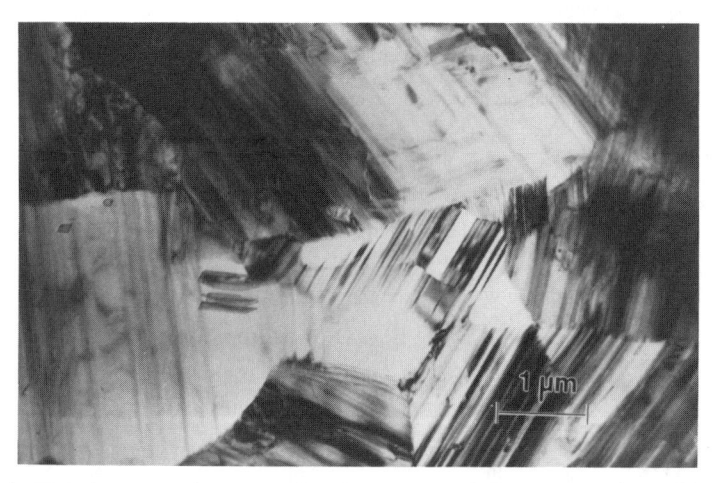

Fig. 6. TEM micrograph of the outer part (>3 μm from the SiC-3/SiC-4 interface) of SiC-4 layer. The SiC grains are greatly coarsened.

As reported earlier [3-6], the outermost fiber coating is a multilayer composite by itself consisting of a carbon matrix reinforced with different densities of β–SiC particulates of varying sizes. The fiber coating can be divided into three sublayers (see Fig. 1): sublayers 1 and 3, with slightly different thicknesses (1.7 μm and 1.3 μm, respectively), have essentially the same microstructure of SiC-reinforced carbon while sublayer 2 is a ~0.1 μm thick carbonaceous region with basically no SiC particulates. Sublayer 2 separates sublayers 1 and 3 and is the weakest region in the coating [6]. After annealing, there is an even clearer distinction between the three sublayers. The reason is that the coarsening of SiC particles in the B zone (the SiC-rich region, see [3]) of sublayers 1 is much greater (from 5 nm to 200 nm) than the coarsening in the A zone (the SiC-poor region, see [3]) (from 30 nm to 100 nm).

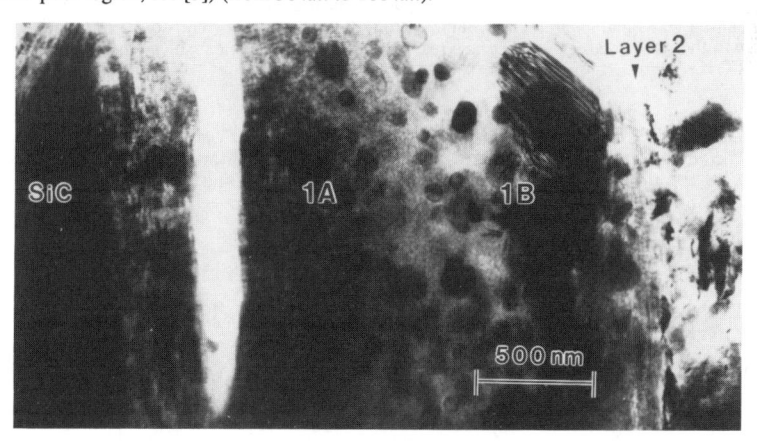

Fig. 7. Cross-sectional TEM micrograph of the outermost coating of the SCS-6 fiber after annealing.

As in the as-received fiber, the density of SiC particles is much higher in the B zones than in the A zones (Fig. 7) and, as a result, the Si/C ratio is higher in the former than in the latter zones. This is also the case in the as-received fiber where the Si/C ratio was determined by PEELS (see Fig. 19 in [3]). The absence of SiC particles in sublayer-2 can be clearly seen in Fig. 7. All the carbonaceous material in the coating is graphitized to different degrees [5] implying that the turbostratic carbon blocks are more ordered as compared to the same regions in the un-annealed fiber. The graphitization is highest in sublayer 2 which could be due to the absence of SiC particles in this layer.

DISCUSSION

The reason for the larger grain growth in SiC-4 (~30 μm) than in SiC-1 (~5 μm) could be the presence of carbon films in the latter. During annealing, the extra carbon must first diffuses along the SiC grain boundaries and segregates at the grain junctions. Subsequent to this, the SiC grains will be interconnected and the process of SiC grain growth can proceed. If, however, the annealing time is not sufficiently long, or the annealing temperature is not sufficiently high, SiC grain growth in the carbon-rich regions will be much slower, or may not even occur. On the other hand, in the stoichiometric SiC-4 region, where no carbon films envelope the SiC grains, grain growth can immediately start on annealing. Thus, the growth rate will be higher in this region than in the carbon-rich regions (SiC-1, SiC-2, and SiC-3).

Because the SCS-6 fiber has a composite microstructure, under tensile loading the outer region, i.e. SiC-4, is the primary element controlling bulk properties such as fiber creep. It would appear then that the enhanced creep resistance after annealing is primarily the result of grain growth in the SiC-4 region. Since the other film regions show much less change after annealing, it would appear that this grain growth is also responsible for the loss in film strength.

CONCLUSION

There are significant changes in the microstructure of the SCS-6 fiber after annealing at 2000°C. The changes are least in the carbon monofilament and the pyrolitic coatings of the fiber. In the SiC-1 region, the SiC grains coarsen and change their morphology from rod-like to equiaxed. In addition, in this region, the carbon films enveloping the SiC grains segregate to the grain junctions. There are little changes after annealing in the morphology and grain size of SiC in the SiC-2 and SiC-3 regions except for the segregation of the extra carbon to the corners of the SiC grains. Except for the innermost 3 μm region, the SiC grains in other parts of SiC-4 layer are coarsened. There is graphitization of the outermost carbonaceous coating after annealing and coarsening of the SiC particles within the coating. The graphitization occurs to different extents in the different regions of the coating. The changes in the fiber microstructure on annealing is accompanied with changes in the mechanical properties of the fiber. The observed decrease in the strength of the fiber and increase in creep resistance appears to be the result of the grain coarsening in the SiC-4 region.

ACKNOWLEDGEMENTS

The authors would like to thank Dr. J. DiCarlo and Mr. G. Morscher for useful discussions and critical comments on the paper. This work was carried out under grant number NCC 3-73 from NASA.

REFERENCES

[1] J. A. DiCarlo, J. Mater. Sci. **21**, 217 (1986).
[2] R. T. Bhatt and D. R.. Hull, Ceram. Eng. Sci. Proc. **12**, 1832 (1991).
[3] X. J. Ning and P. Pirouz, J. Mat. Res. **6**, 2234 (1991).
[4] X. J. Ning, P. Pirouz, K. P. D. Lagerlof and J. DiCarlo, J. Mater. Res. **5**, 2865 (1990).
[5] X. J. Ning, Ph.D thesis, Case Western Reserve University, (1992).
[6] P. Pirouz, G. M. Morscher, and J. Chung, In *Surfaces and Interfaces of Ceramic Materials*, Ed. L. Dufour *et al.*, *pp.* 737-760, Kluwer Academic Publishers (1989).

LOW-TEMPERATURE PACVD SILICON CARBIDE COATINGS

W. HALVERSON,* G.D. VAKERLIS,* D. GARG,** AND P.N. DYER**
* Spire Corporation, One Patriots Park, Bedford, MA 01730-2396
** Air Products and Chemicals, Inc., 7201 Hamilton Boulevard, Allentown, PA 18195

ABSTRACT

Plasma-assisted chemical vapor deposition (PACVD) is used extensively to coat planar (2-dimensional) substrates. In principle, the technique can be used to deposit coatings on 3-dimensional objects. However, extending PACVD to coat 3-dimensional objects uniformly requires careful control of the plasma, substrate temperature, and reactant concentrations over a large volume. A novel low-temperature radio frequency PACVD reactor design was developed to deposit coatings uniformly and reproducibly on 3-dimensional metallic substrates. The design features a temperature-controlled reaction chamber fitted with one or more rf-driven electrodes to generate uniform, large-volume plasma. The reactor was used to develop a series of silicon carbide coatings, which were deposited at or below 500°C. The coatings contain SiC and varying amounts of free silicon and/or amorphous carbon (diamond-like carbon), depending on reagent gas composition and reactor operating parameters. The coatings significantly reduced wear on stainless steel samples in ball-on-disk and abrasive wear tests and provided oxidation protection to molybdenum and titanium alloy.

INTRODUCTION

Chemical vapor deposition (CVD) processes are used routinely to form coatings on complicated 3-dimensional substrates such as cutting tools and rocket nozzles. Plasma-assisted CVD, however, has been largely limited to electronic applications, principally for fabricating electronic devices. PACVD has also been studied for tribological applications, and published papers on the subject go back at least to the late 1970s and early 1980s [1,2]. The deposition technique has several advantages over conventional CVD, because hard, adherent coatings can be formed at relatively low temperatures (typically less than 600°C vs. more than 1000°C for CVD). In PACVD the reagent gases are dissociated, excited, and partially ionized by energetic plasma electrons, which have temperatures of 5,000 to 20,000 K. The activated reagent species combine chemically on all surfaces exposed to the plasma and are not limited by "line-of-sight", as in physical vapor deposition processes.

This paper describes a PACVD technique which deposits hard ceramic coatings on 3-dimensional metallic substrates. The coating material most extensively studied has been silicon carbide; the coatings have varying amounts of free silicon, stoichiometric SiC, and diamond-like carbon (DLC). The composition was controlled by varying the flow ratio of carbon- and silicon-carrying reagent gases and reactor operating parameters; coating hardness was found to be directly related to the C/Si atomic ratio in the deposited material.

REACTOR DESIGN

The capability to coat 3-dimensional components is provided by the PACVD reactor chamber design shown in Fig. 1. Substrates with dimensions up to 8 cm in length and 3 cm

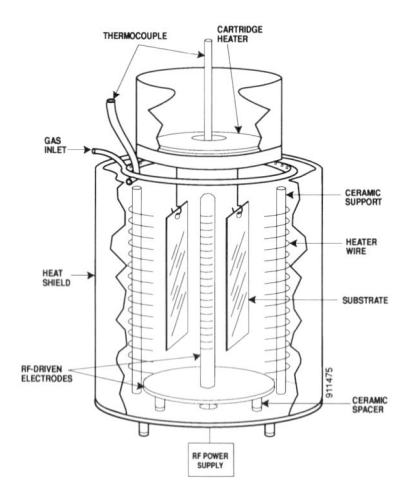

Fig. 1 Diagram of PACVD reactor deposition volume.

wide are hung from a re-entrant cylindrical well; the well is electrically isolated so that dc bias can be applied to the substrates. Reagent gases are injected into the discharge through a perforated stainless steel ring mounted above the nichrome heater assembly. Rf power from an amateur radio transmitter is applied to a 15 cm diameter stainless steel plate, which is mounted by stand-off insulators on a grounded plasma shield. The rf-driven plate is located approximately 12 cm below the end of the re-entrant well. Initial depositions with simple parallel-plate electrodes resulted in poor coating uniformity because of plasma inhomogeneity. The brightest plasma was formed near the rf-driven plate, and deposition occurred principally from the bright plasma. This problem was solved by adding a vertical stainless steel post at the center of the rf-driven plate; the 1.3 cm diameter post is 10 cm long, so that its end is about 2 cm from the re-entrant well and 1 to 2 cm from the substrates to be coated. The plasma is visibly more uniform and the coatings are of nearly constant thickness since the addition of the central electrode post.

SILICON CARBIDE DEPOSITIONS

Silicon carbide was deposited under a broad range of reactor operating conditions and on several substrate materials. Typical process parameters are listed in Table I; the most extensively studied substrate type was 304 stainless steel. As many as 5 rectangular coupons with dimensions of 2.54 by 7.62 by 0.16 cm were hung vertically in the reactor, oriented as shown in Fig. 1. Parameters varied during the studies included reagent gas flow rates and pressure, drive frequency, substrate bias, and deposition temperature.

Table I *Typical deposition parameters for silicon carbide coatings.*

Pressure	65 to 265 Pa
Temperature	450 to 500°C
Reagent gas flow rate:	
SiH_4	5 to 17.5 sccm
CH_4	10 to 15 sccm
Ar	15 sccm
Rf excitation:	
Frequency	1.8 to 14 MHz
Power	250 watts
Dc bias	0 to -200 volts

After deposition, coatings were analyzed for thickness, hardness, elemental composition, crystallinity and crystalline phase. Although all reactor operating parameters had some effect on SiC coating composition and properties, the most striking were caused by variations of the methane/silane concentration ratio. X-ray diffraction analysis of coatings from depositions in which the C/Si ratio in the feed gas was varied between 1 and 3 (determined by the flow rates of CH_4 and SiH_4) revealed that the coatings contained low-crystalline SiC or a mixture of Si and SiC. Laser Raman spectroscopy confirmed the presence of Si and SiC and also showed diamond-like carbon (DLC) in coatings formed with C/Si ratios higher than about 1. Several samples analyzed by the nitrogen nuclear reaction technique were found to have between 21 and 26 atomic percent of hydrogen, with higher values corresponding to films with high carbon concentration. The coating color ranged from a light metallic grey in Si-rich to a dull grey-brown for C-rich material.

X-ray photoelectron spectroscopy and Vickers hardness testing under 0.25 N load showed strong correlation between C/Si atomic ratio in the feed gas and coating composition and hardness. Fig. 2 is a plot of C/Si atomic ratio in the coating as a function of CH_4/SiH_4 molecular ratio in the reactor feed gas; in the regime studied, there is a nearly linear relationship between coating composition ratio and C/Si atomic ratio in the reagent gas. Fig. 3 shows the relationship between C/Si atomic ratio in the coating and hardness; also shown are the approximate regimes in which Si plus SiC, Si plus SiC plus DLC, and SiC plus DLC were found in the coatings. Clearly, the harder coatings were those containing high carbon concentrations and DLC.

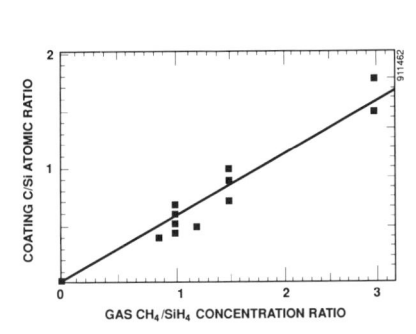

Fig. 2. Deposited C/Si atomic ratio as a function of CH_4/SiH_4 reagent gas concentration ratio. Line is least-squares fit to data.

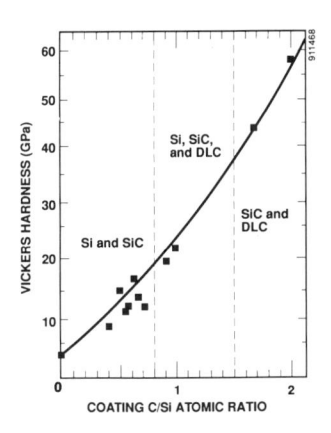

Fig. 3. Vickers hardness (0.25 N load) of coating *vs.* deposited C/Si atomic ratio. Also shown are approximate regimes which coatings consist of Si plus SiC, Si plus SiC plus DLC, and SiC plus DLC. Curve shows only trend of data.

COATING EVALUATION

A series of depositions was performed on AM-350 stainless steel wear disks, abrasive wear coupons, molybdenum, and titanium alloy oxidation coupons for coating evaluation studies; Table II summarizes the resulting coating properties.

Friction and Wear Tests

Friction and wear tests on coated AM-355 precipitation-hardened stainless steel wear discs were performed at the Centre Suisse d'Electronique et de Microtechnique (CSEM), Neuchatel, Switzerland, using a ball-on-disk tribometer. All tests were performed on 2.54 cm diameter,

Table II *Properties of PACVD coatings for evaluation studies.*

Coating Code	Composition	C/Si Coating Ratio	Vickers hardness (GPa)	Thickness (μm)
A	Si + SiC	0.63	17.8	2.7
B	Si + SiC	0.71	12.5	2.9
C	Si + SiC	0.83	17.5	3.4
D	Si + SiC	0.89	20.1	2.6
E	SiC + DLC	1.7	45.5	3.4

0.32 cm thick wear disks, with a 0.6 cm diameter, uncoated, hardened 52-100 chromium steel ball. Testing continued for a minimum of 4,000 revolutions at a constant load of 5 N and a relative surface speed of 10 cm/s. The tests were performed unlubricated, in air at 1% relative humidity, and lubricated with Mobil SHC 626 oil. An uncoated control disk and coated specimens from three PACVD runs were lightly polished and cleaned with dry ethanol prior to testing. Friction coefficient was calculated from the applied contact force and measured drag force; wear rate per unit contact force was calculated from the measured volume loss at the end of the test.

Table III shows measurements of friction coefficients and wear rates for dry and lubricated environments. Although the friction coefficient appears to be little influenced by the coatings, it is clear that the wear rate in the unlubricated dry and lubricated environments was significantly reduced by the coating containing SiC plus diamond-like carbon. This was particularly true for the test with oil lubricant, where no measurable wear was noted on the coated disks after 1.6×10^6 revolutions.

Table III *Ball-on-disk friction and wear testing results.*

	Dry		Lubricated	
Coating Code	Friction Coeff.	Wear Rate* (10^{-15} m^2 N^{-1})	Friction Coeff.	Wear Rate* (10^{-15} m^2 N^{-1})
Uncoated	1.12	322	0.06	0.06
A	0.96	74	0.07	1×10^{-5}
B	0.95	96	0.07	1×10^{-5}
E	0.82	23	0.07	1×10^{-5}

* Tabulated wear rates are the sum of disk and ball wear measurements.

Abrasive Wear Testing

A Model 503 Teledyne Taber tester was used to study the abrasive wear properties of SiC coatings from runs B, C, and E. The coated specimens and an uncoated control were tested using a mild CS-17 abrasive wheel at a constant load of 9.8 N. The test procedure involved (a) abrading uncoated and coated test specimens for a pre-specified number of revolutions; (b) wiping specimens with a cloth lightly wetted with methanol to remove debris; (c) examining the specimens for weight loss and coating failure; and (d) repeating steps (a), (b), and (c) until coating failure.

The coated specimens tested with the mild abrasive CS-17 wheel showed considerably lower wear rates than the uncoated control, which showed a weight loss of 18 mg after 10,000 revolutions. The weight loss of the coatings from runs B, C, and E was 1.3, 1.7 and 5.1 mg, respectively. Apparently the softer, more ductile coatings containing a mixture of Si and SiC had better resistance to abrasive wear than the harder coating containing Si, SiC, and DLC.

Oxidation Testing

Molybdenum and Ti/6Al/4V specimens coated in run D were tested to determine their oxidation behavior at temperatures equal to or less than 650°C. An uncoated and the SiC-coated Mo specimens were heated in a muffle furnace to 650°C in flowing air; the samples were removed periodically from the furnace and inspected for coating integrity and weight change. Test results, summarized in Table IV, showed a significant weight gain by the uncoated Mo specimen after 12 hour exposure to 650°C air. The 2.6 µm thick coating provided excellent resistance to oxidation; the sample's weight increased insignificantly during the 180 hour test. Scanning electron microscopy (SEM) after 48 hours of testing showed that the coating was developing surface cracks, although no weight change was measured. Energy dispersive spectroscopy identified no Mo or other foreign elements in and around the surface cracks.

Uncoated and coated Ti/6Al/4V samples were heated in the muffle furnace in air, for 24 hours at 550°C and at 600°C for the remainder of the test. The specimens were removed periodically from the furnace for inspection of coating integrity and weight change. As shown in Table IV, little weight variation occurred for the initial 24 hour period, at which time the temperature was increased to accelerate oxidation. The small weight gain of the coated Ti/6Al/4V specimen gives a clear indication that the SiC coating provided oxidation protection.

Table IV *Weight gain (mg) of specimens during oxidation testing in air.*

Coating Code	8 hrs	12 hrs	24 hrs	36 hrs	108 hrs	180 hrs	192 hrs
Molybdenum Specimens							
Uncoated	152	239*	--	--	--	--	--
D	0.4	0.6	--	1.0	1.3	1.9*	--
Ti/6Al/4V Specimens							
Uncoated	0.6	--	1.5	--	9.6	--	16.6
D	-0.1	--	-0.4	--	0.4	--	0.8

* Test discontinued.

CONCLUSIONS

Uniform, adherent SiC coatings have been deposited at low temperature by PACVD on "3-dimensional" substrates of several different metals. The coatings have a wide range of composition and physical properties, depending principally on the composition of the reactant gas mixture. The coatings were shown to improve friction and wear, abrasive wear, and oxidation behavior of several substrate metals and offer potential for applications such as bearing races, cutting tools, thread guides, and parts in hot engines. The PACVD process is compatible with many other coating materials, for which a wide range of applications should be found.

ACKNOWLEDGEMENT

This research was funded by the Corporate Science and Technology Center of Air Products and Chemicals, Inc.

REFERENCES

1. F.J. Hazlewood and P.C. Iordanis, *Proc. Int. Conf. on Advances in Surface Coating Technology, 1978* (Welding Institute, Cambridge, England, 1978), p. 147.

2. N.J. Archer, Thin Solid Films **80**, 221 (1981).

THIN FILM PROPERTIES OF LPCVD TiN BARRIER FOR SILICON DEVICE TECHNOLOGY

RAMA I. HEGDE, ROBERT W. FIORDALICE, EDWARD O. TRAVIS AND PHILIP J. TOBIN

Advanced Products Research and Development Laboratory,
Motorola, Inc.,
3501 Ed Bluestein Boulevard,
Austin, Texas 78721

ABSTRACT

Thin film properties of LPCVD TiN barriers deposited on Si(100), using $TiCl_4$ and NH_3 as reactants, were investigated as a function of deposition temperature between 400 °C and 700 °C. The TiN film chemistry and film composition were studied by AES and RBS techniques, while the microstructural properties (grain size, lattice parameter and texture) were evaluated by XRD. The TiN deposition rates and film resistivities were also determined. Finally the film properties of the TiN barriers as determined by surface analysis were related to the process parameters.

INTRODUCTION

In the microelectronics industry titanium nitride (TiN) is primarily used as a barrier layer to prevent interdiffusion of silicon and aluminum metallization, and as an adhesion layer prior to LPCVD blanket tungsten deposition [1 - 8]. Conventionally, TiN films are deposited by reactive sputtering or, by thermal nitridation of sputtered titanium layers. As integrated circuit (IC) feature sizes shrink to deep submicron dimensions, films produced by these techniques suffer from poor step coverage, especially in submicron contact holes with high aspect ratios, causing barrier and adhesion properties to degrade. However, TiN films produced by low pressure chemical vapor deposition (LPCVD TiN) provide excellent step coverage and good diffusion barrier characteristics [1 - 8].

EXPERIMENTAL

LPCVD TiN depositions were carried out in a rapid thermal chemical vapor deposition (RTCVD) system, which is a load locked, single-wafer cold wall reactor, with linear cassette to cassette wafer transfer. The substrate is radiantly heated from the backside. A thermocouple in contact with the center of the substrate's backside provides closed loop temperature control within ± 1% of set point. TiN depositions on Si<100> were carried out between 400 °C and 700 °C using TiCl$_4$ and NH$_3$ reactants with Ar as a carrier gas. TiN films were characterized by Auger electron spectroscopy (AES), Rutherford backscattering (RBS), x-ray diffraction (XRD) and transmission electron microscopy (TEM). Oxygen and carbon contents were determined from AES. Chlorine levels were determined by both AES and RBS. Sheet resistance of the films was measured with a Prometrix four point probe. Film resistivities were derived from the sheet resistances and film thickness. Film thicknesses were measured by x-ray fluorescence and calibrated with cross-sectional TEM. Deposition rates were calculated from the thickness of deposited TiN film.

RESULTS AND DISCUSSION

Film chemistry

An Auger spectrum (AES) of an LPCVD TiN film deposited at 650 °C is shown in Figure 1. The Auger peak intensities for Ti+N (380 eV), Ti (418 eV), O (503 eV), Cl (181 eV), and Si (92 eV) in TiN films formed at 400 °C and 700 °C are shown in Figures 2A and 2B, respectively.

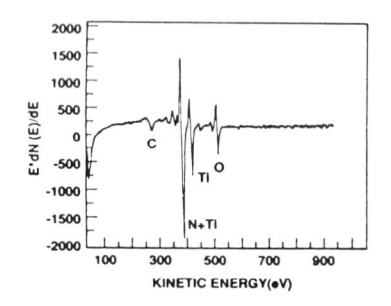

Figure 1

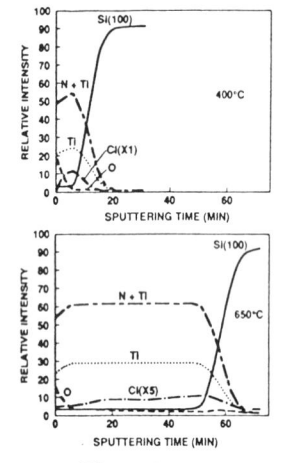

Figure 2

Titanium nitride (TiN) films are formed at all deposition temperatures, and have uniform composition across entire film. Increase in the TiN film thickness was observed with increasing deposition temperature. Depending on deposition temperature, varying amounts of chlorine (Cl), and oxygen (O) impurities are found in the TiN films. No significant level of carbon was detected in any of the TiN films. For the films deposited at 650 °C and 700 °C, the bulk of the TiN layer is oxygen free. The O content is relatively low in all the films (7.7 at% to 2.9 at%), and they are comparable to the levels typically found in reactively sputtered TiN films [4].

In contrast to oxygen behavior, the Cl content of these films show a correlation with the deposition temperature. With increasing deposition temperature a decreasing Cl content was measured. Films deposited at 700 ºC contained as little as 1 at% Cl. A similar trend in the chlorine behavior has been reported by several workers [2, 3, 5 - 9].

Film composition

RBS was conducted to accurately determine the stoichiometry of the TiN. AES measurements were compared to the composition determined from RBS. The [Ti+N]$_{380}$ eV : [Ti]$_{418}$ eV Auger peak height ratio was used as a relative measure of the TiN stoichiometry.

It is apparent from Table 1 that the stoichiometry of all films is the same regardless of deposition temperature. The RBS measurements revealed that the N:Ti ratio is 1.1 ($\pm$0.05) for all the films examined, which is in good agreement with the AES data. Thus no substantial variation in the stoichiometry of the films was found.

Table I

LPCVD TiN$_x$ FILM COMPOSITION

DEPOSITION TEMPERATURE (ºC)	(N+Ti)/Ti [AES]	N/Ti [RBS]
400	2.27 ± .06	1.15 ± .05
550	2.25	1.05
650	2.26	1.10
700	2.24	

Film growth rate

In the present single wafer RTCVD reactor, higher deposition rates were achieved compared to deposition rates of 30 - 65 Å/min obtained in a hot-wall system reported earlier [2]. At lower temperature (400 ºC), the deposition rate is ≈ 80 Å/min. It rises to ≈ 420 Å/min in the higher temperature range.

Film growth kinetics

The typical LPCVD TiN growth rate for the single wafer RTCVD system is shown in Figure 3. As shown the TiN deposition rate is a strong function of deposition temperature. This implies that the kinetics of TiN deposition process is primarily controlled by the surface temperature.

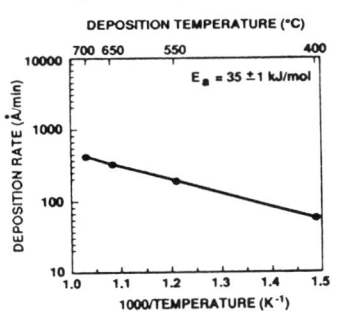

Figure 3

Figure 4 shows the excellent conformality of a TiN film in a 0.6 µm contact, a feature normally attributed to surface controlled reactions. From the slope of the Arrhenius plot (*log* of deposition rate versus reciprocal temperature) in Figure 3, an activation energy of 35 ± 1 kJ/mol was determined. Srinivas et al have reported an activation energy of 40 kJ/mol for TiN deposition from TiCl$_4$ and NH$_3$, which is in close agreement with our results [3]. The reported value of the activation energy for TiN deposition from TiCl$_4$, NH$_3$ and H$_2$ source was 61 kJ/mol [5].

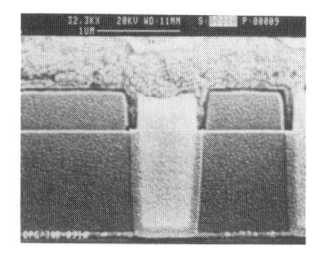

Figure 4

Film resistivity

The measured resistivity of TiN films as a function of deposition temperature is displayed in Figure 5. With increasing deposition temperature, a decreasing resistivity was measured. A similar relation was observed for the Cl content of TiN films. TiN film resistivity on the order of 85 µΩ-cm is obtained for a 670 °C deposition process as a result of optimized annealing (filled square). In our previous study, a value of 47 µΩ-cm was obtained for the LPCVD TiN deposited at 750 °C using a quartz hot wall reactor [2]. Yokoyama et al reported LPCVD TiN resistivity value of 80 µΩ-cm for 700 °C deposition process in good agreement with the value reported here [8].

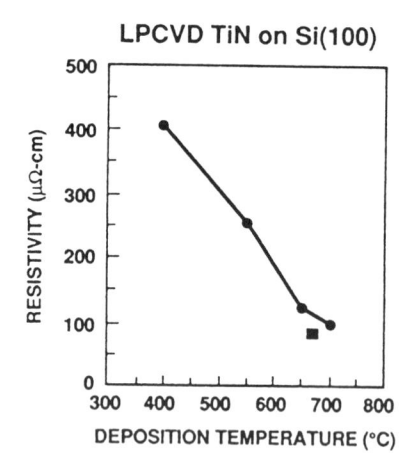

Figure 5

Film structural properties

Thin film XRD patterns of TiN films formed at 400 °C and 700 °C are shown in Figure 6A and 6B, respectively. Crystalline ∂-TiN phase (NaCl type, fcc) was observed regardless of deposition temperature.

The XRD results also revealed that these polycrystalline TiN films exhibit <100> preferred orientation. The variation in average grain size as determined from the half-width value of <200> X-ray line, and using plan view TEM images is shown in Figure 7.

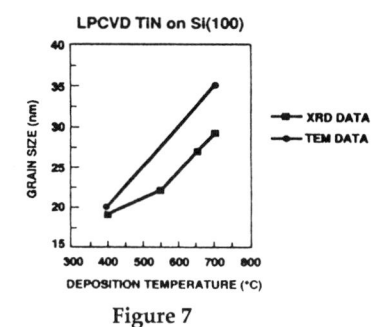

Figure 6

An increase of about 50% in the average grain size was observed in the TiN film deposited at 700 °C. An increase in the lattice parameter with increasing temperature was observed. The value for the TiN film deposited at 700 °C was about 0.9% smaller that of bulk TiN (4.24 Å). This is presumably due to stress in the film [10].

Figure 7

Film surface roughness

Figure 8 is a cross-sectional TEM of the TiN film deposited at 700 °C showing strong columnar growth.

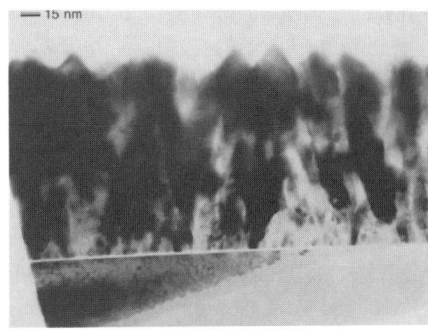

Figure 8

The structure of the columns is fcc-like titanium nitride with predominantly <100> texture. The TiN surface roughness is on the order of 15 - 20 nm for the film deposited at 700 °C, while the film deposited at 400 °C has much smoother surface. Thus, increasing the deposition temperature produces larger TiN grains and rougher surface morphology.

CONCLUSIONS

Stoichiometric TiN films were deposited using an RTCVD, single-wafer reactor system. Composition was uniform in the TiN film. Depending on deposition temperature, varying amounts of chlorine and oxygen impurities are found in the TiN films. The TiN deposition process is surface reaction controlled with an activation energy energy of 35 kJ/mol. Excellent TiN film conformality was observed. The electrical resistivity of the TiN films was found to decrease with increasing deposition temperature. While increasing the deposition temperature increases the TiN film growth rate, it also produces larger TiN grains and rougher surface morphology. The TiN films have crystalline ∂-TiN phase (NaCl type, fcc) with predominantly <100> orientation.

ACKNOWLEDGMENTS

We would like to thank Fabio Pintchovski, Peter Gill, Lou Parrillo, Ron Pyle and Cotton Hance for their support. Special thanks to Vidya Kaushik for his assistance in TEM analysis and Rich Gregory for RBS analysis.

REFERENCES

1. E. O. Travis, W. M. Paulson, F. Pintchovski, B. Boeck, L. C. Parrillo, M. L. Kottke, K. -Y. Fu, M. J. Rice, J. B. Price, E. C. Eichman, IEDM (1990) 47.
2. F. Pintchovski, T. White, E. Travis, P. J. Tobin, and J. B. Price, in Tungsten and Other Refractory Metals for VLSI Applications IV, edited by R. S. Blewer and C. M. McConica (Mater. Res. Soc. Proc. Pennsylvania 1989) pp. 275 - 282.
3. D. Srinivas, J. T. Hillman, W. Triggs and E. C. Eichman, Advanced Metallization for ULSI Applications, MRS (1992) (to be published).
4. R. I. Hegde, R. E. Jones, Jr., V. S. Kaushik, and P. J. Tobin, Appl. Surface Science 52 (1991) 59.
5. M. J. Buiting, A. F. Otterloo and A. H. Montree, J. Electrochem. Soc. 138 (1991) 265.
6. A. Sherman, J. Electrochem. Soc. 137 (1990) 1892.
7. I. J. Raaijmakers and A. Sherman, IEEE 1990 219.
8. N. Yokoyama, K. Hinode and Y. Homma, J. Electrochem. Soc. 13 (1991) 190.
9. S. R. Kurtz and R. G. Gordon, Thin Solid Films, 140 (1986) 277.
10. H. Z. Wu, T. C. Chou, A. Mishra, D. R. Anderson, J. K. Lampert and S. C. Gujrathi, Thin Solid Films, 191 (1990) 55.

PART IV

Chemical Vapor Infiltration

ADVANCES IN MODELING OF THE CHEMICAL VAPOR INFILTRATION PROCESS

THOMAS L. STARR
Georgia Institute of Technology, Georgia Tech Research Institute, Atlanta, GA

ABSTRACT

The technology of chemical vapor infiltration (CVI) has progressed dramatically over the past twenty-five years and stands now as the leading process for fabrication of high temperature structures using ceramic matrix composites. Modeling techniques also have advanced from extensions of catalyst theory to full 3-D finite element code with provision for temperature and pressure gradients. These modeling efforts offer insight into critical factors in the CVI process, suggest opportunities for further advances in process technology and provide a tool for integrating the design and manufacture of advanced components.

Early modeling identified the competition between reaction and diffusion in the CVI process and the resulting trade-off between densification rate and uniformity. Modeling of forced flow/thermal gradient CVI showed how the evolution of material transport properties provides a self-optimizing feature to this process variation.

"What-if" exercises with CVI models point toward potential improvements from tailoring of the precursor chemistry and development of special preform architectures.

As a link between component design and manufacture, CVI modeling can accelerate successful application of ceramic composites to advanced aerospace and energy components.

INTRODUCTION

As with any advanced structural material, incorporation of ceramic composites into mechanical systems seems frustratingly slow to materials scientists. To the design engineer, however, several factors still limit such applications. One of these is fabricability - the ability to reliably produce a finished component at a reasonable cost. Chemical vapor infiltration (CVI) offers considerable promise toward this end with fabrication of a number of near-net-shape components and implementation of commercial-scale production facilities in Europe and the U.S. Modeling has been a key element in development of CVI processing to this point and will be even more important in the future as CVI is used for fabrication of components of increasing size and complexity.

BRIEF HISTORY OF CVI MODELING

There are two variations of CVI in current practice[1], as illustrated in Figure 1. In isothermal CVI (ICVI) a gas mixture is introduced into a furnace containing one or more fiber preforms. Reactant diffuses into the pore spaces of the preform, reacting on the fiber surfaces to form the matrix material. While this variation of the process requires relatively simple equipment and can accommodate a number of large, complex shapes, it suffers from some limitation on component thickness and requires very long processing times, i.e. several hundred hours. In forced flow/thermal gradient CVI (FCVI) the reactant gas mixture is forced to flow through the preform and an applied temperature gradient controls the progress of densification in order to avoid premature pore closure near the gas inlet. While equipment for temperature and gas control is more complex, FCVI is effective for thick components and offers an order of magnitude improvement in processing time.

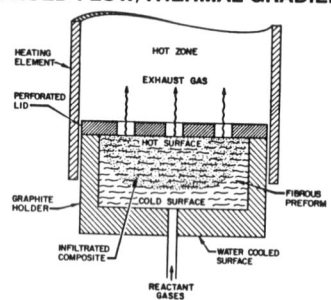

Figure 1. Two variations of the CVI process are in current practice.

The earliest reference to CVI is Bickerdike[2] in 1962 with application of the isothermal process to densification of porous carbon. This early work recognized the importance of modeling and adapted catalyst theory[3] to better understand the balance between deposition rate and diffusion that controls the infiltration process. In the early 1970's Fitzer's group at Karlsruhe began extended investigation of SiC CVI for densification of fiber reinforced composites[4]. This work again adapted existing catalyst models to understand the effect of process conditions on infiltration efficiency. While this work was successful in predicting initial infiltration behavior, there was no attempt to follow the process of densification. Also in the 1970's Naslain's group in Bordeaux started their development of ICVI. Their model[5] included the effect of pore closure and attempted to predict densification time for particular process conditions and pore geometry. Although process development continued through commercialization at Societe Europeenne de Propulsion (SEP) and duPont, modeling efforts lay relatively dormant until the late 1980's.

Like the isothermal process, FCVI was demonstrated in the early 1960's[6]. However, development of the process for ceramic composites did not begin in earnest until the work by Lackey at Oak Ridge National Laboratory in 1984[7] and no model for this process was published until 1987[8]. Development of the process has continued, though with only limited commercial implementation, and modeling efforts have expanded considerably over the past several years.

PROGRESS IN CVI FUNDAMENTALS

A mass balance equation lies at the heart of all CVI models. This equation, based on fundamental conservation of matter, has the form:

$$D\frac{d^2C}{dX^2} + U\frac{dC}{dX} = R \times S \qquad (1)$$

where C is the concentration of a chemical species, D is its diffusion coefficient, U is the gas velocity in the X direction, R is the molar deposition rate per unit of solid surface area and S is the solid surface area per unit volume. The two terms on the left hand side of this equation represent the net diffusive and convective flux of the species into a volume and the term on the right hand side represents the production or depletion of this species through reaction within that volume. In order to evaluate this equation, additional "balance" equations for energy and momentum may be necessary to model temperature and flow. Progress in CVI fundamentals is gauged by how well the mathematical expressions used in solution of these balance equations represent the physical reality of the CVI process.

The source term on the right hand side of eq. (1) is fundamental to all variations of CVI and it value corresponds to the local densification rate for the composite. It is important to note that the rate, R, and the surface area, S, are equally important in determining the magnitude of this term. In general, the rate term depends on temperature, pressure and gas composition. Most CVI models assume that this rate is first-order in a single reactant species and exponential in temperature. In one case, a two step reaction path has been considered where the initial reactant reacts in the gas phase to produce an intermediate chemical species that then reacts on pore surfaces to form the matrix material[9]. More recently, for deposition of SiC from methyltrichlorosilane (MTS), the concentration of reaction product, HCl in this case, has been included as an inhibitor in the deposition rate expression[10].

The surface area term in eq. (1) depends on the microstructure of the fiber preform and its development during densification. Several expressions for this term have been developed for short, randomly oriented fibers or for continuous, parallel fibers[11]. These depend on the fiber diameter and initial packing density in the preform, and differ principally in how quickly the surface area falls toward zero as the density increases. Unfortunately the most useful types of fiber preform, cloth lay-up or 3-D weave, correspond to neither of these ideal cases. For both of these a significant fraction of the porosity is associated with relatively large, low surface area regions between fiber tows (bundles) or between cloth layers.

Our work with one particular cloth lay-up preform suggests a microstructure model with approximately 70% of the volume in "tow" regions with tightly packed, parallel fibers, 25% in "channel" regions between cloth layers and 5% in "hole" regions running through the cloth layers. This leads to a surface area function as shown in Figure 2. The surface area, initially very high, drops to a much lower value as the porosity associated with the "tow" region fills. (Some residual, closed porosity remains but does not contribute to the available surface area.) Additional densification occurs by filling in the "channel" and "hole" regions. Even with a constant deposition rate the densification rate is much lower at this point. This characteristic of decreasing densification rate as density increases is very important as this tends to produce uniform density throughout a component.

In the absence of pressure driven flow, the left hand side of eq. 1 reduces to the first term, reactant transport via diffusion. In this case the CVI problem is identical to the problem of reaction and transport in porous catalysts, and existing models were easily adapted to CVI. Given the proper reaction rate and surface area expressions, only the value of the diffusion coefficient is needed to calculate the densification rate as a function of position within the preform. In early modeling efforts the "effective" diffusion coefficient was estimated from the gas phase diffusion coefficient, the fraction porosity and a tortuosity factor. However, with small pores and at reduced pressure - typical conditions for CVI - Knudsen flow is a more appropriate description of reactant transport, and this has been included in recent models. Unfortunately, it is very difficult to test these transport

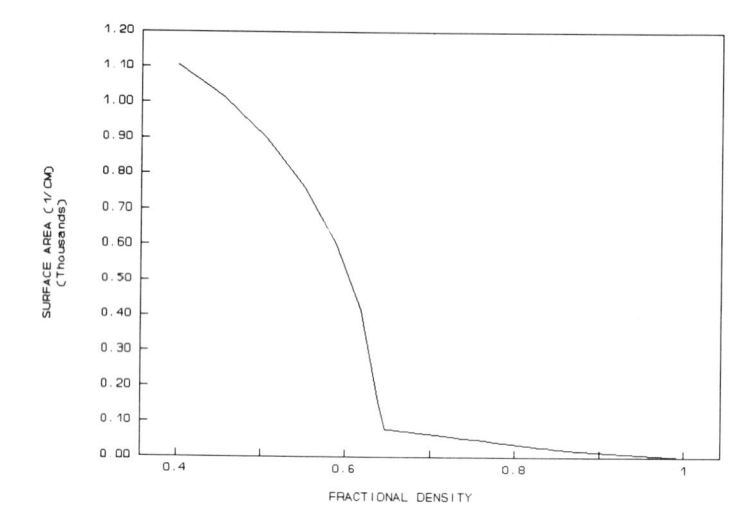

Figure 2. Surface area of cloth lay-up preform decreases as density increases.

models with experimental measurements and no successfully effort has been reported. Recent Monte Carlo computer simulations of transport within fiber structures do offer a somewhat independent test of the diffusion models and generally support their validity[12].

For forced flow CVI the second term in eq. (1), convective flow, dominates mass transport. For a given pressure differential the flow velocity depends on the gas viscosity and on the permeability of the preform. The gas viscosity can be estimated from its composition using standard methods. Permeability of porous materials is the subject of a large body of work and much of this can be applied to CVI modeling. In particular, the Kozeny equation

$$K = \frac{(1-d)^3}{c \times S^2} \qquad (2)$$

relates the gas permeability, K, to the surface area, S, and porosity, $(1-d)$[13]. The constant, c, is a geometric factor that depends on pore shape and orientation. Since we already have found values for S as a function of density, it would seem that permeability may be estimated easily. Unfortunately, for preforms of greatest interest, cloth lay-up and 3-D weave, the porosity is distributed over two very distinct size ranges and the Kozeny equation cannot be used for the preform as a whole. For a cloth lay-up preform, in particular, most of the flow passes through the "channels" and "holes" between the cloth layers. The resulting bulk permeability is anisotropic but can be estimated using Poiseuille's formula, channel dimensions and a semi-empirical ratio of parallel and series flow[14].

For forced flow/thermal gradient CVI (FCVI), a thermal model is needed to derive the local temperature within the preform for a given set of boundary conditions. Conduction is the dominant mode of heat transport in typical experimental configurations although gas convection can be a significant factor[15]. Published thermal conductivity data for CVI SiC/Nicalon™ composite over a range of densities are available[16]. Data for other CVI materials are lacking, as are methods for estimating thermal conductivity from microstructure information for these complex, multiphase composites.

MODELING FOR PROCESS OPTIMIZATION AND CONTROL

The fundamentals of CVI modeling can be combined with advanced numerical modeling techniques to produce a computer simulation of the CVI process[17]. With this approach CVI modeling can be used as a tool for process optimization in much the same way as finite element modeling is used for mechanical optimization of a structural component.

Early work with the isothermal CVI process utilized a simple reaction and diffusion model to obtain "in-depth" infiltration conditions for preforms with a given pore size[18]. In this case, infiltration uniformity improves for conditions that produce a lower deposition rate, i.e. low temperature and pressure. "Optimum" conditions are simply a trade-off between uniformity of infiltration and total processing time.

For forced flow/thermal gradient CVI the number of process parameters is large (orientation and magnitude of applied temperature gradient, flow rates for at least two gases, location of gas inlet and outlet, etc.) and effective densification of a particular component preform may depend critically on choosing the proper set of process conditions. An ongoing collaborative effort between the Oak Ridge National Laboratory and the Georgia Institute of Technology has produced quantitative validation of a "finite volume" process model that predicts local temperature, pressure, reactant concentration and density as a function of time over the infiltration process[19]. This model can be used to select optimum process conditions for large, complex shapes, avoiding a great deal of "cut and try" experimental effort. This has been shown for a hypothetical rotor component (Figure 3) where thick and thin sections densify at different rates[20]

Beyond process optimization, CVI modeling can aid process monitoring and control. For the component described above, the CVI model predicts temperature rise as a function of time at various points (Figure 4). Continuous measurement of these temperatures provides a monitor of the process and can localize problems that may occur during a run due to preform flaws or variation of process parameters. Such a model also may suggest modifications to the process to recover from such deviations, i.e. "feed-forward" process control.

FUTURE DIRECTIONS

The explosion of interest and activity in CVI modeling over the past five years has produced a good understanding of the factors needed for modeling of the process and a number of sophisticated numerical models. Additional efforts are needed to provide (i.e., measure) quantitative values for the transport and reaction rate parameters involved and, using these, validate the models by direct comparison with experimental densification. Better understanding of the relation between fiber architecture and transport properties is critical since preforms of practical interest will certainly involve a variety of lay-up and weave patterns.

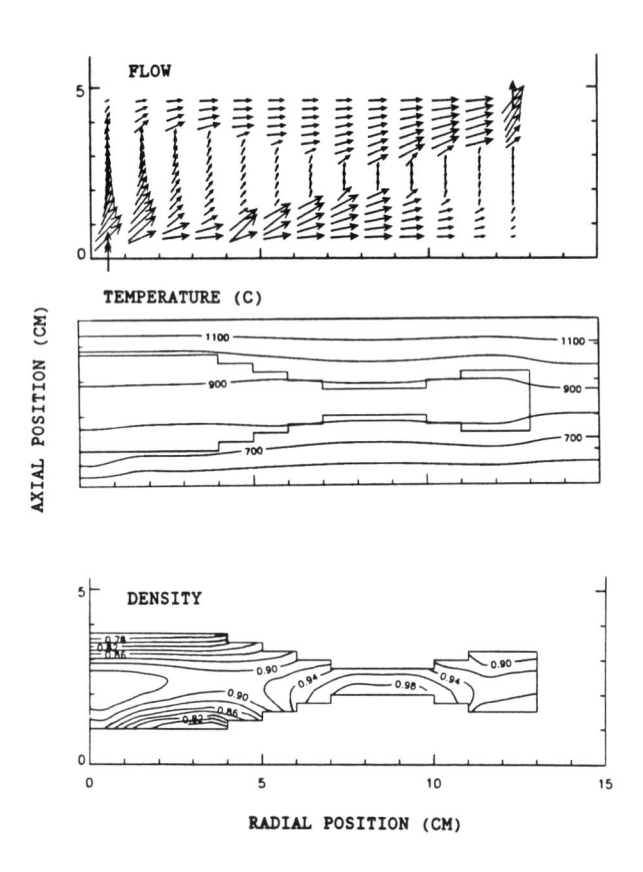

Figure 3. Model simulates infiltration of rotor with thick and thin sections. Cross-sectional views show flow pattern, temperature profile and density variation after partial infiltration[20].

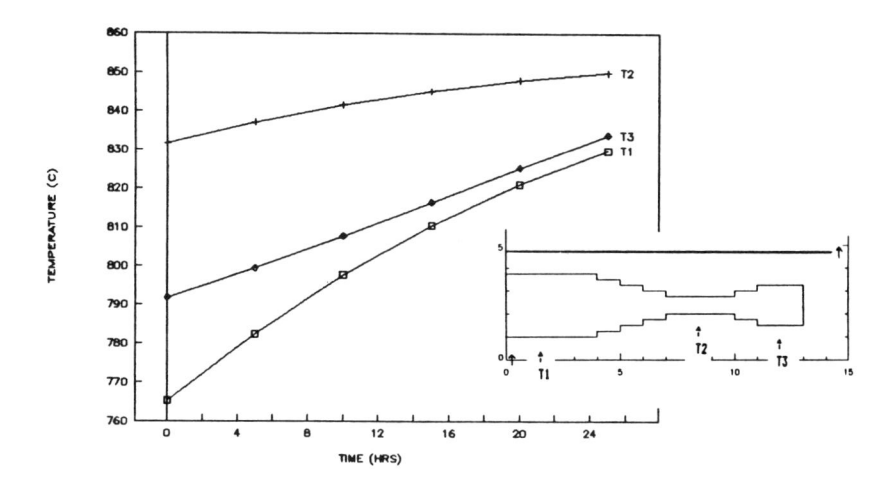

Figure 4. Model predicts rise in local temperatures as infiltration proceeds, offering a method for real-time monitoring of the process[20].

ACKNOWLEDGEMENT

This work has been supported by the U. S. Department of Energy, Fossil Energy Advanced Research and Technology Development Materials Program under Martin Marietta Energy Systems subcontract No. 19X-55901C; and by the U. S. Air Force Wright Laboratory Materials Directorate under Martin Marietta Energy Systems subcontract No.19X-SD324C.

REFERENCES

1. W. J. Lackey and T. L. Starr, in Fiber Reinforced Ceramic Composites: Materials, Processing and Technology, edited by K. S. Mazdiyasni (Noyes Publications, Park Ridge, NJ, 1990), p. 397-450; T. M. Besmann, B. W. Sheldon, R. A. Lowden and D. P. Stinton, Science 253, 1104 (1991).

2. R.L. Bickerdike, A.R.G. Brown, G. Hughes and H. Ranson, in Proceedings of the fifth conference on carbon, (Pergamon press, New York, 1962),pp. 575-582

3. G. Damkohler, Chem. Ing. 3,430 (1937); E.W. Thiele, Ind. Eng. Chem. 31, 91 (1939).

4. E. Fitzer, D. Hegen and H. Strohmeier, in Proceedings of the Seventh International Conference on Chemical Vapor Deposition, edited by T. O. Sedgwick and J. Lydtin, (Electrochemical Society, Incorporated, Pennington, NJ, 1990) pp. 506-512.

5. J.Y. Rossignal, F. Langlais and R. Naslain, in *Proceedings of the Ninth International Conference on Chemical Vapor Deposition*, edited by McD. Robinson, C.H.J. van den Breckel, G.W. Cullen, J.M.J. Blocher and P. Rai-Choudhury (The Electrochemical Society, Incorporated, Pennington, NJ, 1984), pp. 596-614

6. William C. Jenkin, U. S. Patent No. 3 160 517 (8 December 1964).

7. A. J. Caputo and W. J. Lackey, Ceram. Eng. Sci. Pro. 5, 654-667 (1984).

8. T.L. Starr, in *Proceedings of the Tenth International Conference on Chemical Vapor Deposition*, edited by G.W. Cullen (The Electrochemical Society, Incorporated, Pennington, NJ, 1987), pp. 1147-1155

9. S. Middleman, J. Mater. Res. 4, 1515-1524 (1989).

10. T. M. Besmann, B. W. Sheldon, T. S. Moss and M. D. Kaster, J. Amer. Ceram. Soc., in press.

11. T.L. Starr, Ceram. Eng. Sci. Proc. 8, 951-957 (1987); R.P. Currier, J. Am. Ceram. Soc. 73, 2274-2280 (1990).

12. M. M. Tomadakis and S. V. Sotirchos, AIChE J. 37, 1175-1186 (1991).

13. J. Kozeny, Wasserkraft Wasserwirtsch 22, 67 (1927).

14. T. L. Starr and A. W. Smith, in preparation.

15. T. L. Starr and D. P. Stinton, in *Proceedings of the American Society of Composites, Fifth Technical Conference*, edited by L. T. Drzal (Technomic Publishing, Lancaster, PA, 1990) pp. 765-773.

16. H. Tawil, L. D. Bentsen, S. Baskaran and D. P. Hasselman, J. Mat. Sci. 20, 3201-3212 (1985).

17. T. L. Starr and A. W. Smith, in *Chemical Vapor Deposition of Refractory Metals and Ceramics*, edited by T. M. Besmann and B. M. Gallois (Mat. Res. Soc. Sym. Proc. 168, Pittsburgh, PA 1990) pp. 55-60; R.R. Melkote and K.F. Jensen, ibid., pp.506-512; S.M. Gupte and J.A. Tsamopoulos, J. Electrochem. Soc. 137, 3675-3682 (1990); G. Chung and B.J. McCoy, J. Am. Ceram. Soc. 74, 746-751 (1991).

18. E. Fitzer, W. Fritz and R. Gadow, in *Advances in Ceramics*, edited by S. Somiya (KTK Scientific, Tokyo, 1983).

19. T. M. Besmann, T. S. Moss and J. C. McLaughlin, presented at the 16th Annual Conference on Composites and Advanced Ceramics, Cocoa Beach, FL, 1992 (unpublished); T. L. Starr and A. W. Smith, ibid.

20. T. L. Starr, A. W. Smith and G. F. Vinyard, Ceram. Eng. Sci. Proc. 12, 2017 (1991).

X-RAY TOMOGRAPHIC MICROSCOPY OF NICALON PREFORMS AND CHEMICAL VAPOR
INFILTRATED NICALON/SILICON CARBIDE COMPOSITES

M.D. BUTTS[1], S.R. STOCK[1], J.H. KINNEY[2], T.L. STARR[1], M.C. NICHOLS[3],
C.A. LUNDGREN[4], T.M. BREUNIG[1] and A. GUVENILIR[1]

[1]Georgia Institute of Technology, Atlanta GA 30332
[2]Lawrence Livermore National Laboratory, Livermore CA 94550
[3]Sandia National Laboratories, Livermore CA 94550
[4]E.I. DuPont de Nemours, Wilmington DE 19880

ABSTRACT

Following the evolving microstructure of composites through all stages of chemical vapor infiltration (CVI) is a key to improved understanding and control of the process. X-ray Tomographic Microscopy (XTM), i.e., very high resolution computed tomography, allows the microstructure of macroscopic volumes of a composite to be imaged nondestructively with resolution approaching one micrometer. Results obtained with XTM on dense SiC/SiC composites and on woven SiC fiber preforms illustrate how details of the densification process can be followed using this technique during interruptions in processing. Ways in which the three-dimensional microstructural information may be used to improve modeling are also indicated.

INTRODUCTION

Improved control of CVI and decreased densification times are important to achieving widespread use of ceramic composite components fabricated with this process [1]. The assurance that near-critical-sized flaws or that damage-intolerant microstructures have not been incorporated are vital to component designers. More economical production of components, achieved by shortened processing times, will widen the range of feasible applications for ceramic composites.

There are three levels in the hierarchy of porosity which remain after CVI densification of woven preforms: intra-tow microporosity and two types of inter-tow porosity (holes in the weave of the preform and channels between the preform layers) [1]. The channels distribute the gas mixture and play the key role in determining how far and how rapidly the composite densifies. The space between fibers in a tow is quickly eliminated, isolating micropores; and the holes, which are intrinsic to the cloth, remain after the channels to the surface close. Understanding the rate at which the channels narrow, the influence of channel topology and topography and how these relate to the concurrent composite density are important to refining physically-based models of densification. This paper discusses how high resolution x-ray computed tomography (i.e., XTM) can be used to examine the same volume of a preform/composite multiple times during densification and outlines approachs for quantifying the types, sizes and evolution of pores during CVI of Nicalon/SiC composites.

EXPERIMENTS

The XTM used to study the Nicalon preform and Nicalon/SiC composites has been described in detail elsewhere [2,3], and only the barest details are repeated here. The apparatus is based on a two-dimensional, 1320 x 1035 element charge-coupled device (CCD) coupled through a short depth-of-field, variable magnification optical lens system to a single crystal scintillator screen of $CdWO_4$. The sample is viewed with x-rays along a large number of projection directions, and the radiographs are recombined via the filtered back projection algorithm [4,5] into a stack of parallel, two-dimensional, cross-sections (i.e., slices) which map the variation in x-ray attenuation within the sample's interior.

Preforms of stacked, plain weave Nicalon cloth and Nicalon/SiC composites densified using an isothermal CVI process have been examined with XTM. The fiber tows in adjacent layers of cloth are aligned parallel (0° layup), alternate between 0° and 45° orientations (0°/45° layup) or are aligned in the sequence 0°, 30° and 60° (0°/30°/60° layup). The sample described below is cut parallel to the cloth planes from a larger section of a densified, 0° layup composite; its cross-sectional dimensions are approximately 0.8 mm x 1.0 mm. Monochromatic synchrotron x-radiation (18 keV) from Beamline A-2 at CHESS (Cornell High Energy Synchrotron Source) is used to image the sample. This energy allows optimum contrast ($\mu t = 1.5$, where μ is the linear attenuation coefficient and t is the path length) along the longest x-ray path through the sample. The angular increments between views is 0.5°, projections are collected over 180° and the isotropic pixels are 3.4 μm in size in the 88 contiguous slices reconstructed for the sample described below.

RESULTS AND DISCUSSION

Figure 1 shows a typical XTM slice of a densified Nicalon/SiC composite. The darker pixels represent the location of material with lower x-ray attenuation. One well-defined channel, C, curves between tows on the right side of the slice, and an enlarged image of it appears in Fig. 1b. Five layers of Nicalon cloth with fiber axes running vertically are labeled "N" in Fig. 1; the black bands within these tows are elongated micropores. The other layers of the preform have fibers and tows approximately perpendicular to the slice which produces the mottled contrast elsewhere in the composite. This contrast, which is difficult to discern in monochromatic images, arises primarily from differences in attenuation between Nicalon and deposited SiC. Some pixels in these tows are partially occupied by pores as well as Nicalon or SiC.

The calculated attenuation coefficients for Nicalon and SiC are 9.0 and 14.2 cm^{-1}, respectively, based on tabulated densities [6], compositions [7] and

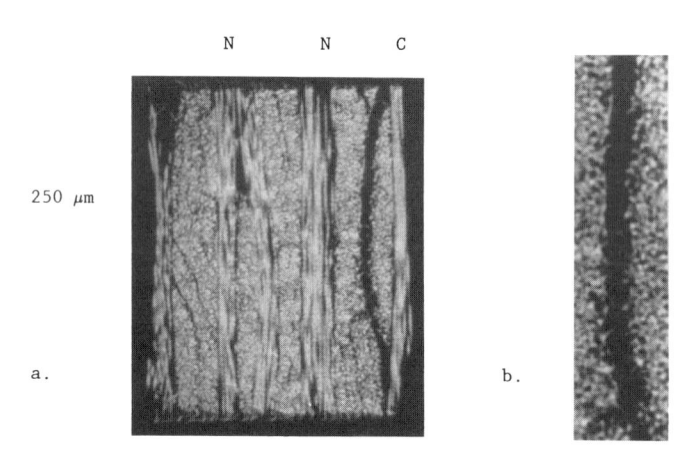

Figure 1. a) XTM slice through a densified Nicalon/SiC composite. A well-defined channel and cloth layers with fiber axes running vertically are labeled C and N, respectively. b) Enlargement of part of the channel.

mass absorption coefficients [8]. A histogram of linear attenuation coefficients from a typical volume of the composite is shown in Fig. 2. The box in the adjacent slice defines which part of Fig. 1 is used to produce the histogram. The continuum of absorption coefficients seen in the histogram is the result of voxels being partially occupied by the different constituents of the composite: Nicalon, SiC and empty space.

Three-dimensional surface renderings are very helpful in visualizing the spatial distribution of porosity. The surface rendering shown in Fig. 3 depicts, from a given viewing position, the surface encompassing all low absorption pixels within a subset of the total volume imaged. Figure 3b shows where the volume intersects the slice shown in Fig. 1, and the surface includes all pixels with values equal to or lower than 25% of the maximum value encountered in the sample (there are 256 gray levels used in the images). Long rods (extended, isolated micropores running parallel to fibers within the tows) are prominent within the image (indicated by "R"). Broad, shallow sheets (labeled "C") show the position of channels between tows.

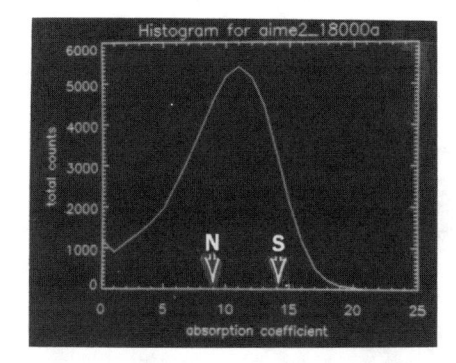

Figure 2. Histogram of linear attenuation coefficients in the slice shown in Fig. 1. The calculated values for Nicalon and SiC are indicated by letters N and S, respectively.

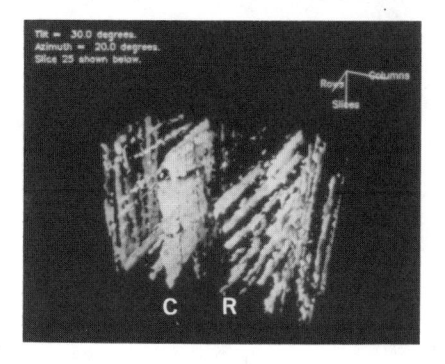

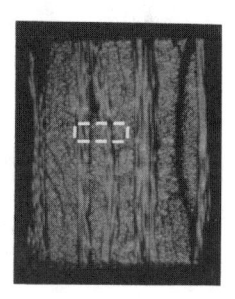

a.

b.

Figure 3. a) Surface rendering showing only those pixels with gray levels below 25 % of the 256 levels of contrast in the pictured above. The position of the volume is shown in b). Extended micropores within tows are labeled R, and channels between tows are labeled C.

The width of the channels is a key variable to quantify in studies relating evolving microstructure to densification during CVI. The digital, volumetric data from XTM is ideal for obtaining this information. First, a volume containing the channel of interest is selected. The channel width is measured across each row (i.e., horizontally across the slice shown in Fig. 1) of the slice, and the measurement is repeated for all slices in the volume. A simple algorithm is used to determine channel width automatically: the program counts the number of consecutive pixels in a row which fall below a defined critical value of the absorption coefficient. The algorithm needs to be robust, in order to avoid unintentionally measuring the widths of micropores which may be included in the volume containing the channel. The simple program first looks in each row for three or more adjacent pixels which fall below the critical level. If these are found, the first pixel is marked as the edge of the channel. Next, the program searches for three pixels in that row which are above the critical level; the first of these is marked as the end of the channel. If the resulting channel width is zero, the same procedure is followed in a search for two adjacent pixels below and then above the critical absorption value.

Figure 4 shows channel widths in the area in Fig. 1b. The critical absorption coefficient value used is 5 cm^{-1} which is much lower than the values for Nicalon and SiC. The variation in channel width is quite pronounced as is the physical irregularity of the channel walls. Apparently, the rough walls result from SiC deposited on fibers snaking from the main body of the tows. Figure 5 shows the variation of channel width within the volume sampled. Variability is similar to that in Fig. 4.

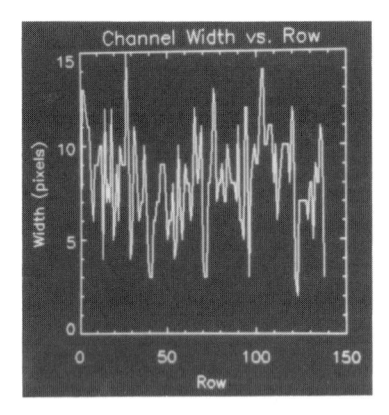

Figure 4. Variation of channel width in the area shown in Fig. 1b. The width is measured along the rows of the image, approximately perpendicular to the channel walls.

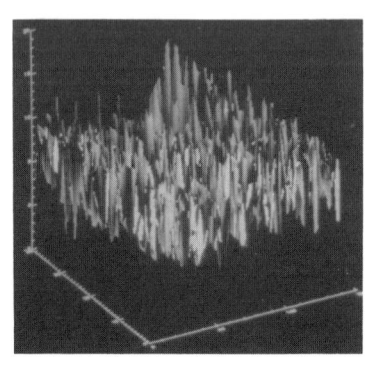

Figure 5. Variation of channel width (vertical axis) for the rows in each slice (horizontal axis) and for all slices (angled axis), i.e., throughout the volume indicated in Fig 1b. All axes are in units of pixels.

This approach should be very valuable in future XTM characterization of samples' progression to complete densification: the change in channel width after each increment of infiltration can be measured precisely and rapidly for the large number of channels in a sample. Availability of channel width data, of changes in channel width during each infiltration step and of channels' proximity to holes as a function of position in the sample allow three-dimensional visualization of the network of pores and holes within the composite. For example, plotting surfaces representing the centers of channels and using a range of colors to indicate local channel width is one promising strategy for analyzing the large amount of data generated by XTM. Better estimates of permeability and local deposition rates will result and should lead to a new generation of numerical models for CVI and perhaps to design rules for optimum composite preform architecture.

CONCLUSIONS

An approach for improving modeling of CVI of composites is based on data collected nondestructively from sample interiors using XTM. Results from a densified Nicalon/SiC ceramic matrix composite are used to indicate how channel widths and microporosity can be studied. Of particular interest is the possibility of examining the same volume of material after each stage of infiltration. Obtaining such XTM data should lead to better understanding of the CVI process and allow improved modeling.

ACKNOWLEDGEMENTS

This research was supported through grant W-7405-ENG-48 by US Department of Energy, Advanced Industrial Materials Program, Advanced Industrial Concepts Division. The work was done partially at the Cornell High Energy Synchrotron Source which is supported by the National Science Foundation.

REFERENCES

1. T.M. Besmann, B.W. Sheldon, R.A. Lowden and D.P. Stinton, Science, 253, 1104 (1991).
2. U. Bonse, Q.C. Johnson, M.C. Nichols, R. Nusshardt, S. Krasnicki and J.H. Kinney, Nucl. Instrum. Methods, A246, 644 (1986).
3. J.H. Kinney, Q.C. Johnson, U. Bonse, M.C. Nichols, R.A. Saroyan, R. Nusshardt, R. Pahl and J.M. Brase, Mat. Res. Soc. Bull., XIII, 13 (1988)
4. A.C. Kak and M. Slaney, Principles of Computerized Tomographic Imaging, (IEEE Press, New York, 1987).
5. R.H. Huesman, G.T. Gullberg, W.L. Greenberg and T.F. Budinger, Users Manual: Donner Algorithms for Reconstruction Tomography, (PUB-214, Lawrence Berkeley Laboratory, University of California, 1977).
6. K.K. Chawla, Composite Materials Science and Engineering, (Springer-Verlag, New York, 1987) p. 54.
7. Data sheets, Dow Corning Corporation, Midland, Michigan.
8. International Tables for X-ray Crystallography (Kynoch Press, Birmingham, England, 1972), Vol. 4, p. 61.

EFFECTS OF FIBER ORIENTATION AND OVERLAPPING ON KNUDSEN, TRANSITION, AND ORDINARY REGIME DIFFUSION IN FIBROUS SUBSTRATES

Manolis M. Tomadakis and Stratis V. Sotirchos
Department of Chemical Engineering
University of Rochester
Rochester, NY 14627

ABSTRACT

We present effective diffusion coefficients of gases in porous media whose structure can be represented as an assemblage of cylindrical fibers, such as the media used as substrates in chemical vapor infiltration. Structures consisting of non-, partially, or freely overlapping fibers of various orientation distributions are considered, and effective diffusion coefficients are computed by means of a Monte Carlo simulation scheme. In order to be able to examine the interrelation of ordinary, transition, and Knudsen diffusivities and tortuosities, computations are carried out over the whole diffusion regime, $i.e.$, from bulk to Knudsen. Our simulation results are compared with variational bounds and experimental values of tortuosity of fibrous beds reported by other investigators.

INTRODUCTION

Knowledge of the mass transport characteristics of fibrous structures used as preforms in chemical vapor infiltration (CVI) helps us better understand the mechanism of the CVI process. This in turn enables us to identify operating conditions and procedures for improving the process, both in its conventional isothermal, diffusion-driven form [1] and modified, temperature-pressure gradient (ORNL process [2]) and pulse-CVI [3,4] versions. However, only a few diffusivity measurements for dilute beds in the slip [5] and the ordinary diffusion regime [6,7] are available in the literature, with the Knudsen and transition regimes left totally unexplored. Moreover, theoretical work in this area has primarily been focused on the derivation of bounds using variational principles or other methods [8-11].

Simulation results for the variation of the effective diffusion coefficient of random fiber structures with the porosity in the whole diffusion regime, from bulk to Knudsen, are presented in this work. Effective diffusivities are computed by using a Monte Carlo simulation scheme to compute the mean square displacement of test molecules travelling in the pore space [12]. We consider fiber structures formed by randomly overlapping cylindrical fibers distributed randomly in d (d = 1, 2, or 3) directions (d-directional, random fiber structures), that is, with their axes parallel to a line ($d = 1$), parallel to a plane ($d = 2$), or oriented randomly in the three dimensional space ($d = 3$), and structures with fibers grouped into d (d = 1, 2, or 3) bundles of parallel, randomly overlapping fibers, with the bundles arranged in mutually perpendicular directions (d-directional, parallel fiber structures). Since the fibers in an actual preform do not overlap with each other, structures of freely overlapping fibers can be used as models of actual preforms only for relatively high porosities. However, the initial porosity of the fibrous structures that are used as preforms in composite fabrication by chemical vapor infiltration varies in the relatively broad range 0.4-0.85 [13,14]. For this reason, we also present results for structures consisting of unidirectional nonoverlapping or partially overlapping fibers, the latter representing the evolving states of a preform of parallel fibers as it is being densified by chemical vapor infiltration.

CONSTRUCTION OF FIBROUS STRUCTURES AND COMPUTATION OF EFFECTIVE DIFFUSIVITIES AND TORTUOSITY FACTORS

The construction of finite samples of porous structures consisting of unidirectional, nonoverlapping fibers of uniform size is accomplished by means of a scheme based on the Metropolis Monte Carlo method [15]. The fibers are initially positioned in a cubic unit cell at the sites of a regular triangular lattice. Random structures of nonoverlapping fibers are obtained through a large number of random sequential moves of the fibers from their initial positions. Partially overlapping fiber structures are produced from the nonoverlapping ones by increasing the fiber size by a certain amount. A partially overlapping fiber structure resulting from a nonoverlapping one after the fibers of the latter are let to grow by 25% is shown in Fig. 1. For a structure undergoing densification by chemical vapor infiltration, the rings added around the fibers of the original nonoverlapping fiber structure, displayed using solid black pattern, correspond to the material deposited within the preform during the process. The structure shown in Fig. 1 resembles, in almost every important detail, photomicrographs of partially densified fiber-reinforced composites [16,17].

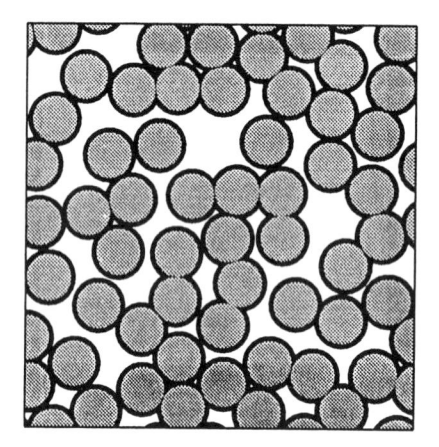

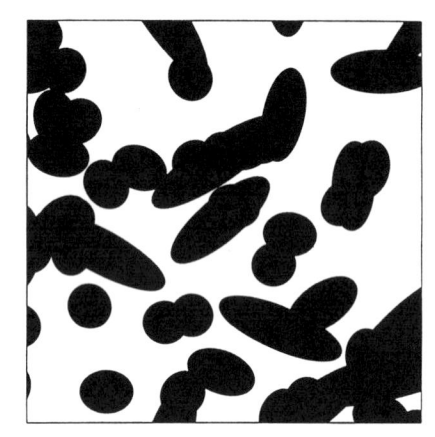

Figure 1. Cross section of a partially overlapping unidirectional fiber structure of 21% porosity. $\varepsilon_0 = 45\%$.

Figure 2. Section of a randomly overlapping tridirectional fiber structure of 50% porosity.

A porous structure consisting of mutually perpendicular bundles of parallel, randomly overlapping fibers is constructed by generating a random distribution of points on a face of the cubic finite sample, treating the points as the traces of the axes of the fibers that are perpendicular to the face, and repeating for all d bundles, on d mutually perpendicular faces of the cell. Structures of randomly oriented fibers are constructed by distributing randomly in 2 or 3 directions the axes of the fibers, according to the mean free path-randomness (μ-randomness) mechanism [18]. The section of a 50%-porosity structure of fibers oriented randomly in three directions is shown in Fig. 2. For fibers of 1 μm in radius, Fig. 2 depicts a 20 μm × 20 μm section of the fiber structure.

Effective diffusivities are computed using the mean square displacement, $<\xi^2>$, of molecules travelling in the void space of the porous medium for adequately large travel times τ, using the formulas [12]

$$D_e = \frac{<\xi^2>}{6\tau}; \quad D_{ej} = \frac{<\xi_j^2>}{2\tau} \quad (1a,b)$$

where D_e stands for the orientationally averaged effective diffusivity, and D_{ej} for the diffusivity in direction j. The computation of the mean square displacement is accomplished by following the trajectories of a large number of molecules, introduced randomly and travelling independently in the unit cell. The procedure used for computing the trajectories of molecules undergoing diffusion in a porous, capillary or fibrous, structure is described in detail in other publications [19-22].

The effective diffusivity results may be used to estimate a tortuosity factor for the porous medium, η, using the equations

$$D_e = \frac{\varepsilon}{\eta}D; \quad D_{ej} = \frac{\varepsilon}{\eta_j}D; \quad \frac{1}{D} = \frac{1}{D^b} + \frac{1}{D^K} \quad (2a,b,c)$$

The reference diffusivity, D, is the self-diffusion coefficient of the diffusing species in a cylindrical pore of radius equal to the average pore radius of the fibrous structure under the same conditions of pressure and temperature (*i.e.*, same mean free path, $\bar{\lambda}$, and mean thermal speed, $\bar{v}$, for the gas molecules). It is computed using the reciprocal additivity approximation (Bosanquet formula) for transition regime diffusion in a cylindrical tube [23], according to which (see eq. (2c)) the transition regime diffusivity is approximated closely by the harmonic mean of the continuum self-diffusion coefficient, D^b, and Knudsen diffusion coefficient, D^K.

RESULTS AND DISCUSSION

Effects of Fiber Overlapping on Diffusivities

Computer simulation results for Knudsen tortuosities corresponding to diffusion in non-, partially, or freely overlapping unidirectional, unimodal fiber structures are shown in Fig. 3. The tortuosities in directions perpendicular to the fibers are seen to be lower for structures consisting of nonoverlapping fibers than for structures of freely overlapping fibers of the same porosity. As the porosity increases, however, the extent of fiber overlapping in freely overlapping (fully penetrable) fibers decreases, and thus, the tortuosities for the two cases get closer to each other and eventually coincide as the porosity approaches unity. In the vicinity of 100% porosity, the tortuosity factors for both cases approach the lower bound ($\eta = 1.747$) for diffusion perpendicularly to fully overlapping fibers [10]. The tortuosity factor in directions perpendicular to the fibers becomes infinite (*i.e.*, the effective diffusivity becomes zero) at 0.0931 porosity (the porosity of a triangular array of closely packed solid cylinders) for structures of nonoverlapping fibers, while for freely overlapping fibers it approaches infinity at a much higher porosity (~ 0.33). The tortuosity factor for diffusion parallel to the fibers depends weakly on porosity and extent of fiber overlapping, having an average value of 0.549.

Fig. 3 also presents tortuosity results for partially overlapping structures for three values of initial (hard-core) porosity, ε_0. If the partially overlapping fiber structures of the figure had been obtained through a densification process, ε_0 would correspond to the initial porosity of the fibrous preform. Thus, notice that the curves for the partially overlapping structures start on the corresponding curve for nonoverlapping fibers at the hard core porosity. The results of Fig. 3 reveal that partially overlapping fiber structures exhibit behavior intermediate to those of the two extreme cases. As the starting porosity (ε_0) increases, the tortuosity vs. porosity curve moves closer to that for freely overlapping fibers. The percolation threshold of a partially overlapping fiber structure lies between the values for non- and freely overlapping fibers, that is, in the range [0.0931, 0.33]. It should be pointed out that the solid curves shown in this figure, as well as in all other figures, were obtained through cubic spline approximation to the simulation data, using software provided with the graphics package employed. The dashed curves give the predictions of a correlation that we will discuss later.

Effect of Fiber Orientation and Knudsen Number on Mass Transport

Tortuosity vs. porosity results on Knudsen diffusion in randomly overlapping fibrous structures of various directionalities are presented in Fig. 4. The results shown indicate that the percolation thresholds of 2-d and 3-d fibrous structures are much lower than that of a structure of unidirectional fibers, decreasing in the direction of increasing directionality. Specifically, 3-d structures percolate at about 4% porosity, while the percolation threshold of 2-d structures is around $\varepsilon=0.11$. Since the percolation threshold determines the lowest porosity that can be achieved during densification, the above observation suggests that higher densities can be achieved when 3-d structures are used as preforms in chemical vapor infiltration instead of cloth layups. Because of the anisotropy of 2-d structures, the effective diffusivities in directions parallel and perpendicular to the fiber mat are different, but their difference is much smaller than the difference between the diffusivities in directions parallel and perpendicular to the fibers of a unidirectional structure.

Tortuosity factors in the three regimes of diffusion are plotted in Fig. 5 as functions of porosity for diffusion perpendicularly to the fibers of 2-d random fiber structures. The values of the bulk tortuosity factor, η^b, were obtained for $Kn = 0.02$, while those of the Knudsen tortuosity factor, η^K, for $Kn = 100$. The Knudsen number, Kn, is defined as $Kn = \bar{\lambda}/\bar{d}$ with $\bar{d}$ being the mean intercept length of the porous medium ($= 4 \times$ porosity / surface area). Values of the transition regime tortuosity are given for $Kn = 1$, that is, for the case where the mean free path of the molecules and the mean distance between successive molecule-wall collisions are equal. Simulations for 9 different realizations of the fibrous structure were carried out at each porosity to get the results shown in Fig. 5.

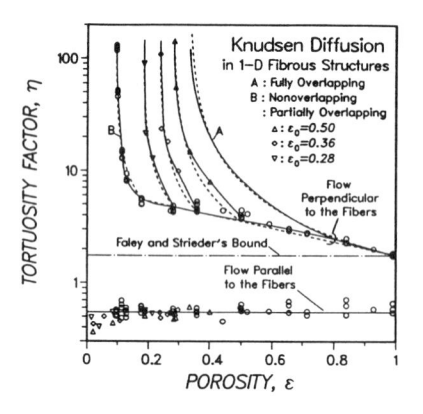

Figure 3. Variation of the tortuosity factor with the porosity for Knudsen diffusion in unidirectional fiber structures.

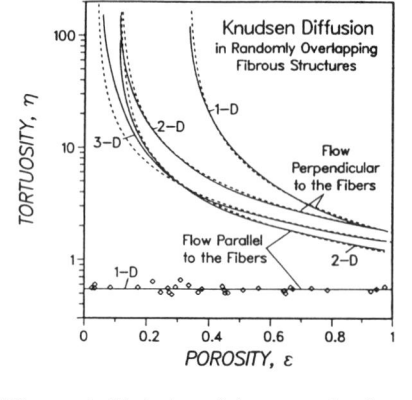

Figure 4. Variation of the tortuosity factor with the porosity for Knudsen diffusion in randomly overlapping fiber structures.

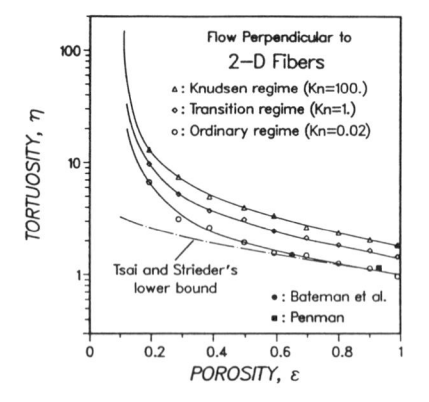

Figure 5. Variation of the tortuosity factor with the porosity for diffusion in bidirectional fiber structures.

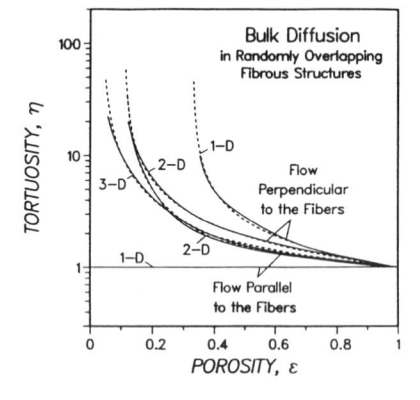

Figure 6. Variation of the tortuosity factor with the porosity for ordinary diffusion in randomly overlapping fiber structures.

Comparison of our numerical results for the tortuosity factor in the bulk diffusion regime with the predictions of a variational bound by Tsai and Strieder [9] (see Fig. 5) showed that the latter gives a very good approximation to the actual tortuosity for porosities greater than about 60%. Similar conclusions were reached from the results for bulk diffusion in 3-d and 1-d fiber structures. Figure 5 also shows that the numerically computed bulk tortuosities for 2-d structures are in very good agreement with two experimental data points by Bateman et al. and Penman [6,7]. Bateman et al. [6] measured the bulk tortuosity for diffusion perpendicularly to the fibers of a cellulosic filter of 65% porosity. Penman [7] measured the bulk tortuosity for a similar configuration of flow in steel wool of $\varepsilon = 0.93$.

Fig. 6 presents tortuosity vs. porosity curves for bulk diffusion through fibrous beds of various directionalities. Several observations can be made by comparing the results of Fig. 6 with the analogous results presented in Fig. 4 for Knudsen diffusion. The tortuosities for bulk diffusion are lower than the corresponding values for Knudsen diffusion for all fiber structures and diffusion directions, except for flow parallel to the fibers of unidirectional fiber structures. The differences among diffusivities in different directions for anisotropic structures (1-d and 2-d) are smaller for bulk diffusion. Finally, as the solid fraction becomes zero, the bulk tortuosity approaches unity for all cases, while the Knudsen diffusivity approaches a limit that in general depends on the type of the fibrous structure and the direction of diffusion.

In applications, it is customary to compute the effective diffusion coefficient in the transition regime from the Bosanquet formula (eq. (2c)) by using the effective values of bulk and Knudsen diffusivities in the place of the single pore values, that is, by writing:

$$\frac{1}{D_{ej}} = \frac{1}{D_{ej}^b} + \frac{1}{D_{ej}^K} \qquad (3a)$$

Using eq. (2c) and the definition of the tortuosity factor (eq. (2b)), eq. (3a) leads to the following relation between the transition, Knudsen, and bulk diffusion regime tortuosities:

$$\eta_j = \frac{\eta_j^b + \eta_j^K Kn}{1 + Kn} \qquad (3b)$$

with subscript j denoting the direction of diffusion. Application of eq. (3b) to our simulation data in the Knudsen and bulk diffusion regimes gave tortuosities that were almost identical to those obtained by independent simulations in the transition regime for all cases, with the exception of diffusion parallel to fibers of a unidirectional structure. In the last case, eq. (3b) was found to underpredict the transition regime tortuosity, by as much as 20% in the vicinity of $Kn = 1$. To the best of our knowledge, this is the first time that the validity of the reciprocal additivity correlation has been demonstrated for a class of random porous media.

Tortuosity-Porosity Correlations

In order to render our simulation results readily usable by people working in areas involving diffusion in fibrous beds, such as the fabrication of ceramic composites by chemical vapor infiltration, we searched for simple, one-parameter correlations that would provide satisfactory approximations to the computed tortuosities, using the known structural properties of the fibrous structures. Various one-parameter correlations were tested, and the one that appeared to work the best for all cases examined in this study was of the form

$$\eta_j = \eta_j(\varepsilon_0) \left(\frac{\varepsilon_0 - \varepsilon_p}{\varepsilon - \varepsilon_p} \right)^\alpha \qquad (4)$$

with ε_p being the percolation threshold of the fiber structure in the direction of diffusion. The predictions of eq. (4) are given by the dashed curves of Figs. 3-6.

In the case of partially overlapping unidirectional fiber structures, ε_0 denotes the hard-core porosity of the fibers, i.e., the porosity at the point where the tortuosity factor curves for partially overlapping structures leave the curve for nonoverlapping fibers, and $\eta_j(\varepsilon_0)$ the corresponding tortuosity factor. The dependence of ε_p on ε_0 was investigated in detail in a previous study [21] and can be approximately followed using the data given in Fig. 3. Parameter α was found to depend linearly on ε_p. Specifically, $\alpha = 2.28\varepsilon_p + 0.35$ with 99.93% correlation coefficient. The Knudsen tortuosity factor of nonoverlapping fibers was correlated for $\varepsilon \geq 0.18$ using the equation $\eta = 1.747 exp[1.4(1 - \varepsilon)]$. For lower porosities, we used eq. (4) with $\varepsilon_0 = 0.18$, $\varepsilon_p = 0.0931$, and $\alpha = 0.72$.

For diffusion in randomly overlapping fiber structures, $\varepsilon_0 = 1$ and $\eta_j(\varepsilon_0) \equiv \eta_{j\,min}$. The values of α are listed in Table 1 along with the minimum values of Knudsen tortuosity for each structure and direction of diffusion. For bulk diffusion, the minimum value of tortuosity is equal to 1 for all cases. It should be noted that the values of α of Table 1 for bulk diffusion in a 3-d structure or parallel to the fiber mat of a 2-d structure are good for $\varepsilon > 0.4$ only. For the low porosity region, better results are obtained using eq. (4) with $\varepsilon_0 = 0.4$, $\alpha = 0.872$ for the 2-d structure, and $\alpha = 0.965$ for the 3-d structure, with $\eta_j(0.4)$ computed from eq. (4) and Table 1.

Table 1. Parameters used in eq. (4)

FIBROUS STRUCTURE AND DIRECTION OF DIFFUSIONAL FLOW		ε_p	KNUDSEN REGIME		BULK REGIME
			$\eta_{j\,min}$	α	α
1-D	parallel	0	0.549	0	0
	perpendicular	0.33	1.747	1.099	0.707
2-D	parallel	0.11	1.149	0.954	0.521
	perpendicular		1.780	1.005	0.785
3-D	all directions	0.037	13/9	0.921	0.661

REFERENCES

1. D.P. Stinton, T.M. Besmann, and R.A. Lowden, Am. Ceram. Soc. Bull. 67, 350 (1988).
2. D.P. Stinton, A.J. Caputo, and R.A. Lowden, Am. Ceram. Soc. Bull. 65, 347 (1986).
3. K. Sugiyama and T. Nakamura, J. Mat. Sci. Let. 6, 331 (1987).
4. K. Sugiyama and Y. Ohsawa, J. Mat. Sci. Let. 7, 1221 (1988).
5. J.C Brown, TAPPI 33, 130 (1950).
6. B.R. Bateman, J.D Way and K.M. Larson, Sep. Sci. Tech. 19, 21 (1984).
7. H.L. Penman, J. Agri. Sci. 30, 437 (1940).
8. G.W. Milton, J. Appl. Phys. 52, 5294 (1981).
9. D.S. Tsai and W. Strieder, Chem. Eng. Comm. 40, 207 (1986).
10. T.L. Faley and W. Strieder, J. Appl. Phys. 62, 4394 (1987).
11. T.L. Faley and W. Strieder, J. Chem. Phys. 89, 6936 (1988).
12. A. Rahman, Phys. Rev. 136, A405 (1964).
13. A.J Caputo, W.J. Lackey and D.P. Stinton, Ceram. Eng. Sci. Proc. 6, 694 (1985).
14. R. Naslain, F. Langlais and R. Fedou, J. de Physique 50, C5-191 (1989).
15. N. Metropolis, A.W. Rosenbluth, M.N. Rosenbluth, A.H. Teller and E. Teller, J. Chem. Phys. 21, 1087 (1953).
16. M.A. Meyers and K.K. Chawla, Mechanical Metallurgy Principles and Applications (Prentice Hall, New Jersey, 1984)
17. G.R. Hopkins and J. Chin, J. Nucl. Mat. 143, 148 (1986).
18. R. Coleman, J. Appl. Prob. 6, 430 (1969).
19. V.N. Burganos and S.V. Sotirchos, Chem. Eng. Sci. 44, 2451 (1989).
20. M.M. Tomadakis and S.V. Sotirchos, AIChE J. 37, 74 (1991).
21. M.M. Tomadakis and S.V. Sotirchos, AIChE J. 37, 1175 (1991).
22. M.M. Tomadakis and S.V. Sotirchos, (in preparation) (1991).
23. W.G. Pollard and R.D. Present, Phys. Rev. 73, 762 (1948).

CONTRIBUTION OF GAS–PHASE REACTIONS TO THE DEPOSITION OF SiC BY A FORCED–FLOW CHEMICAL VAPOR INFILTRATION PROCESS

CHING–YI TSAI* and SESHU B. DESU**
*Department of Engineering Science and Mechanics
**Department of Materials Engineering
Virginia Polytechnic Institute and State University, Blacksburg, VA, 24061

ABSTRACT

A model, incorporating both gas–phase and surface reactions, for simulating thickness profile of SiC, deposited from trichloromethylsilane (TMS), along the longitudinal direction of a single pore is presented in this paper. The transport mechanisms considered include both forced–flow and diffusion. With the nonlinear nature of this model, a finite element model was developed to solve the problem numerically. Simulation results were in good agreement with the reported experimental data by Fedou et al. (1990). Effects of critical parameters, such as deposition temperature, ratio of sticking coefficients of TMS and intermediate species, and forced–flow, on the deposition thickness profile were investigated. Forced–flow effect was found to be small for the chemical vapor infiltration (CVI) processes at high deposition temperatures.

INTRODUCTION

Ceramic materials have long been considered as the ideal materials for high temperature applications because of their good stability, resistance to corrosion, and high strength. However, the brittle nature of ceramics has precluded their use in applications where significant levels of toughness are required.

Primary goal of many ceramic materials researchers has been to alter the properties of ceramic materials through composition, design, and processing in order to minimize the effects due to the brittle nature and retain other desirable properties. To achieve this goal, researchers have developed the ceramic matrix composites (CMCs), which basically consists of a ceramic matrix reinforced with high performance fibers.

Recently, chemical vapor infiltration (CVI) processes have received a considerable attention as a strong candidate for the fabrication of CMCs because of its versatility in creating all major families of ceramic matrices by a single, continuous deposition step, low processing temperature feature, and geometry–preserving properties. The development of forced–flow CVI processes makes the CVI processes more compatible because they ensure better uniformity of the deposit and reduce the processing time significantly [1].

Extensive work is being done in the area of CVI process modeling in an effort to find an optimum relationship between the processing conditions and product properties [2–11]. Geometrical considerations of these models range from single–pore [2–5], overlap pore [6], unit–cell configuration [7], to effective medium [8–9], which considers the porous medium as pore networks. Diffusion is assumed to be the main transport mechanism of the gas species [2–8]. Depending on the CVI process to be modeled, convective flow (forced–flow) was also used [9–10].

In general, chemical reactions involved in the CVI processes are quite complex and can be broadly classified as gas–phase and surface reactions. Most CVI models proposed so far considered only the heterogeneous surface reactions, with the exception of two papers published recently by Middleman [4] and Sheldon [11], which also considered the homogeneous gas–phase decomposition reactions. Both these models [4 & 11] assume that the formation of the intermediate species on the surface sites comes entirely from the adsorption of the gas phase intermediate species. In another words, the sticking coefficient of the precursors used was assumed to be zero. However, the intermediate species on the surface sites might also originate from the decomposition of the adsorbed parent species.

Here we present a detailed kinetic model for the deposition of SiC from trichloromethylsilane (TMS) along a single pore. Both homogeneous and heterogeneous reactions were included in the deposition mechanism of SiC from TMS precursor. Furthermore, allowance is also made for the possibility that both adsorbed parent species and gas phase intermediate species led to the formation of intermediate species on the surface sites, which are responsible for the SiC deposition.

The deposition thickness profile was obtained by solving the transport–reaction equations, which included both forced–flow and diffusion transport mechanism.

Parameters, such as forced–flow, deposition temperature, ratio of the sticking coefficients of TMS and intermediate species, were studied in order to understand their influence upon the deposition thickness profile.

THEORY

(1) Deposition mechanism

Our kinetic model of the deposition of SiC from CH_3SiCl_3 (TMS) is based on the assumption that the deposition rate is kinetically controlled. The gas phase decomposition of TMS into certain intermediate species is also assumed. Schematic diagram of the assumed kinetic model for the deposition of SiC from TMS is shown in Fig.1. Shaded areas represent species on the surface sites, and plain rectangular areas represent gas phase species.

Assume that the surface concentration of TMS (θ_{tms}) and intermediate species (θ_{ip}) are in steady state condition, then the following relationship could be obtained:

$$\theta_{ip} = \frac{k_2 C_{ip} + k_3 C_{tms} - k_{-3}\theta_{tms}}{k_{-2} + k_5} \qquad (1)$$

where C_{tms} and C_{ip} are the gas phase concentration of TMS and intermediate species respectively.

Thus the surface growth rate (r_5) could be expressed as,

$$r_5 (mol/cm^2 sec) = \frac{k_2 C_{ip} + k_3 C_{mts}}{1 + k_{-2}/k_5} \qquad (2)$$

with the assumption that $k_{-3}\theta_{tms} \ll k_2 C_{ip} + k_3 C_{tms}$.

From the kinetic theory of gases, k_2 and k_3 could be expressed in terms of the sticking coefficient of the gas species as [12],

$$k_2 = 0.25 (Sc\ Vt)_{tms}\ ,\ k_3 = 0.25 (Sc\ Vt)_{ip} \qquad (3)$$

with
Sc_i = Sticking coefficient of the gas species i
Vt_i = The thermal mean velocity of the gas species i = $(8kT/\pi M_i)^{0.5}$

From equation (2) it is clear that the deposition thickness profile along the pore direction depends on the concentrations of both TMS and intermediate species. These concentration can be determined by considering the simultaneous transport mechanism and chemical reactions within the single pore geometry, which will be discussed in the following section.

(2) Transport–reaction equations

The governing transport–reaction equation in a steady state one dimensional flow for a general species (i) can be written as [13,14]

$$-D_i \frac{\partial^2 C_i}{\partial X^2} + U_i \frac{\partial C_i}{\partial X} = q_i \quad (4)$$

where D_i, U_i, and C_i are the diffusion coefficient (cm²/sec), flow velocity (cm/sec), and the concentration (mol/cm³) of the species (i). Also q_i (mol/cm²/sec) is the rate of generation (or depletion) of the species (i) per unit length of the single pore.

With the application of the assumed deposition mechanism and equation (4), transport–reaction equations for the species TMS and intermediate can be written as,

$$-D_{tms} \frac{\partial^2 C_{tms}}{\partial X^2} + U_0 \frac{\partial C_{tms}}{\partial X} + (k_1 + Sc_1 Vt_1/d)C_{tms} - k_{-1}C^2_{ip} = 0 \quad (5a)$$

$$-D_{ip} \frac{\partial^2 C_{ip}}{\partial X^2} + U_0 \frac{\partial C_{ip}}{\partial X} + (k_{-1}C_{ip} + Sc_2 Vt_2/d)C_{ip} - k_1 C_{tms} = 0 \quad (5b)$$

where d is the diameter of the pore.

The associated boundary conditions are,

$$\text{at } X = 0$$
$$C_{tms} = C_{tms0}, \ C_{ip} = C_{ip0} \quad (6a)$$

$$\text{at } X = L/2 = L_0$$
$$\frac{\partial C_{tms}}{\partial X} = \frac{\partial C_{ip}}{\partial X} = 0 \quad (6b)$$

Analytical solutions are difficult to obtain for the nonlinear coupled transport–reaction equations (5a) and (5b). In the following section, the finite element method (FEM) was used to solve the transport–reaction equations numerically.

FINITE ELEMENT MODELING

The following group of dimensionless parameters were used for the dimensional analysis of the transport–reaction equations (5a,b), and the associated boundary conditions (6a,b):

$$X/L_0 = X^*, \ C/C_{tms0} = C^*, \ \text{Ratio} = D_{tms}/D_{ip}$$

Finite element formulations were obtained by multiplying the dimensionless governing equations with weighting functions $\Psi_1(X)$ and $\Psi_2(X)$ respectively, then integrating over the entire domain [15]. The star symbol is dropped in the subsequent expressions for clarity.

Finite Element Formulations

$$\begin{bmatrix} K_{11} & K_{12} \\ K_{21} & K_{22} \end{bmatrix} \begin{bmatrix} C_{tms} \\ C_{ip} \end{bmatrix} = \begin{bmatrix} 0 \\ 0 \end{bmatrix} \quad (7a)$$

Boundary Conditions
$$\text{at } X = 0$$
$$C_{tms} = 1.0, \ C_{ip} = C_{ip0}/C_{tms0} \quad (7b)$$
$$\text{at } X = 1.0$$
$$\frac{dC_{tms}}{dX} = \frac{dC_{ip}}{dX} = 0.0 \quad (7c)$$

with

$$K_{11}^{(e)} = \int_{\Omega_e} \{\frac{\partial N_i}{\partial X} \frac{\partial N_j}{\partial X} + \text{Peclet } N_i \frac{\partial N_j}{\partial X} + (\text{Gas}_1 + \text{Surf}_1) N_i N_j\} dX$$

$$K_{12}^{(e)} = -\int_{\Omega_e} \{\text{Gas}_{-1} C_{tms0} C_{ip} N_i N_j\} dX$$

$$K_{21}^{(e)} = -\int_{\Omega_e} \{\text{Gas}_1 \text{Ratio } N_i N_j\} dX$$

$$K_{22}^{(e)} = \int_{\Omega_e} \{\frac{\partial N_i}{\partial X} \frac{\partial N_j}{\partial X} + \text{Pectlet Ratio } N_i \frac{\partial N_j}{\partial X} +$$
$$(\text{Gas}_{-1} C_{ip} C_{tms0} \text{Ratio} + \text{Surf}_2) N_i N_j\} dX$$

with
  Peclet = $U_0 L_0 / D_{tms}$ = Peclet Number of TMS
  Gas$_1$ = $k_1 L_0^2 / D_{tms}$ = Forward Gas Phase Damkohler Number of TMS
  Gas$_{-1}$ = $k_{-1} L_0^2 / D_{tms}$ = Backward Gas Phase Damkohler Number of TMS
  Surf$_1$ = $Sc_1 Vt_1 L_0^2 / d / D_{tms}$ = Surface Damkohler Number of TMS
  Surf$_2$ = $Sc_2 Vt_2 L_0^2 / d / D_{ip}$ = Surface Damkohler Number of Intermediate Species

Two—node linear elements are used to discretize the domain. The evaluation of K_{12} and K_{22} terms requires prior knowledge of C_{ip} and renders the equations nonlinear. Also because of the existence of convection terms in the formulations, the stiffness matrix is unsymmetrical. The direct iteration scheme was used to solve the nonlinear finite element problem [15].

RESULTS AND DISCUSSIONS

(1) Effect of the deposition temperature upon the thickness profile

Fedou et al. [5] studied the chemical vapor infiltration process of SiC from TMS precursor along single pores with 1cm length and 34μm diameter. The deposition temperatures were 1223°K, 1323°K, and 1373°K, with the deposition pressure maintained around 20 kPa. The ratio of H$_2$ and TMS was about 5.

The data used to simulate this SiC CVI process is presented in Table I. Degree of dissociation of TMS molecules (α) was used to estimate the equilibrium constant K_g. The model predictions were found to be in good agreement with the reported experimental results [5], as shown in Fig.2.

(2) Sticking coefficient ratio effect

As mentioned above, there are two contributions to the surface concentration of intermediate species: (1) from the adsorption of intermediate gas phase species and (2) from the adsorption of TMS itself, which then decomposes into the intermediate product after being adsorbed.

With the same entrance deposition thickness, different sticking coefficient ratio (Sc_1/Sc_2) resulted in different deposition profiles especially for low deposition temperature (1223°K), as illustrated in Fig.3a/3b.

(3) Forced—flow effect

The forced—flow effects upon the deposition thickness profile is very small for Peclet numbers up to 1.0 for both low and high deposition temperatures, as shown in Fig.4a/4b. For low deposition temperature (1223°K) with high Peclet numbers (10.0), the convection flow can carry the reactive gas species deep into the pore, thus resulting in a more uniform deposition profile. On the other hand, the influence of the forced—flow upon the deposition profile at high deposition temperatures (1373°K) is minute even for high Peclet number (10.0).

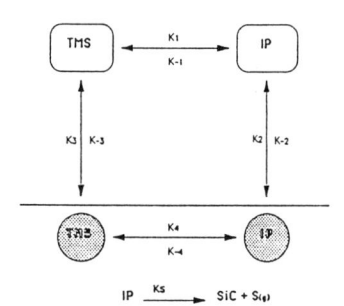

Table I Simulation data for SiC deposition from TMS

T	α	DR	k_{-2}/k_3	Vt_1	Vt_2	D_1	D_2	C_1	C_2
°K	%	nm/min		cm/sec		cm^2/sec		mol/cm^3 10^{-4}	10^{-8}
1223	2.0	70	0.01	41547	71962	21.57	26.98	0.321	0.656
1323	5.0	150	5.0	43212	74846	23.39	29.56	0.288	1.515
1373	8.0	200	18.0	44021	76247	24.29	30.84	0.269	2.336

T	K_g	k_1	Sc_1	Sc_2
°K	mol/cm^3 10^{-9}	1/sec	10^{-4}	10^{-4}
1223	0.1339	1	0.342	0.6848
1323	0.7974	30	3.140	3.140
1373	2.0313	100	9.950	9.950

Fig.1 Schematic diagram of deposition mechanism of SiC from TMS

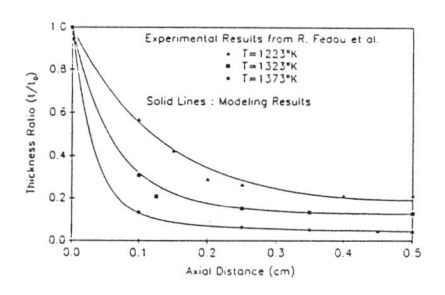

Fig.2 Thickness profiles of SiC deposited from TMS.

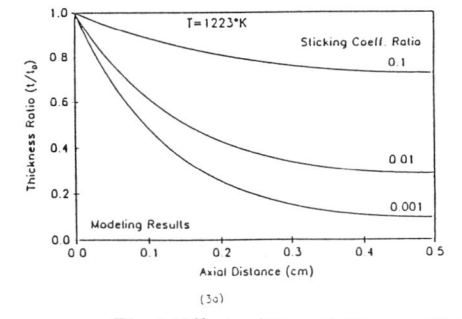

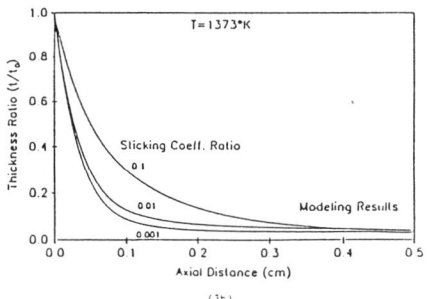

Fig.3 Effect of the sticking coefficient ratio on the thickness profile

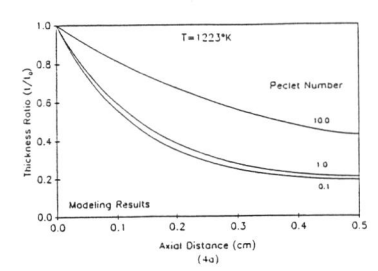

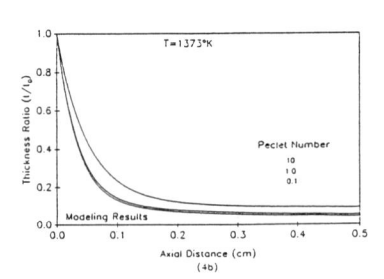

Fig.4 Forced–flow effect on the deposition profile

CONCLUSIONS

This paper presents a new model for the deposition of SiC from TMS along a pore geometry. Both homogeneous and heterogeneous chemical reactions were considered in this model. Simulation results were in good agreements with the reported experimental data by Fedou et al. [5].

The effect of critical parameters such as, deposition temperature, sticking coefficient ratio, and forced–flow, on the deposition thickness profile were investigated in this paper. Influence of the sticking coefficient ratio on the thickness profile was found to be large for the low deposition temperatures. Forced–flow effect is very small unless the Peclet Number is larger than 1.0.

Since the aspect ratio of the infiltrated composite preform is usually very small, premature pore blockage is expected for isothermal CVI processes of SiC deposited from TMS. The application of the forced–flow can not wholly solve the problem unless the Peclet number is very high and the deposition temperature is low enough. This illustrates the importance of the thermal gradient CVI processes in lowering the final porosity of the infiltrated parts.

ACKNOWLEDGEMENT

The authors are indebted to Professor J.N. Reddy for his help concerning the Finite Element Method. Thanks also to the Center for Composite Materials and Structures of Virginia Polytechnic institute and State University for providing financial support to one of the authors (Ching Yi Tsai).

REFERENCES

1. A.J. Caputo, W.J. Lackey and D.P. Stinton, Ceram. Eng. Sci. Proc., 6, 694 (1985).
2. Nyan–Hwa Tai and Tsu–Wei Chou, J. Am. Ceram. Soc., 72, 414 (1989).
3. S.M. Gupte and J.A. Tsamopoulos, J. Electrochem. Soc. ,136, 555 (1989).
4. Stanley Middleman, J. Mater. Res., 4, 1515 (1989).
5. R. Fedou, F. Langlais and R. Naslain, Proc. of the 11th Inter. Confer. on Chemical Vapor Deposition, 513 (1990).
6. R.P. Currier, J. Am. Ceram. Soc., 73, 2274 (1990).
7. T.L. Starr, Ceram. Eng. Sci. Proc., 8, 951 (1987).
8. S.M. Gupte and J.A. Tsamopoulos, J. Electrochem. Soc. ,137, 1626 (1989).
9. S.M. Gupte and J.A. Tsamopoulos, J. Electrochem. Soc. ,137, 3675 (1990).
10. Nyan–Hwa Tai and Tsu–Wei Chou, J. Am. Ceram. Soc. 73, 1489 (1990).
11. Brian W. Sheldon, J. Mater. Res., 5, 2729 (1990).
12. Kazunori Watanabe and Hiroshi Komiyama, J. Electrochem. Soc. ,137, 1222 (1990).
13. Surya R. Kalidindi and Seshu B. Desu, J. Electrochem. Soc. ,137, 624 (1990).
14. R.B. Bird, W.E. Steward, and E.N. Lightfoot, Transport Phenomena, (John Wiley & Sons, New York, 1960).
15 J.N. Reddy, An Introduction to the Finite Element Method, (McGraw–Hill, New York, 1984)

FIBER-REINFORCED TUBULAR COMPOSITES BY CHEMICAL VAPOR INFILTRATION*

D. P. Stinton, R. A. Lowden, and T. M. Besmann
Oak Ridge National Laboratory, Oak Ridge, Tennessee 37831

ABSTRACT

A forced-flow thermal-gradient chemical vapor infiltration process has been developed to fabricate composites of thick-walled tubular geometry common to many components. Fibrous preforms of different fiber architectures (3-dimensionally braided and filament wound) have been investigated to accommodate components with different mechanical property requirements. This paper will discuss the fabrication of tubular, fiber-reinforced SiC matrix composites and their mechanical properties.

INTRODUCTION

Fiber-reinforced SiC-matrix composites appear promising for gas turbine applications because of their high strength at elevated temperature, light weight, thermal shock resistance, damage tolerance, and oxidation and corrosion resistance. However, incorporation of continuous ceramic fibers into ceramic matrices without significant damage to the fibers is difficult. Hot-pressing of fiber-reinforced composites is impractical because the extremes of temperature and pressure weaken the continuous fibers. Cold-pressing and sintering routes to the fabrication of composites are of no value because typical sintering temperatures greatly exceed the temperature limit of the fibers [1,2]. Because of these limitations several novel impregnation processes have been developed. One such process impregnates a fibrous preform with liquid precursors that transform to ceramic materials on heat treating [3-5]. A second process impregnates a fibrous preform with extremely fine metallic silicon which converts to silicon nitride when reacted with nitrogen gas at elevated temperatures [6,7]. The greatest success has been achieved by vapor-phase processing, leading to a class of techniques termed chemical vapor infiltration (CVI). Two distinctly different vapor phase processes have been used to fabricate matrices: isothermal CVI, [8-11] which is used commercially, and forced CVI under development at Oak Ridge National Laboratory (ORNL) [12-15].

Fiber-reinforced composites are being considered for combustors, burner tubes, heat exchangers, headers, hot-gas filters and even rotors for stationary gas turbine engines. Unfortunately, neither of the CVI processes described above has demonstrated the ability to fabricate thick-walled tubular shapes appropriate for turbine engines. Isothermal CVI is ideal for the fabrication of thin-walled structures including complex shapes, however, infiltration times become extremely long for thick cross sections. The forced CVI process has been developed for thick-wall plates, however, very few tubular shapes have been infiltrated. Therefore, the focus of this investigation was the development of the forced CVI process for the fabrication of thick-walled tubular composites.

FIBROUS PREFORMS

Nicalon fiber preforms of a tubular geometry (2.5 to 3.8 cm ID; 0.6 cm wall thickness; ≈15 cm long) were fabricated with different fiber architectures. Nicalon is a polymer derived SiC fiber that is microcrystalline or amorphous in nature and contains significant amounts of silica [16,17].

---

*Research sponsored by the U.S. Department of Energy, Fossil Energy AR&TD Materials Program, under contract DE-AC05-84OR21400 with Martin Marietta, Inc.

Filament winding of fiber tows was used to fabricate components that require high hoop or radial stengths, but relatively modest axial strengths. A fiber architecture of this type would be ideal for combustors or headers. Three dimensional braiding was used to fabricate components for applications such as burner tubes or heat exchangers that require high axial strengths but only modest hoop strengths. Preforms were also fabricated by wrapping layers of cloth around a mandrel so that half the fibers were in the hoop direction and half in the axial direction. Filament wound preforms formed on graphite mandrels were prepared by the K-25 plant in Oak Ridge. The graphite mandrels contained hundreds of holes to permit ready access of reactant gases to the fibrous preform. The initial preforms were unidirectionally wound with the fibers less than 1° off the hoop direction. Satisfactory tubes with fiber contents of 45 to 55 vol % were produced, but because of the small angle, expected strengths in the axial direction are minimal. Therefore, tubes were produced with the fibers 10° off the hoop direction to increase the axial strength (Fig. 1). Initially, space was left between fiber tows, however, large pores became aligned on top of each other making thorough infiltration difficult. The winding technique was then modified to place adjacent tows in contact. This winding technique produced a density of about 47 vol % and prevented the formation of large pores in the preform.

Fig. 1 Preforms with different fiber architectures; (left) filament wound with fibers 10° off the hoop direction, (center) same except adjacent tows are touching, (right) 3-D braid.

Quadrax Corporation* prepared 3D braided preforms on graphite mandrels for this study. The initial tubes were braided very loosely and contained only about 15 vol% Nicalon fibers (Fig. 1). Near the ends of the preforms, the braided fibers were bound to the mandrel with graphite yarn to prevent the braid from unraveling. Infiltration of such a low density preform proved to be impossible with the forced CVI process. Therefore, researchers at Quadrax Corporation modified their process in order to fabricate tubes containing up to 40 vol % Nicalon fibers.

In another approach, process was developed to fabricate preforms for hot-gas filters that combines continuous and chopped fibers. The continuous fiber produces the strength, damage tolerance, and thermal shock resistance while the chopped fibers control the permeability and filtering efficiency of the filter surface. Preforms of this type are fabricated by fiber molding chopped fibers into the open pores of a braided substrate. Hybrid preforms have proven to be ideal for hot gas filter applications in work being performed at The 3M Company. One further advantage of the process is that expensive Nicalon fibers can be replaced by less expensive

---

*Quadrax Corporation - 300 High Point Avenue, Portsmouth, R. I.

chopped alumina or Nextel fibers. Because of the reduced cost, hybrid preforms may also find application in heat exchangers or regenerators where strength requirements are quite modest.

FIBER-MATRIX INTERFACE

The mechanical properties of fiber-reinforced ceramic composites are controlled by the amount of fiber in each orientation and the strength of the bond between the fibers and the matrix [18]. Experience has shown brittle composites result when the SiC matrix is applied directly onto the Nicalon fibers. Fracture surfaces for these materials are flat and smooth with no evidence of fiber pullout that results in toughening. To control the interfacial bond strength and protect the fibers from the corrosive, HCl laden CVI atmosphere, a thin layer of pyrolytic carbon (0.1 to 0.5μm thick) is deposited onto the Nicalon fibers. The laminar structure of the graphitic carbon coating protects the fibers from HCl attack and promotes the movement or slip of the fiber within the matrix.

CVI PROCESSING

Forced chemical vapor infiltration as developed at the Oak Ridge National Laboratory utilizes a thermal gradient and a pressure gradient to efficiently infiltrate simple flat plate preforms of thick cross-section. The forced CVI process was recently modified so that tubular preforms could be infiltrated by creating a thermal gradient from the outside of the tube to the inside [19]. While the outside of the preform is heated, cold water is circulated through the stainless steel injector to cool the inside diameter of the fibrous preform (Fig. 2). The gaseous reactants enter the furnace through tubing that runs within the water cooling passage. Reactants flow from the tubing in the cooling passage into a graphite gas distributor and are dispersed along the length of the preform through parallel slots in a graphite gas distributor. Reactants then proceed uniformly through holes in the graphite mandrel into the preform (Fig. 3).

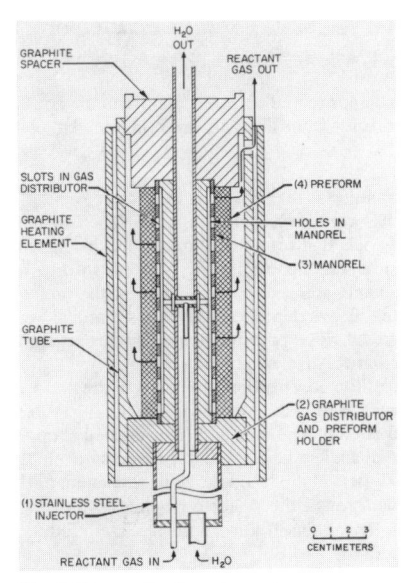

Fig. 2 Schematic of equipment used to infiltrate tubular preforms.

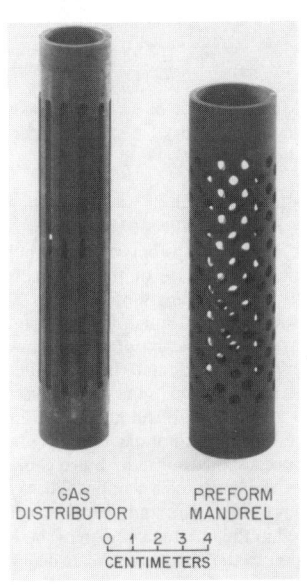

Fig. 3 Photo of graphite gas distributor and mandrel for infiltrating tubular composites.

Densification of the tubular preforms occurs when hydrogen and methyltrichlorosilane (CH_3SiCl_3 or MTS) flow through the mandrel. Decomposition of the MTS and deposition of SiC occurs as the gases approach the higher temperature regions near the outer diameter of the preform. Deposition of SiC within the hot region of the preform increases the density and thermal conductivity of the material. Therefore, the deposition zone moves from the outer diameter, hotter regions toward the inner diameter, cooler regions.

Filament wound preforms were effectively infiltrated from the outset using this system. The high fiber content of the preform produced a sufficient backpressure to disperse the reactants along the length of the tube allowing them to flow uniformly through the walls of the preform. The microstructure of filament wound composites (Fig. 4a) demonstrates that the limited amount of porosity is distributed uniformly through the thickness. In contrast, flat plate preforms produced by cloth layups result in a very different microstructure (fig. 4b). Note the greater amount of porosity and the tendency for voids to line up regardless of the layup (0°-90° vs 30°-60°-90° etc.).

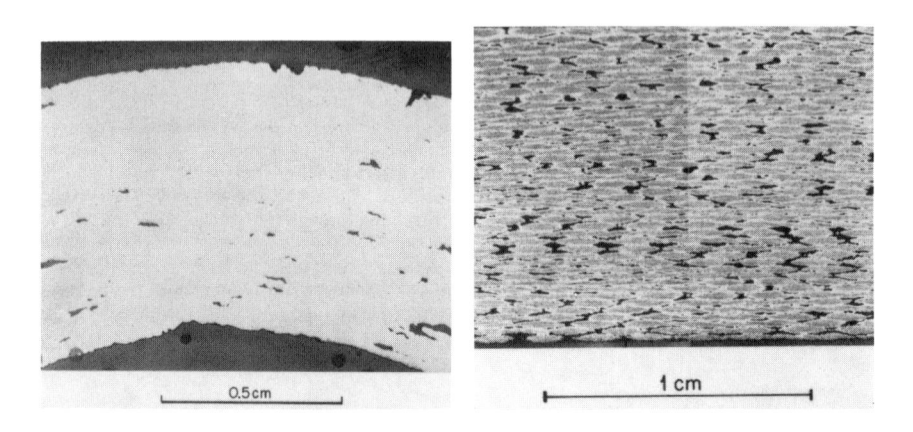

Fig. 4 Microstructure of a) filament wound tubular composite and b) a flat plate type composite (cloth layup preform). Note the interconnected porosity from layer to layer in the flat plate composite.

Braided preforms were much more difficult to infiltrate than filament wound preforms. Braided preforms with a fiber content of only 15 vol% were investigated initially. The low fiber content and large gaps between fiber bundles created voids that extended through the thickness of the preform. Because of these voids, no backpressure was created by the preform to disperse reactants along the length of the tube. Therefore, reactant gases entered the center of the preform and flowed directly through the walls so that very little SiC was deposited. Preforms that had been braided to a fiber loading of 33 vol % were also infiltrated. The porosity created by this braid was more uniformly distributed, however the permeability was relatively high and reactants moved through the preform so that little deposition occurred near the ends of the preform. Braided preforms with still higher densities will be required to obtain proper infiltration. Preforms fabricated for use as hot-gas filters containing braided continuous fibers and fiber molded chopped fibers were also investigated. Since chopped fibers are much less permeable to gases than cloth, the hybrid preform created a significant backpressure that dispersed the reactants along the length of the tubular preform. The uniform movement of reactants through the preform resulted in SiC being deposited in the hot outer region of the tube. After sufficient densification occurred, the infiltration proceeded to the center and inner diameter of the preform.

MECHANICAL PROPERTIES

The mechanical properties of SiC matrix composites reinforced with Nicalon fibers have been characterized when fabricated as simple disks [19]. Forced CVI composites exhibit an average flexure strength of ≈380 MPa and an apparent fracture toughness of ≈23 MPa•m$^{1/2}$. To date, the mechanical property characterization of tubular shaped composites has been very limited. The mechanical properties of one filament wound 3.8 cm diameter (6 mm wall thickness) tubular composite have been measured. The 16 cm long tubular composite was cut into 6mm thick C-ring specimens. Room temperature compression testing of three of these rings ranged from 777 MPa to 842 Mpa. Testing at 1000°C resulted in slightly lower strengths that ranged from 574 to 603 MPa. Composites reinforced with braided or filament wound preforms are currently being fabricated for tension/torsion testing at room and elevated temperature.

CONCLUSIONS

Fiber-reinforced SiC matrix composites of tubular geometry are desired for application in advanced gas turbines because of their oxidation resistance, damage tolerance, and thermal shock resistance. A process has been developed that can efficiently fabricate thick-walled tubes in short times without damaging the reinforcing fibers. Tubular preforms were obtained with different fiber architectures to satisfy different strength requirements. Densification of filament wound tubes was accomplished without difficulty because the tubular preforms were sufficiently permeable to disperse the reactants along the length of the preform. Infiltration then proceeded as desired from the hot face to the cool face of the preform. Densification of braided preforms was considerably more difficult. Braided preforms were less dense and had porosity that extended through the walls. Therefore, the preforms were so permeable that reactants failed to disperse properly along the length of the preform. After denser, much less permeable braided preforms were obtained, densification of tubular preforms was achieved. C-ring strengths of filament wound composites were found to be quite high (≈800 MPa) at room temperature but decreased to ≈600 MPa at 1000°C.

REFERENCES

1. G. Simon and A. R. Bunsell, "Creep Behavior and Structural Characterization at High Temperatures of Nicalon SiC Fibers," J. Mater. Sci. 19, 3658 (1984).

2. P. J. Lamicq, G. A. Bernhart, M. M. Dauchier, and J. G. Mace, "SiC/SiC Composite Ceramics," Am. Ceram. Soc. Bull. 65 (no. 2), 336 (1986).

3. F. I. Hurwitz, P. J. Heimann, J. Z. Gyekenyesi, J. Masnovi, and X. Ya Bu, "Polymeric Routes to Silicon Carbide and Silicon Oxycarbide CMC," Cer. Eng. Sci. Proc., 12(7-8), 1292 (1991).

4. R. P. Boisvert, and R. J. Diefendorf, "Polymeric Precursor SiC Matrix Composites," Cer. Eng. Sci. Proc., 9(7-8), 873 (1988).

5. R. Lundberg, and P. Goursat, "Silicon Carbo-Nitride Ceramic Matrix Composites by Polymer Pyrolysis," Developments in the Science and Technology of Composite Materials, ed. A. R. Bunsell, P. Lamicq, and A. Massiah, Elsevier Applied Science, London, p. 93, (1989).

6. A. Lightfoot, L. Ewart, J. Haggerty, Z. Q. Cai, J. Ritter, and S. Nair, "Processing and Properties of SiC Whisker- and Particulate-Reinforced Reaction Bonded Si$_3$N$_4$, "Cer. Eng. Sci. Proc., 12(7-8), 1265 (1991).

7. T. L. Starr, J. N. Harris, G. B. Freeman, and D. L. Mohr, "Reaction Sintered Silicon Nitride Composites with Continuous Fiber Reinforcement," Proceedings of the Fifth Annual

Conference on Fossil Energy Materials, ORNL Report CONF-9105184, 47 (1991).

8. F. Christin, L. Heraud, J. J. Choury, R. Naslain, and P. Hagemuller, "In-Depth Chemical Vapor Deposition of SiC within Porous Carbon-Carbon Materials," Proc. 3rd Eur. Conf. on Chem. Vapor Dep., 154 (1980).

9. J. W. Warren, "Fiber- and Grain-Reinforced CVI Silicon Carbide Matrix Composites," Cer. Eng. Sci. Proc., 6(7-8), 64 (1985).

10. R. E. Fisher, C. V. Burkland, and W. E. Bustamante, "Ceramic Composites Based on Chemical Vapor Infiltration," Proc. of the Metal and Ceramic Matrix Composite Processing Conf., Battelle Columbus Labs., Columbus, Ohio, Nov. 14, 1984.

11. M. H. Headinger, and M. J. Purdy, "Design of a Production-Scale Process for Coating Composite Structures," Surface and Coat. Tech., 33, 433 (1987).

12. D. P. Stinton, A. J. Caputo, and R. A. Lowden, "Synthesis of Fiber-Reinforced SiC Composites by Chemical Vapor Infiltration," Am. Ceram. Soc. Bull., 65[2], 347-50 (1986).

13. A. J. Caputo, and W. J. Lackey, "Fabrication of Fiber-Reinforced Ceramic Composites by Chemical Vapor Infiltration," Cer. Eng. Sci. Proc., 5(7-8), 654-67 (1984).

14. T.M. Besmann, B. W. Sheldon, R. A. Lowden, and D. P. Stinton, "Vapor-Phase Frication and Properties of Continuous-Filament Ceramic Composites," Science, Vol. 253, 1104-9 (1991).

15. D. P. Stinton, T. M. Besmann, and R. A. Lowden, "Advanced Ceramics by Chemical Vapor Deposition Techniques," Am. Ceram. Soc. Bull., 67(2), 350-55 (1988).

16. S. Yajima, et al., "Synthesis of Continuous SiC Fibers with High Tensile Strength, "J. Am. Ceram. Soc., 59(7-8), 324-27 (1976).

17. S. Yajima, et al., "Anomalous Characteristics of the Microcrystalline State of SiC Fibers," Nature 27(21), 706-7 (1979).

18. R. A. Lowden, D. P. Stinton, and T. M. Besmann, "Characterization of Fiber-Matrix Interfaces in Ceramic Composites," Conference Proceedings, Whisker- and Fiber-Toughened Ceramics, ASM International, 253-64, 1988.

19. D. P. Stinton, R. A. Lowden, and R. H. Krabill, "Mechanical Property Characterization of Fiber-Reinforced SiC Matrix Composites," ORNL Report No. ORNL/TM-11524, April 1990.

EFFECT OF BN INTERFACIAL COATING ON THE STRENGTH
OF A SILICON CARBIDE/SILICON NITRIDE COMPOSITE

KIRK P. NORTON AND HOLGER H. STRECKERT
General Atomics, P.O. Box 85608, San Diego, CA 92186-9784

ABSTRACT

Boron nitride has been identified as a good fiber/matrix interface coating for SiC fibers in a Si_3N_4 matrix. Boron nitride thin films prepared from BCl_3 and NH_3 at 1000°C and from B_2H_6 and NH_3 at 550°C were deposited on SiC (Nicalon®) cloth in a low pressure chemical vapor deposition reactor. The coating thickness was varied from 0.1 μm up to 4 μm. Six layers of cloth for each coating thickness were incorporated into a chemical vapor infiltrated matrix of amorphous Si_3N_4.

Flexural strength data and composite toughness were obtained from three-point bend tests. Scanning electron microscopy and metallographic analysis of fracture surface morphology were used to determine the failure mode.

INTRODUCTION

The development of ceramic matrix composite fabrication technology has progressed to the point where fiber-matrix interface coatings will be required to achieve satisfactory fracture resistance [1,2]. Interface coatings applied are known to principally serve two functions. The interface coating may provide a barrier against diffusion between the fiber and the matrix [3,4]. Secondly, to transfer the load from the matrix to the fiber, thereby enhancing the strength of the material and, if the interface bonding is not too strong, enhance the fracture toughness of the composite through fiber pullout [1-5].

Boron nitride has been identified as a good barrier coating for silicon carbide fibers. Indentation studies, where the force to push the fiber through the matrix is measured, show a low frictional stress of ~6.6 MPa in a Silicon Carbide matrix [6] and 16 MPa in a Si_3N_4 matrix [5]. Low matrix shear stress results in greater fiber pullout during failure.

EXPERIMENTAL PROCEDURE

The coatings discussed here were made on Nicalon® CG SiC cloth. The cloth was cut into strips, 5 cm by 20 cm and placed inside a cylindrical graphite susceptor. The susceptor was 5 cm ID by 25 cm long, supported inside a quartz tube. The graphite was inductively heated. Coating gases were fed through the top of the quartz tube and evacuated through the bottom.

Amorphous boron nitride coatings were prepared by two different CVD routes[7]. A high temperature BN, was prepared from boron trichloride and ammonia in hydrogen.

The following first order reaction is assumed:

$$BCl_3 + NH_3 \Rightarrow BN + 3HCl.$$

This reaction for deposition of BN on Nicalon® cloth is carried out at 1000°C and a pressure of 3 to 4 torr. The reaction proceeds rapidly at these conditions. A deposition rate on the order of 0.15 μm/min is not unusual. The second, a dual BN coating was prepared in two steps. The first step consists of reacting diborane and ammonia at 550°C in hydrogen according to the following reaction:

$$B_2H_6 + 2NH_3 \Rightarrow 2BN + 6H_2.$$

This low temperature BN was found to protect the fibers from chemical attack such as Cl- from the BCl_3 reaction. The second step in the dual coating is the deposition of high temperature BN as previously described.

Composite Fabrication

Composite coupons were made from the coated Nicalon® fabric and chemically vapor infiltrated with silicon nitride. The coupons consisteded of six layers of cloth. The BN coated strips of cloth were made into preforms by binding the layers together with polystyrene. Preforms made in this manner have the advantage that fiber architecture and fiber volume fraction can be more accurately controlled. These preforms were then cut into 4 cm x 4 cm squares. The thickness of the coupons was influenced by the thickness of the coating on the fibers and ranged from 2.8 mm to 3.8 mm. In addition to the coupons made with coated Nicalon®, one sample was made from Nicalon® which received no coating.

The composite infiltration was performed using the 33 cm ID coater, designed at GA for that purpose. The coater can be used for chemical vapor infiltration in several configurations such as isothermal, thermal gradient, or thermal gradient with forced flow. For making the composites described here, a thermal gradient with forced flow was used. The samples were constrained in a graphite fixture and radiantly heated from above and actively cooled from below. The coating gases were forced through channels in the graphite fixture and the layers of cloth. As the reactant gases pass through the thermal gradient, reactions take place which result in the deposit being formed at the hot surface. As the process continues, the deposition front moves downward through the cloth raising the local temperature. The deposition of Si_3N_4 is achieved by reacting $SiCl_4$ with NH_3 at 750 to 800°C and a pressure of 100 torr.

Composite Testing

The composite coupons were carefully removed from the graphite fixture. To avoid weakening the coupons by prying them away from the fixtures, any graphite remaining on the coupons was oxidatively removed by heat treating overnight at 650°C in an air furnace. Each coupon was sectioned into six (6) strips using a diamond blade sectioning saw. The average dimensions for the test beams were; 6mm X 3mm and 40mm long.

RESULTS

A short beam 3-point bend test was performed on three samples from each coupon. Figure 1 summarizes the average flexural strengths for the samples tested at room temperature. The composites made from the high temperature BN coated cloth showed the highest strength with an average strength of 212 MPa. The composites made from the cloth with a dual coating of BN exhibited a moderate strength with an average of 132 MPa.

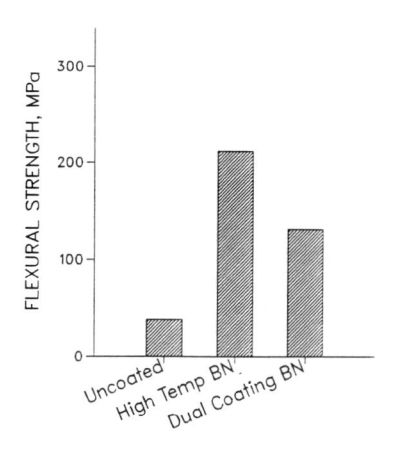

Figure 1. Results of 3-point bend tests for composite coupons with and without BN coating on fibers.

Scanning electron microscopy was used to examine the fracture surfaces of the broken samples. The samples in which the fibers were coated with BN showed considerable fiber pullout. The fracture surface of the sample made from uncoated fibers was relatively smooth and indicates an undesirable brittle failure. The stress strain curves for these samples support the same result.

Figure 2 displays an optical micrograph of two of the broken samples, magnified 8x. In the figure, the sample made from the dual BN coated Nicalon® is on the top (flexural strength, 132 MPa), and the sample made from the high temperature BN coated Nicalon® is below (flexural strength, 212 MPa). While both show the fiber pullout characteristic of tough composites, the degree of pullout in the bottom sample helps to explain the difference in strength between them.

Fiber Coating Optimization

The high temperature BN coated Nicalon® composite exhibited the greatest flexural strength in this study. Not immediately clear, whether the difference in strengths seen between the high temperature BN and the dual BN coated Nicalon® composites is due to the difference in the chemistry of the BN or some other parameter, such as coating thickness. The variation in coating thickness is readily apparent since the high temperature BN was on the order 3 to 4 µm thick while the dual BN was approximately 0.75 µm thick.

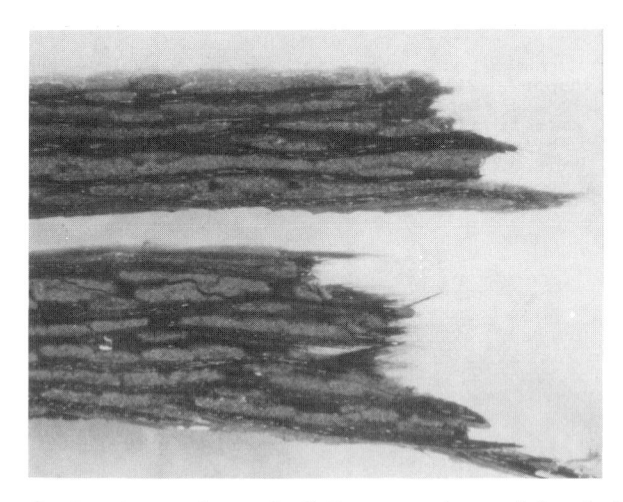

Figure 2. Fracture edges of Si_3N_4 composites with; dual coating of BN in the top sample, and high temperature BN in the bottom sample. (magnification 8X)

However, examination of the fracture surfaces for the two interface coatings shows that debonding occurred at the fiber/coating interface. The scanning electron micrograph in Figure 3 shows the debonding and crack deflection at the fiber/coating interface. The granular-like appearance of the surface in this figure is an artifact of the polishing procedure.

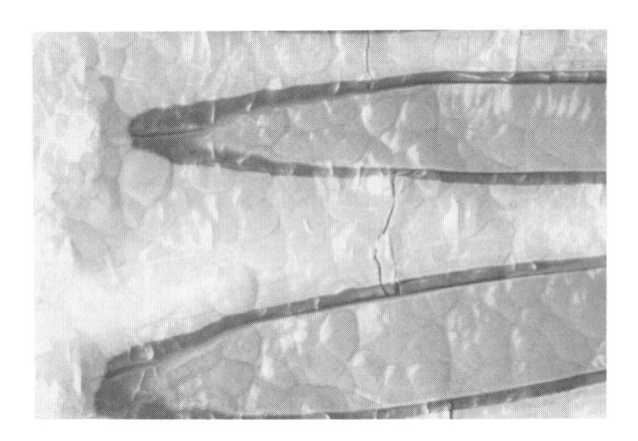

Figure 3. Scanning electron micrograph photo showing crack deflection at the fiber/coating interface. (magnification 1500X)

If equal or perhaps greater strength in the composite can be obtained by a thinner coating, this would enhance not just the mechanical properties, but would speed the processing as well. In order to address this issue, additional composites were made with different fiber coating thicknesses. Four coupons were made with high temperature BN coatings, targeting thicknesses of 0.1, 0.5, 1, and 4 µm thick on Nicalon® HVR. The actual thickness achieved were 0.12, 0.55, 1.07, and 5.6 µm. One coupon of Nicalon® HVR with 1.0 µm thick coating of low temperature BN was included. The infiltration was repeated as described previously with Si_3N_4. The coupons were sectioned and tested using a 3-point bend test. These results are shown in Figure 4. This figure illustrates that for the high temperature BN the composite strength is independent of fiber coating thickness for thicknesses between 0.55 and 5.6 µm. However, there is a significant reduction in strength when the coating thickness is reduced to 0.12 µm. Also, when comparing the 0.16 µm thick low temperature BN with the 0.12 µm thick high temperature BN, there is a distinct contrast between the effectiveness of the two coatings, with the high temperature BN composite being stronger, though thinner.

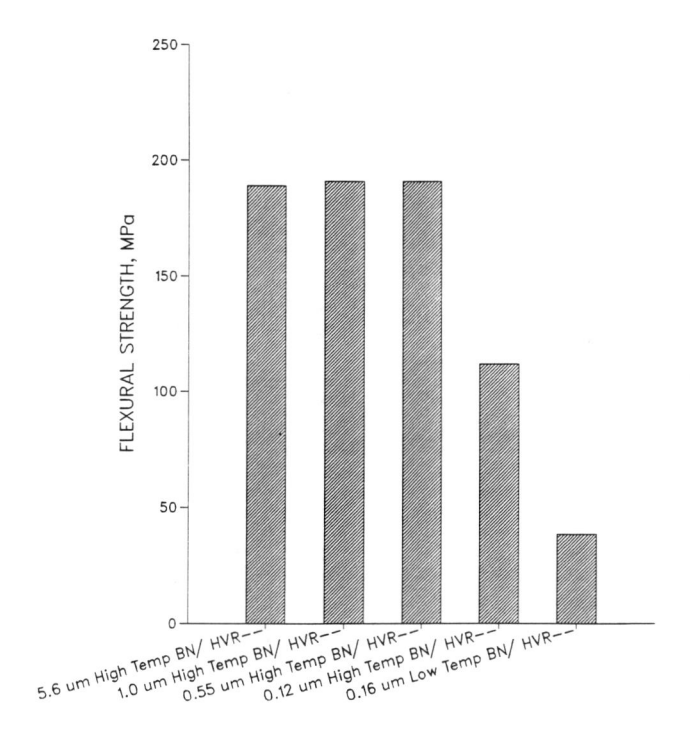

Figure 4. Results of 3-point bend tests for Si_3N_4 matrix composites with varying thickness of BN interface coatings.

CONCLUSION

Boron nitride has shown to be a good interface coating for SiC fibers in a amorphous Si_3N_4 matrix. A BN coating prepared from BCl_3 at high temperature (1000°C) results in a much stronger and tougher composite than does a coating prepared at low temperature (550°C) from B_2H_6. The optimum fiber coating thickness for the high temperature BN is greater than 0.12 µm but no more than 0.55 µm for SiC fibers in a Si_3N_4 matrix composite.

ACKNOWLEDGMENTS

The authors express their thanks to D. R. Wall for scanning electron microscope work and M. Jackson and R. Harrington for composite preparation and testing.

This research was supported by the Advanced Research Projects Agency of the Department of Defense and was monitored by the Air Force Office of Scientific Research under Contract No. F4920-89-C-0078. The United States Government is authorized to reproduce and distribute reprints for governmental purposes notwithstanding any copyright notation hereon.

REFERENCES

1. K. M. Prewo," FIBER-REINFORCED CERAMICS: NEW OPPORTUNITIES FOR COMPOSITE MATERIALS," **68** [2] 395-400 (1989)

2. R. J. Kerans, R. S. Hay, N. J. Pagano, and T. A. Parthasarathy," THE ROLE OF THE FIBER-MATRIX INTERFACE IN CERAMIC COMPOSITES," *Ceramic Bulletin,* **68** [2] 429-42 (1989)

3. R. N. Singh and M. K. Brun," EFFECT OF BORON NITRIDE COATING ON FIBER-MATRIX INTERACTIONS," *Advanced Ceramic Materials,* **3** [3] 235-37 (1988)

4. D. Lewis III," STRENGTH AND TOUGHNESS OF FIBER-REINFORCED CERAMICS AND RELATED INTERFACE BEHAVIOR," Proceedings of " Whisker-and Fiber-Toughened Ceramics," Ed. by R. A. Bradley, etal., Oak Ridge, Tenn., 265-273 (June 1988)

5. M. K. Brun and R. N. Singh," EFFECT OF THERMAL EXPANSION MISMATCH AND FIBER COATING ON THE FIBER/MATRIX INTERFACIAL SHEAR STRESS IN CERAMIC MATRIX COMPOSITES," *Advanced Ceramic Materials,* **3** [5] 506-509 (1988)

6. R. A. Lowden and K. L. More," THE EFFECT OF FIBER COATINGS ON INTERFACIAL SHEAR STRENGTH AND THE MECHANICAL BEHAVIOR OF CERAMIC COMPOSITES," MRS Symposium Proceedings, Vol 170 (1990)

7. T. D. Gulden, D. A. Hazlebeck, K. P. Norton, and H. H. Streckert," CERAMIC FIBER COATING BY GAS-PHASE AND LIQUID-PHASE PROCESSES," *Ceram. Eng. Sci. Proc.* **11** [9-10] 1539-53 (1990)

MICROWAVE ASSISTED CHEMICAL VAPOR INFILTRATION

D.J. Devlin, R.P. Currier, R.S. Barbero, B.F. Espinoza, and N. Elliott
Materials Science and Technology Division
Los Alamos National Laboratory
Los Alamos, NM 87545

ABSTRACT

A microwave assisted process for production of continuous fiber reinforced ceramic matrix composites is described. A simple apparatus combining a chemical vapor infiltration reactor with a conventional 700 W multimode oven is described. Microwave induced inverted thermal gradients are exploited with the ultimate goal of reducing processing times on complex shapes. Thermal gradients in stacks of SiC (Nicalon) cloths have been measured using optical thermometry. Initial results on the "inside out" deposition of SiC via decomposition of methyltrichlorosilane in hydrogen are presented. Several key processing issues are identified and discussed.

INTRODUCTION

Many materials used in high temperature service, including monolithic ceramics, tend to be brittle and susceptible to catastrophic failure through crack propagation. This has lead to concentrated interest in the fabrication of fiber reinforced ceramic matrix composites (CMCs). CMCs consist of a fibrous backbone, or substrate, whose void spaces are filled with a "matrix" material. The chemical composition of the matrix may or may not be the same as that of the fibers. While a random pile of individual fibers may be used as reinforcement, the tougher CMCs typically consist of continuous fiber bundles, or yarn, woven together to form either two or three dimensional "textile" cloths.

Processing techniques for CMCs differ primarily in the way matrix materials are infiltrated into the porous substrate. Infiltration may involve molten liquids, sol-gels, polymeric materials, powders, or vapors. Most of these techniques require final densification steps, e.g. sintering, involving matrix shrinkage. Large residual stresses can develop when a matrix shrinks around non-shrinking fibers. Chemical vapor infiltration (CVI) is an attractive alternative for matrix deposition since it avoids stressing the fibrous backbone during processing. Also, relatively low temperatures are used in CVI which limits adverse chemical attack on the fibers. Conventional CVI processes may be either isothermal or involve intentionally imposed thermal gradients. Deposition may be reaction or diffusion limited, or may rely on forced and/or pulsed reactant flows. The various configurations have recently been reviewed by Besmann et al [1]. To varying degrees, conventional CVI processes are typically subject to some or all of the following drawbacks: preferential deposition in the substrate's outer regions leading to pore blockage; long processing times with intermittent machining operations; non-uniform composite density; high residual porosity; and limitations on substrate geometry. However, despite inherent limitations commercial CVI operations are now in place.

The idea of using electromagnetic radiation, in particular microwaves, to heat substrates during CVI has recently been explored numerically [2]. The potential advantage in using microwave heating is the ability to heat the substrate internally, giving rise to "inverted" thermal gradients. With the internal region of the substrate hot, cool reactant gases could penetrate inward prior to the onset of reaction. Consequently, deposition could occur from the inside-out. A successful microwave CVI process could offer several advantages over conventional technologies. First, constraints on substrate geometry would be removed. Second, more spatially uniform, high

This work is performed under the auspices of the U.S. Department of Energy Office of Industrial Technology, Advanced Industrial Concepts (AIC) Materials Program.

density composites should be attainable. Third, relatively short processing times should be possible. Fourth, machining operations to reopen closed pores should not be necessary since densification would occur from the inside-out. Success will of course require proper management of heating and cooling rates. The electric field within the preform governs the local heating rate. Interaction of an electromagnetic field with a porous preform is complicated. In general, the local heating rate is proportional to the square of the electric field strength and to the effective dielectric constant and loss factors for the growing composite. On the other hand, substrate heat losses occur primarily by way of radiation and convection. One expects these losses to depend on geometry, temperature, flow rates, and to a lesser extent, on weave architecture and lay-up pattern. The present work is an experimental investigation into the feasibility of microwave assisted CVI. Emphasis is on quantifying induced thermal gradients, combining CVI and microwave heating to demonstrate inside-out densification, and examination of materials and processing issues.

EXPERIMENTAL RESULTS

In order to quantify the microwave induced thermal gradients, optical thermometry experiments were conducted. Measurements were made in cylindrical SiC cloth lay-ups subject to 700 W of 2.45 GHz microwave energy. Measurements were conducted in a commercial General Electric multimode oven. Temperatures were measured along both the cylinder axis and mid-plane radius using quartz fiber optic cables woven directly into the substrate. The optical cables were connected via photodiodes to amplifier circuits from which voltages were read. Each optical thermometer was first calibrated using stacks of SiC cloth placed in a conventional high temperature furnace. The thermal profiling data shown in Figure 1 indicates that steep inverted thermal gradients can be established in SiC cloth lay-ups.

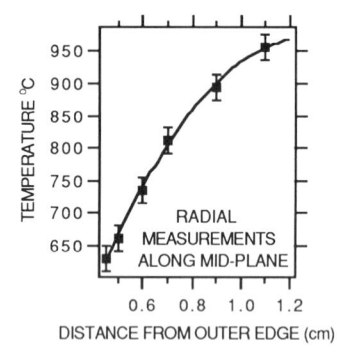

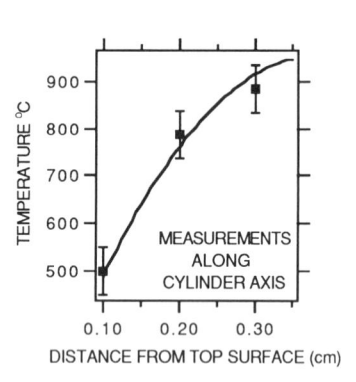

FIGURE 1. Inverted thermal gradients in SiC cloth lay-ups.

Internal temperatures in excess of 1000°C were observed in stagnant air along with gradients on the order of several hundred °C/cm. Similar gradients are seen in atmospheres other than air, e.g. H_2, CH_4, and C_3H_6. The error bars shown in the figures include contributions from positioning cables within the preform and variations in the voltage readings, as deduced from the calibration.

Figure 2 is a schematic of the gas delivery system and microwave CVI cavity. The latter consisted of a modified 2.45 GHz, 700 W General Electric commercial multimode oven. A circular hole was laser cut through the oven floor through which a Pyrex bell jar reactor was inserted. The interior floor of the reaction vessel was a metallic plate with a circular hole in the center used for reactant feed and product gas removal. A metallic sleeve was utilized to reseal the cavity.

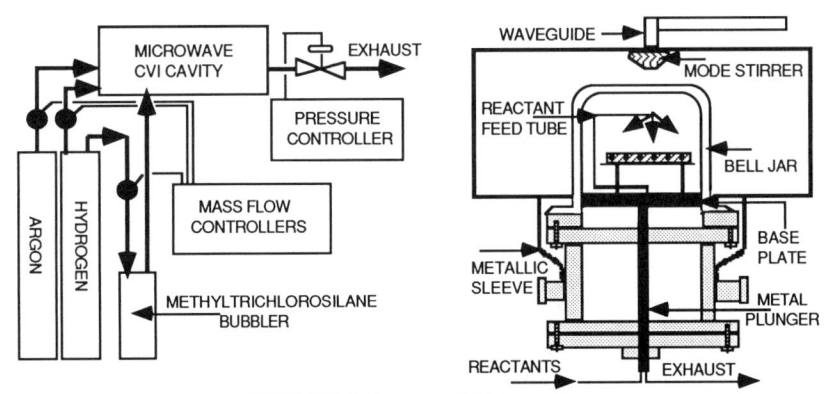

FIGURE 2. Microwave CVI Apparatus.

Provisions were made for sampling of reactant and exhaust gases for analysis by FTIR and mass spectroscopy. Substrates consisted of a pile of ten circular Nicalon (SiC) cloth sections with a 7 cm diameter. A quartz ring 2.5 cm high was used to positioned these off the reactor floor. Initial experiments involved deposition of SiC from methyltrichlorosilane (MTS) and hydrogen through the reaction $CH_3SiCl_3 + H_2 = SiC + 3HCl$. Two sets of reaction conditions were considered:

Case	Total Pressure	Partial Pressure MTS	Hydrogen Flow	Rate (mg/min)
I	300 Torr	1 Torr	4000 sccm	≈ 0.5
II	600 Torr	20 Torr	500 sccm	≈ 7.0

These infiltration conditions were chosen to represent both low and high SiC deposition rates. Figure 3 shows photographs of cloth layers from the stack center under Case I conditions.

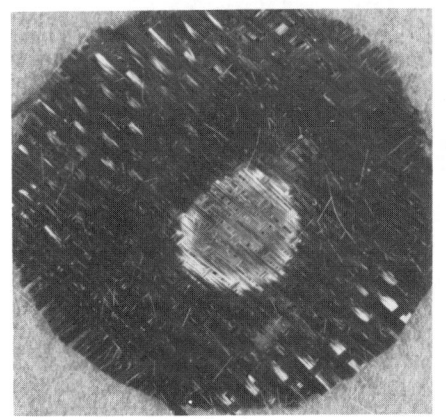

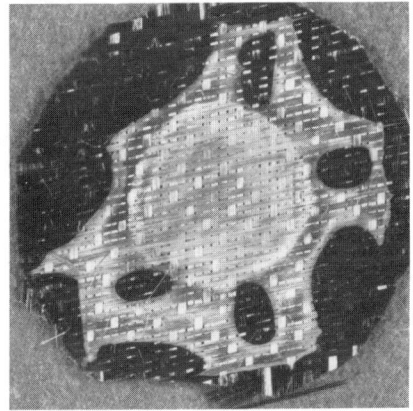

(a)　　　　　　　　　　　　(b)
FIGURE 3. Partially densified cloth. (a) Inside-out densification. (b) Nonuniform heating pattern.

With radiation and convective heat losses at the surface, the highest temperatures were at the preform center. CVI against the thermal gradient resulted in preferential deposition in the hot region. The coated region, which appears white, is clearly seen in Figure 3(a). The size of this infiltrated region decreases in diameter moving along the stack axis toward either the top or bottom surface in general agreement with the thermal profile data shown in Figure 1. However, with the longer infiltration times the substrate center appeared to cool and patterns such as that seen in Figure 3(b) typically evolved. Several factors may contribute to this phenomenon. First, regions may be selectively cooled by reactant gas flow patterns. If this is the case, then a systematic study using various cloth lay-up patterns, weave architectures, and gas delivery configurations is called for. However, initial experiments do not suggest a strong sensitivity to lay-up pattern or gas flow rate. Another possible factor contributing to inhomogeneous heating patterns is that the microwave cavity employed in these experiments is less-than-perfect. It is widely known that commercial multimode ovens of these dimensions do not heat uniformly. A related issue is that as densification occurs it may be necessary to continuously increase the microwave power input above the currently available 700 W to maintain the desired heating rate. A more robust microwave CVI reactor should eliminate these difficulties. Finally, the possibility that the deposit affects microwave susceptibility must be considered. If filaments are coated with a material which does not couple well with the electric field to produce heating, then the heating rate may diminish once the coating thickness exceeds the skin penetration depth:

$$\text{Skin depth} = 1 / (\pi f \mu \sigma)^{1/2} \tag{1}$$

where f the frequency, μ is the permittivity, and σ the dc conductivity. If this occurs, then an initially hot region may be partially densified with subsequent cooling. The microwaves may then more effectively couple with a different region of the substrate. The impact of microwave-material interactions during SiC CVI is discussed in more detail in the following section. A combination of the above mentioned factors is most likely responsible for the phenomenon seen in Figure 3(b).

Figure 4 shows an end-on view of a densified fiber bundle taken from the preform center obtained using Case II CVI conditions. Also shown is the cloth total weight gain versus time data.

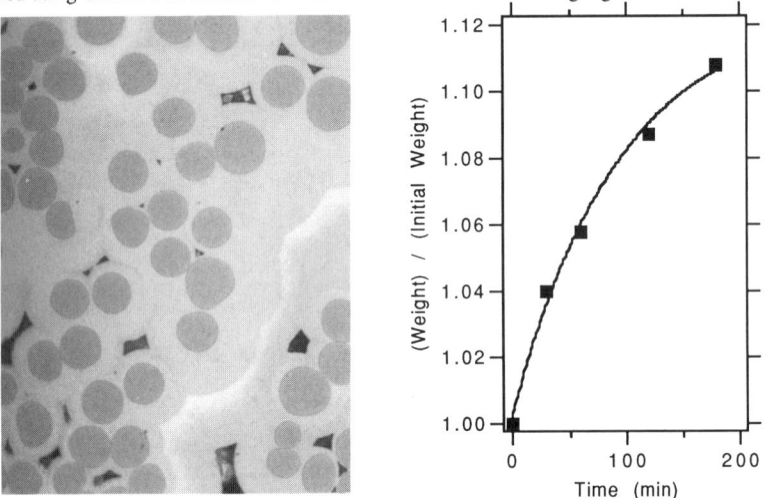

FIGURE 4. (a) Infiltrated bundle. (b) Weight gain data.

Under Case II infiltration conditions, heating patterns remained circular on a given cloth layer. The radius of the hot spot increased with time and there was no apparent cooling of the interior regions as deposition occurred. Fiber bundles in the interior of preform were reasonably dense, as seen in Figure 4. Also, the weight gain data for the stacked cloth layers showed a response typical for CVI. However, x-ray diffraction patterns for these samples indicated the deposit was rich in Si. Within the densified region some zones were clearly denser than others. This again may be due to the inhomogeneous electric fields produced with the present cavity. It should be noted that the change in composition from beta-SiC to SiC/Si resulting from changes in reactant composition and pressure is in general agreement with recent findings on the MTS-H_2-SiC system [3].

DISCUSSION

Case II conditions resulted in regions of high density consisting of a SiC/Si mixture while under Case I conditions substrates were effectively extinguished after one hour of SiC deposition. Pure beta-SiC coatings of more than a few microns could not be achieved with either set of CVI conditions. These observations suggest the coating may adversely affect absorption of microwave energy. For example, if the conductivity of the semiconducting beta-SiC coating is greater than that of the Nicalon cloth, then an increase in reflected energy from the coated surface is anticipated. Reported room temperature conductivities for beta-SiC range from 10^{-2} to 10^2 (ohms-cm)$^{-1}$ [4], with a band gap of approximately 2.0 eV. Given these values, one would expect beta SiC to be a good conductor at elevated temperatures. Manufacturers report the conductivity of Nicalon cloth as 10^{-3} (ohms-cm)$^{-1}$ and do not expect it to vary significantly with temperature. Considering the difference in properties between the fiber and and the coating, a drop in temperature is reasonable. It is interesting to note that with Case I conditions, the initial heating occurs in the preform center for several minutes, a time sufficient to deposit on the order of a micron of SiC. With subsequent heating, the hot zone spreads radially and visually dulls. This condition persists for an additional 15 minutes and again the hot region dulls and heating is observed only at the outer edges of the preform. It would appear that each time a critical coating thickness is reached, a new set of modes is established and therefore a different heating pattern occurs. Eventually no glow is observed and deposition ceases. For Case II however, deposition of both SiC and Si occurred with no apparent decrease in temperature. For these conditions, namely, a temperature of approximately 1000°C, high pressure, and low residence time, a reduction in temperature could result in the preferential deposition of Si over SiC. The room temperature conductivity for silicon is on the order of 10^{-5} (ohms-cm)$^{-1}$, while at 1000°C it is approximately 100 (ohms cm)$^{-1}$. Presumably, the conductivity is low enough to maintain heating at a slightly reduced temperature. Figure 5 shows an interfacial

FIGURE 5. End-on view of fiber bundle for Case II. Note region surrounding fibers.

layer on the order of a micron thick between the Nicalon fiber and the Si rich matrix. This is consistent with the idea that beta-SiC is initially deposited resulting in a subsequent decrease in temperature favoring the deposition of Si.

Initial results suggest the infiltration of beta-SiC with our present limitations on power will be difficult. It also points out the need for careful control of the matrix properties during CVI. For semiconductors, values of the conductivity in the extrinsic range can vary dramatically, covering many orders of magnitude for small variations in impurity concentration. Fortunately the CVD technique is well suited for control of these factors. The possibilities of co-deposition or doping of suitable phases to alter the conductivity and dielectric properties of the matrix material needs to be explored. This of course is only possible to the extent that it does not degrade the required mechanical properties of the composite. The effect of interfacial layers, such as carbon or boron nitride which are used to enhance the composite toughness, on subsequent microwave heating of the fibers also needs to be explored.

For future experiments, consideration should be given to both single and multimode microwave cavities. Multimode cavities are inexpensive and easy to operate. Given a uniform electric field and control of substrate cooling, multi-mode cavities should be capable of promoting inverted gradients in complex shaped substrates. In contrast, a single mode cavity involves regions of high field concentration and localized hot-spots. The single mode cavity is usually more energy efficient and can localize heating. However, using a single mode cavity during densification of complex shaped substrates by CVI will likely require continuous and non-trivial tuning schedules and power modulation schemes. Of primary concern should be the ability to manipulate and control the thermal gradients, which in turn govern the densification process. For example, if the localized heating shown in Figure 3(a) is such that the hot region is essentially isothermal with a steep thermal gradient at the outer edge then problems typically associated with convention CVI, such as pore blockage in the outer region, are possible under certain conditions. Flexibility can be maximized through use of in-situ diagnostics, variable input power, and cavity tuning.

Microwave radiation, in combination with sufficiently reduced pressures, will result in a glow discharge due to electrical breakdown in the vapor. Low operating pressures may be either intentionally set or be encountered during specific phases of a cyclical pulsed-pressure CVI process [5]. The presence of a discharge could have beneficial effects on CVI. For example, in systems where vapor phase reactions are involved, chemical kinetics may be substantially enhanced and new reaction pathways may become available. However, since it is not clear that a glow discharge can be maintained within porous preforms, plasma assisted CVI may involve only the infiltration of neutral species formed in the discharge. Furthermore, the power consumption involved in maintaining the discharge can be appreciable and thus limit the ability to heat the substrate as desired. Possible damage to both fiber integrity and fiber-matrix interfaces from ion bombardment must also be considered. CVI in the presence of glow discharges will be explored in a future work.

In summary, microwave-induced inverted thermal gradients and the possibility of inside-out densification by CVI have been demonstrated. Several issues central to further process development have been identified. This includes control over microwave power input, flow rates, and cavity tuning. In addition, it appears that control over the matrix composition may prove necessary in order to maintain a suitable level of microwave heating, or conversely, to avoid a thermal runaway.

REFERENCES

1. Besmann, T.M., Sheldon, B.W., Lowden, R.A. and Stinton, D.P. Science 253: 1104 (1991).
2. Gupta, D. and Evans, J.W., J. Mater. Res. 6(4): 810 (1991).
3. Motojima, S. and Hasegawa, M., Thin Solid Films 186: L39 (1990).
4. Pohl, R.G. in Silicon Carbide: A High Temperature Semiconductor. Proc. Conf. on Silicon Carbide. O'Conner, J.R. and Smiltens, J., Eds., p. 318, Pergamon Press. Boston, (1960).
5. Sugiyama, K. and Nakamura, T., J. Mater. Sci. Letters 6: 331 (1987).

CVD OF SILICON CARBIDE ON STRUCTURAL FIBERS: MICROSTRUCTURE AND COMPOSITION

Lisa C. Veitch, Francis M. Terepka, Suleyman A. Gokoglu, NASA Lewis Research Center, Cleveland, OH

ABSTRACT

Structural fibers are currently being considered as reinforcements for intermetallic and ceramic materials. Some of these fibers, however, are easily degraded in a high temperature oxidative environment. Therefore, coatings are needed to protect the fibers from environmental attack.

Silicon carbide (SiC) was chemically vapor deposited (CVD) on Textron's SCS6 fibers. Fiber temperatures ranging from 1350 to 1500 °C were studied. Silane (SiH_4) and propane (C_3H_8) were used for the source gases and different concentrations of these source gases were studied. Deposition rates were determined for each group of fibers at different temperatures. Less variation in deposition rates were observed for the dilute source gas experiments than the concentrated source gas experiments. A careful analysis was performed on the stoichiometry of the CVD SiC-coating using electron microprobe. Microstructures for the different conditions were compared. At 1350°C, the microstructures were similar; however, at higher temperatures, the microstructure for the more concentrated source gas group were porous and columnar in comparison to the cross sections taken from the same area for the dilute source gas group.

INTRODUCTION

Ceramic fibers are of interest as reinforcing materials for advanced ceramic and intermetallic matrix composites for aerospace applications. For many of these commercially existing fibers, additional coatings will be needed to protect the fibers from oxidation. These coatings will also serve to enhance the fiber/matrix interfacial properties.

Silicon carbide coating has been considered to protect the outer carbon-rich coatings of Textron Specialty Materials' SCS6 silicon carbide fiber and can be synthesized by chemical vapor deposition (CVD) techniques. A number of different precursors, conditions and reactors have been discussed in the literature [1-3]. In this work, the deposition rates, stoichiometry and microstructure of the CVD SiC on SCS6 was examined.

EXPERIMENTAL

For the deposition experiment, a vertical batch fiber reactor was used to coat 30 cm long fibers [4]. Deposition temperatures ranged from 1350 to 1500 °C, and SiH_4 and C_3H_8 were used as the source gases with H_2 as the carrier gas. A 1:9 Si:C atom ratio of source gases was held constant for all deposition experiments. Two distinct sets of experiments were conducted, based on different source gas concentrations. Group A experiments were performed using 3 mole% SiH_4 and 9 mole% C_3H_8. Group B experiments were conducted using 0.3 mole% SiH_4 and 0.9 mole% C_3H_8. The temperature of the fibers during the CVD process was monitored at the center (15 cm) of the fiber with a two-colored optical pyrometer.

RESULTS AND DISCUSSION

The CVD SiC coating thickness on the fibers was measured using a split image microscope. SEM micrographs of polished cross sections were also used from several locations along each fiber to confirm the optical microscope readings. Figures 1 and 2 show the deposition rate variation along the fibers for different temperatures for each group. The error bars indicate the standard deviation in deposition rate measurements. A large variation in SiC thickness and, hence, deposition rate was evident for the Group A fibers. For the dilute source gas case for Group B fibers, less variation was seen. It is possible that the Si atoms are able to migrate over a longer range on the surface of the fiber for dilute flows before becoming incorporated in the surface as SiC. This would give a more uniform profile along the fiber.

It is difficult to conclude what are the dominant mechanisms for this particular SiC CVD process. The flat profile for Group B fibers may be evidence of diffusion controlled process. The large rates at the beginning of the skewed profile for the Group A fibers may be explained by a thermally controlled chemical kinetic process and, then, farther along the fiber, the process seems to have become more diffusion and depletion controlled. However, for both groups, there are a number of different decomposition reactions occurring during the CVD process and, what species are available and at what concentration is not known. The uncertainties in the fiber temperature measurement will also contribute to the observed scatter in the deposition rate data [4].

Electron microprobe analysis was used to determine the stoichiometry of the SiC coating. CVD single crystal beta-SiC was used as a standard. The fiber samples analyzed were taken from the center position of the fibers where the temperature was monitored. Figures 3 and 4 show the variation in stoichiometry for the different temperatures for each group. For Group A fibers, the SiC deposited at 1350 °C was only 3 microns thick and resulted in an unreliable chemical analysis. The measurement error for the electron microprobe analysis is ~10%. From Figures 3 and 4, all of the fiber coatings fall within the electron microprobe error. For this reason, all of the SiC coatings can be considered close to stoichiometric.

Several places along the fibers from each of the groups were also analyzed. At 1400 °C, the Si-rich SiC coating became more stoichiometric from the 17 cm position to the end of the fiber (30 cm), and by 1500 °C, the CVD SiC coating became stoichiometric as early as the 7 cm position (close to the gas inlet). A similar trend was observed for the Group B fibers also. One can explain this by by the fact that C_3H_8, which is harder to crack than silane, starts to dissociate faster at higher temperatures into more reactive species, such as CH_3, C_2H_5, C_2H_4 and C_2H_2, and becomes competitive with Si-containing species for the formation of SiC [3].

SEM micrographs (backscatter and secondary electron) revealed that the microstructures of the higher temperature CVD SiC for Group A fibers was columnar and less dense than that of the higher temperature cases for Group B fibers. Figure 5 shows the 1450 °C cases for each group (the cross sections are taken from the center position of the fibers). This type of growth can be attributed to the fact that the Group B fibers have a much slower deposition rate than the Group A fibers. Similar morphologies with other precursors have been reported for these temperatures [1,3]. Close to the gas inlet position, both fiber groups exhibited a very porous and columnar structures for temperatures between 1400 and 1500 °C.

Figure 6 compares the microstructures for the 1350 °C cases. Here, the microstructures appear similar. It seems that temperature, rather than deposition rates, is the critical variable in determining the microstructure evolving at 1350°C.

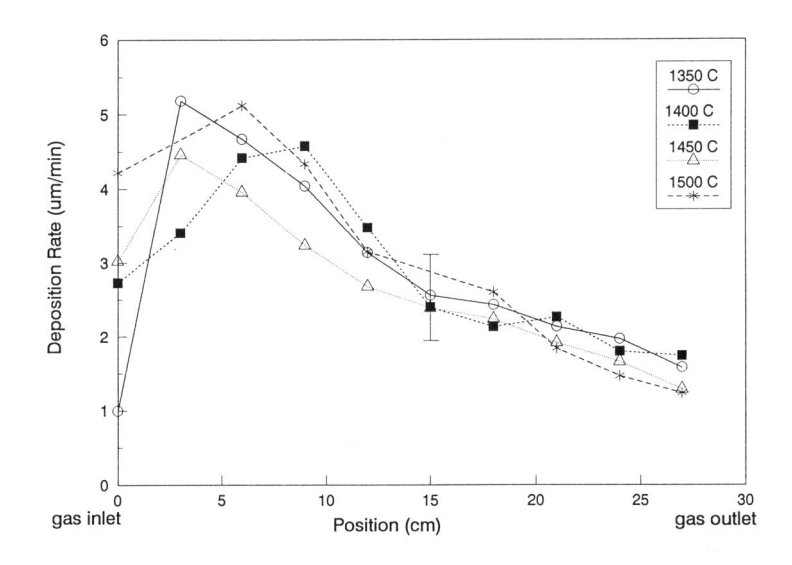

Figure 1. Deposition rate vs. position on fiber for various temperatures for Group A.

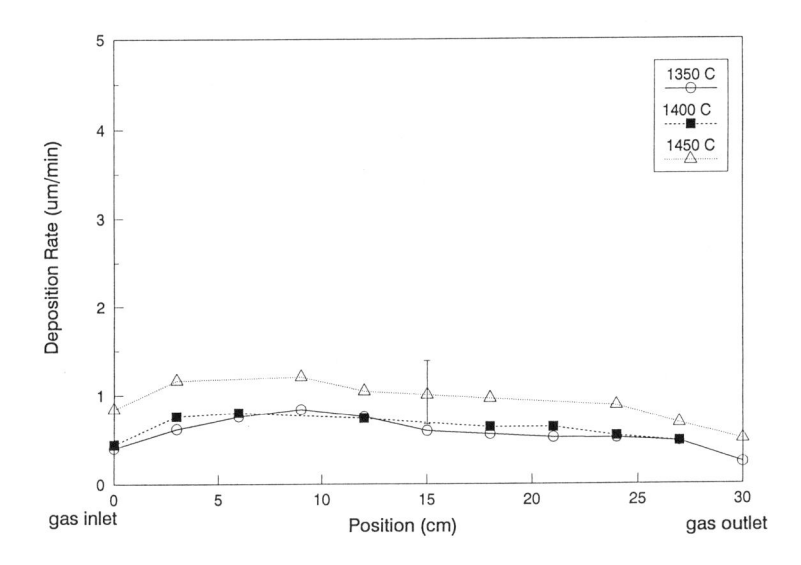

Figure 2. Deposition rate vs. position on fiber for various temperatures for Group B.

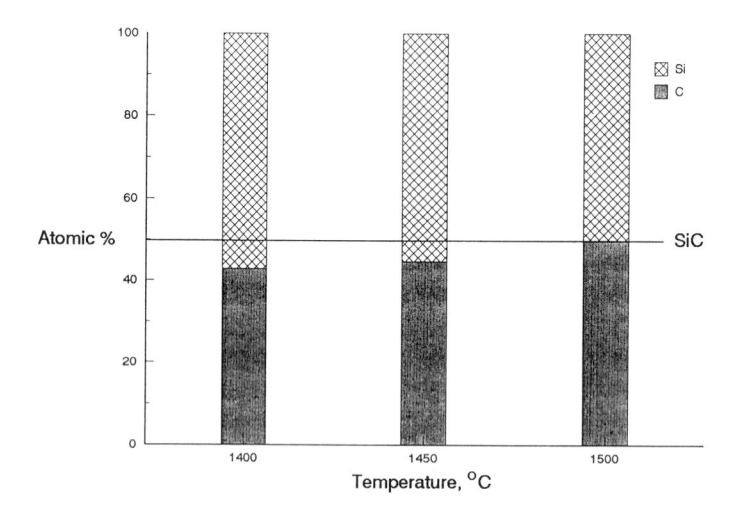

Figure 3. Electron microprobe analysis of CVD SiC for Group A.

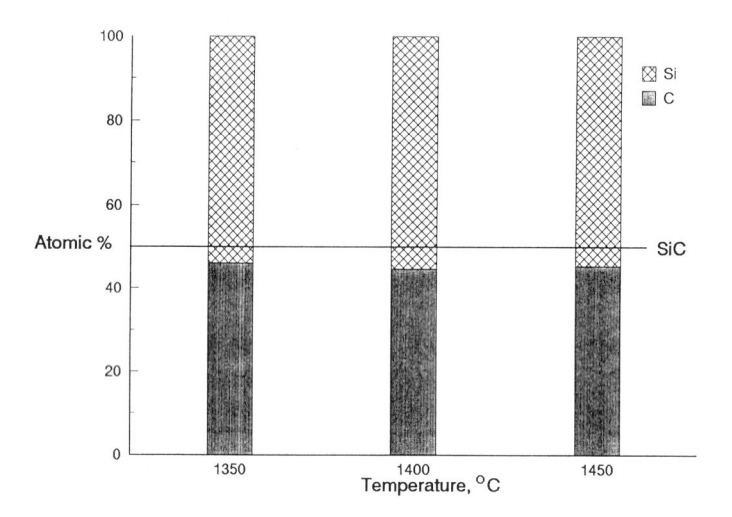

Figure 4. Electron microprobe analysis of CVD SiC for Group B.

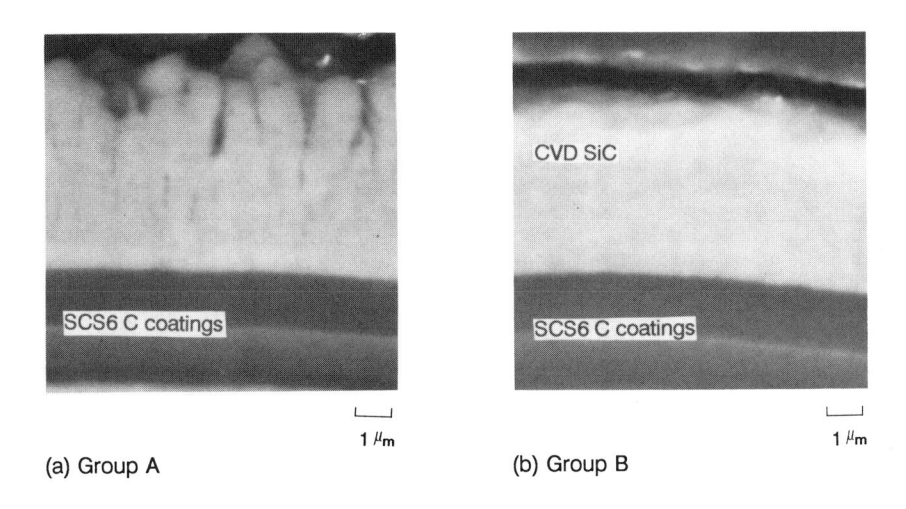

Figure 5. SiC coatings on SCS6 deposited at 1450°C (a) Group A and (b) Group B.

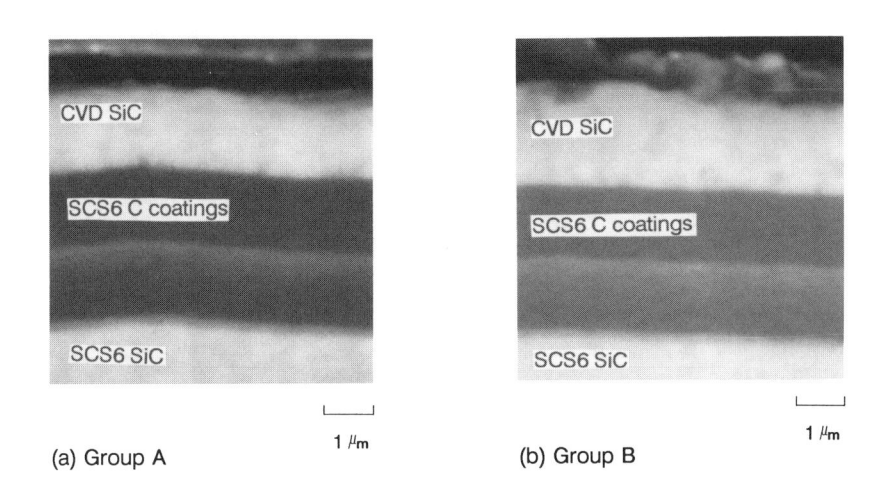

Figure 6. SiC coatings on SCS6 deposited at 1350°C (a) Group A and (b) Group B.

SUMMARY

Silicon carbide was deposited on Textron's SCS6 fiber using SiH_4 and C_3H_8 precursors. The deposition rates for temperatures ranging from 1350 to 1500 °C and different source gas concentrations were determined. Electron microprobe analysis was used to determine the stoichiometry along the fibers. The CVD SiC coatings were close to stoichiometric for all of the temperatures and source gas concentrations studied. Also, at higher deposition temperatures, the SiC coatings were more stoichiometric along longer lengths of the fiber. The microstructure of the SiC coatings were compared. At 1350°C, the microstructures were similar; however, at higher temperatures, the microstructure for the more concentrated source gas group were porous and columnar in comparison to cross sections taken from the same area for the dilute source gas group.

REFERENCES

1. B.J. Choi and D.R. Kim, J. Mater. Sc. Let. 10 860 (1991).

2. A. Saigal and N. Das, Adv. Cer. Mater. 3 (6), 580-583 (1988).

3. K. Nilhara, A. Suda, and T. Harai (Proc. of Int'l. Symp. on Cer. Comp. for Engines, Japan, 1983) pp.480-489.

4. S.A. Gokoglu, M. Kuczmarski, L.C. Veitch, P. Tsui and A. Chait, in Chemical Vapor Deposition, edited by K.E. Spear and G.W. Cullen (The Electrochem. Soc. Proc. 11 Pennington 1990) pp. 31-37.

THE CVD OF CERAMIC PROTECTIVE COATINGS ON SIC MONOFILAMENTS

K. L. CHOY AND B. DERBY
Department of Materials, University of Oxford, Parks Road, Oxford, OX1 3PH, UK

ABSTRACT

The TiB_2, TiC and TiN coatings were studied as potential protective coatings for SiC fibre reinforced titanium alloys. These coatings were deposited by a CVD technique on resistively heated SiC monofilament fibres at reduced pressure in a cold wall reactor. The deposition conditions that are required to coat each of these materials were determined. The morphology, microstructure, phase composition and nature of the coatings were examined. The comparison and evaluation of the effectiveness and potential of these as protective coatings were carried out by incorporating the coated fibre into Ti-6Al-4V using the diffusion bonding method at 1100°C under 10 MPa for 1 hour.

INTRODUCTION

The development of Ti-based composites is hindered largely due to the interfacial reaction between SiC and Ti matrices at elevated temperatures which leads to the formation of a brittle reaction zone consisting of titanium carbides and titanium silicides, which weakens the mechanical properties of the composites [1,2]. This has prompted research into the development of chemically inert protective coatings onto SiC fibres to inhibit the fibre/matrix interaction. The prospective coatings of interest in this study are TiB_2, TiC and TiN. These coatings are all chemically inert.

EXPERIMENTAL

The SiC fibres used in this study were BP Sigma SiC monofilaments (~100 μm diameter with tungsten core). The schematic diagram and general operation of the CVD apparatus used for the deposition has been described elsewhere [3]. A SiC monofilament was drawn into a cold wall reactor. Subsequently it was heated to the deposition temperature. Then a gaseous mixture of the reactants was fed into the reactor, these underwent reactions and deposited the coatings onto the SiC substrate. The reactants and deposition reaction used for the deposition of TiB_2, TiC and TiN, were as follows :

$$TiCl_4(g) + 2BCl_3(g) + 5H_2(g) \longrightarrow TiB_2(s) + 10HCl(g)$$
$$TiCl_4(g) + CH_4(g) \xrightarrow{H_2} TiC(s) + 4HCl(g)$$
$$2TiCl_4(g) + N_2(g) + 4H_2(g) \longrightarrow 2TiN(s) + 8HCl(g)$$

The coating process variables chosen were substrate temperature, deposition pressure and input gas ratio. The experimental parameters were systematically varied to investigate the effect of these parameters on the surface morphology, deposition rates and microstructures of the coatings. Table 1 summarizes the chemical vapour deposition conditions for different types of coatings. The coated fibres were characterized by optical microscope, SEM, TEM, X-ray diffraction and chemical analysis.

Table 1. The chemical vapour deposition conditions for different types of coatings

Coating	Temperature (°C)	Pressure kPa	Flow rate(cm^{3}min^{-1})				
			TiCl$_4$	BCl$_3$	CH$_4$	N$_2$	H$_2$
TiB$_2$	1050-1450	6-23	24-216	72-192	-	-	300
TiC	1000-1350	8-25	30-60	-	30-180	-	0-450
TiN	950-1050	10-25	30-60	-	-	200-400	200-600

The efficiency of the depositions as protective coatings to inhibit the fibre-matrix interfacial reaction were evaluated by incorporating these coated SiC fibres between two sheets of Ti-6Al-4V foils and diffusion bonding under 10 MPa at 1100°C for 1 hour. Thicker coatings ~6μm were used in this evaluation. The bonded specimens were sectioned, mounted and polished, and then were examined using SEM, and EPMA to investigate the composition of the fibre-coating-matrix interface.

RESULTS AND DISCUSSION

TiB$_2$ Coating

Dense and adherent TiB$_2$ coatings can be deposited under the above conditions. A detailed study of TiB$_2$ coatings has been published elsewhere [3] and a summary of our findings is presented here. The typical cross section of the coated fibre is shown in Fig.1. No crack or spalling of the coatings was observed. The fibre fracture surface, Fig. 2 revealed that the adherence of the TiB$_2$ coating was good so that even when the fibre was deliberately fractured, the SiC substrate was delaminated together with the TiB$_2$ coating, and the original surface of the SiC fibre which consists of uniform columnar grains was destroyed. X-ray diffraction analysis showed that the dominant phase that deposited in this range of deposition conditions was TiB$_2$. The other thermodynamically possible phases such as B or TiB, were not detected by the X-ray analysis. The EPMA showed a constant titanium to boron ratio across the coating thickness for the TiB$_2$ coatings deposited at BCl$_3$: TiCl$_4$ = 1 : 1 - 3 : 1. The TEM micrograph showed that there

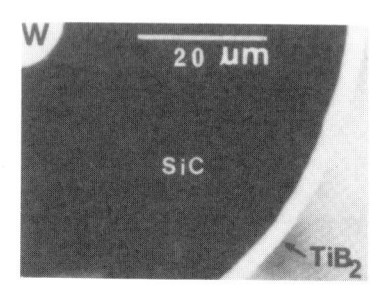

Fig. 1 The typical cross section of the TiB$_2$ coated SiC fibre.

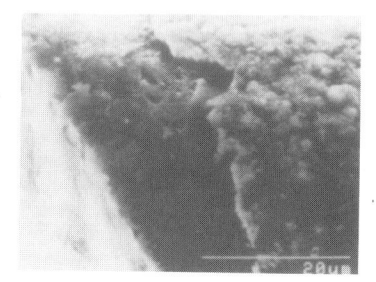

Fig.2 The fractured surface of the TiB$_2$ Coated SiC fibre.

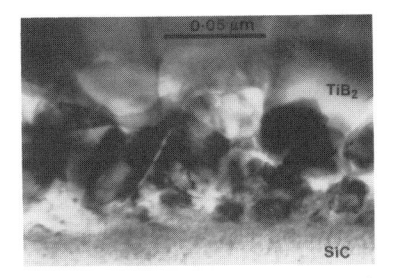

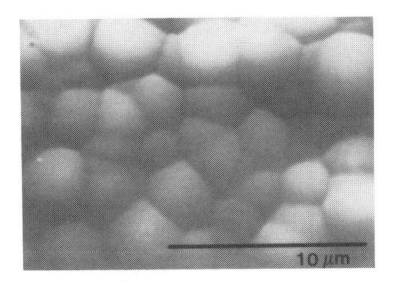

Fig.3 TEM micrograph of the TiB$_2$/SiC microstructure.

Fig. 4 SEM of the TiB$_2$ deposited at 1100°C, 15kPa and TiCl$_4$: BCl$_3$=1:3.

was no interfacial reaction between the TiB$_2$ coating and SiC substrates (Fig.3). The surface morphology of the coatings was very sensitive to the deposition temperature. At low temperatures, fine-grained polycrystals were deposited, however, at high temperatures larger faceted grains were obtained. The deposition rate increases with the increase of temperature. The transition from chemical kinetic limiting to mass transport limiting for TiB$_2$ deposition occurs at ~1350°C. The chemical kinetic limiting mechanism has an activation energy of 144 kJ/mol. The deposition rate also increases as the pressure increases. As the TiCl$_4$: BCl$_3$ input gas ratio increases, the nucleation rate increases, however the growth rate is slow, resulting in a low deposition rate. This is due to the fact that in the reactor, the local equilibrium of TiCl$_4$ and its subchlorides and HCl may easily establish in a homogeneous gas reaction as follows [4] :

$$TiCl_4 + 1/2\, H_2 \longrightarrow TiCl_3 + HCl$$
$$TiCl_4 + H_2 \longrightarrow TiCl_2 + 2HCl$$
$$TiCl_4 + 3/2 H_2 \longrightarrow TiCl + 3HCl$$

The HCl has been reported as having retarding effects in the TiN and TiC system [4,5]. Therefore, it is expected that this HCl also exhibits the same retarding effects in the deposition of TiB$_2$ which reduces the growth rate and hence the deposition rate. In the range of study, TiB$_2$ coatings consisting of smooth surfaces can be obtained at lower temperatures (eg. 1100°C) and higher pressures (15 kPa) and TiCl$_4$: BCl$_3$= 1 : 3 (Fig. 4).

TiC Coating

The deposited coatings were identified by X-ray diffraction analysis to be TiC. Other phases such as Ti, Ti$_2$C and TiC$_2$ were not detected. The temperature has a strong effect on the morphology of the deposited TiC coatings. Fig.5 shows that fine grains were deposited at lower temperatures, and coarser and larger faceted crystals at higher temperatures. For the coating deposited at 1200°C or higher, small grains were observed near the substrate with a transition to a columnar structure with the increase of grain size (Fig. 6). However, a similar observation was not apparent in thin coatings that were deposited at 1150°C or lower. Cracks were easily observed in TiC coatings that were deposited at high temperature, above 1200°C (Fig.5c). Similar cracks were not detected in coating

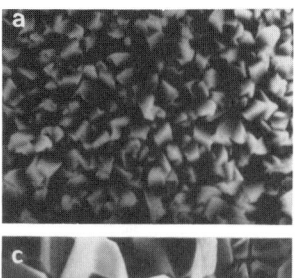

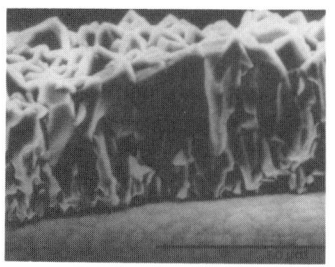

Fig. 5 Scanning electron micrograph of the surface morphology of the surface morphology of the TiC coatings deposited at pressure 15 kPa, flow rates (cm^3/min) : CH$_4$, 75; TiCl$_4$, 30; H$_2$, 300 and deposition temperatures at (a) 1000°C (b) 1150°C (c) 1250°C.

Fig. 6 Typical fracture surface of the TiC coating deposited above 1200°C, which revealed the transition from small grains near the substrate to a columnar structure.

deposited at 1150°C or lower. From SEM micrographs, it can be concluded that the cracking mode of the coating is intragranular. This may be due to the large difference in the thermal expansion co-efficient between TiC and SiC, resulting in the coating cracking readily under thermal shock during cooling. In general the adherence of TiC coatings on SiC monofilament fibres is poorer compared to TiB$_2$ coatings. This is evident from the inability of TiC to withstand external force which causes delamination and fracture of the coatings. This effect is more pronounced in the coatings deposited at high temperatures. This is clearly seen in Fig. 7a which shows the fractured end of the TiC coating deposited at 1300°C. More cracks were generated and migrated along the fracture end, and the coatings at the fracture end spalled (chipped) off severely. However, coatings that were deposited at lower temperatures were more resistant to external force. Fig. 7b shows the fracture end of the TiC coated fibres deposited at 1000°C. The lower the deposition temperature, the fewer cracks were generated and the coatings were not so easily spalled off. From the Arrehenius plot, it was obvious that more than one mechanism operated. The transition from chemical kinetics limiting to mass transport limiting is at ~1250°C. The activation energy for the chemical kinetics limited mechanism was found to be ~178 kJ/mol, which is very different from the reported activation energy of pyroltic graphite deposition from methane which is 431kJ/mol (103 kcal/mol) [6]. Therefore within the surface-reaction kinetics controlled region, it is more likely that the adsorption

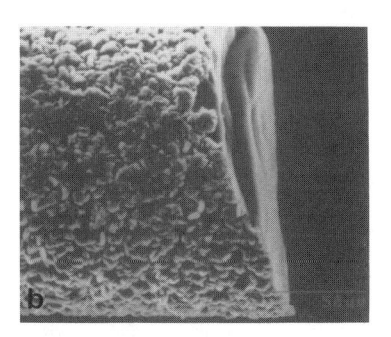

Fig. 7 SEM of the fracture end of the TiC deposited at (a) 1300°C (b) 1000°C.

and desorption of the reactant and the product gases rather than the dissociation of the carbon from the hyrocarbon molecules is the rate-controlling step. As the concentration of methane increases, finer grains were deposited.

However the TiC coatings had an excess of carbon. An equal input ratio of methane to TiCl$_4$ produced larger crystals but with a more stoichiometric TiC coating. An increase in the CH$_4$ concentration also increases the deposition rate, however the deposition rate decreases as the TiCl$_4$ concentration increases. This may be the same as the effect of TiCl$_4$ on the deposition of TiB$_2$ that was discussed above. As the pressure increases, finer and less faceted grains were deposited. The effects of pressure on TiC deposition seems to be similar to that of the TiB$_2$, therefore can be explained by the same mechanism as discussed earlier in [3]. The increase in pressure also caused the deposition rate to increase.

TiN Coating

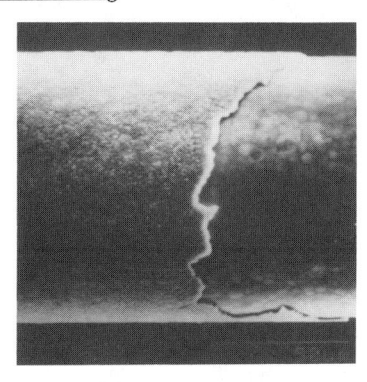

Fig. 8 Scanning electron micrograph of a typical TiN coated SiC fibre which revealed severe cracking and spalling of the coating.

The typical TiN coating on SiC monofilament is as shown in Fig. 8 where severe cracking and spalling occurs during handling. Variation in the deposition condition and thickness of the coating have no effect on the improvement of the adherence of the coating. This may be due to the large thermal expansion coefficient between the coating and SiC. Therefore it is expected that TiN would not provide effective protection for SiC fibres in Ti matrix. As a result, the reinforcement of TiN coated SiC was not carried out.

Evaluation of the effectiveness as protective Coating

TiB_2 coating is stable with the SiC fibre and has effectively inhibited the deleterious interfacial reaction between SiC and Ti-alloy matrix [7]. However, if the TiB_2 was deposited under high BCl_3 : $TiCl_4$ condition, the codeposition of B occurs that leads to the formation of needle like TiB adjacent to TiB_2 coating in the Ti-alloy matrix.

TiC coating was found to react with SiC fibres under the diffusion bonding conditions used [8]. Si from the SiC fibres diffuses into TiC coating and possibly forms a Ti-Si-C and/or titanium silicide compounds, this requires further analysis of its composition. However, no significant reaction between the TiC and the Ti-alloy was observed. Konitzer et al [9] have reported a very limited interaction between the TiC particulate and the Ti-6Al-4V matrix even after heat treatment of 50 hours at 1050°C. Although TiC coating cannot completely eliminate the deleterious interfacial reactions, it helps to retard the 6 μm reaction layer that forms when uncoated SiC fibre reinforced with Ti-6Al-4V is subjected to the same composite fabrication condition.

CONCLUSIONS

A dense and uniform TiB_2 coating can be deposited successfully on SiC fibres by the CVD technique. The presence of this coating has effectively inhibited the SiC fibre/Ti-6Al-4V interfacial reaction. The adherence of TiC on SiC fibres is weaker compared with TiB_2 coating. The TiC coating was found to react with SiC fibres. However, there was no apparent interfacial reaction between TiC and Ti-alloy matrix. TiN cannot be coated successfully onto the SiC monofilament. It may be necessary for the presence of another thin intermediate layer, to improve the adherence of TiN coatings on the fibre. Among these coatings, obviously, TiB_2 is the most promising protective coating for SiC fibres.

ACKNOWLEDGEMENTS

The authors thank SERC for funding, BP Research Lab., Sunbury-on-Thames for provision of the SiC fibres and C. J. Salter for the chemical analysis.

REFERENCES

1. P. R. Smith and F. H. Froes, J. Metals, 36, 19 (1984).
2. P. Martineau, R. Pailler, M. Lahaya, R. Naslain, J. Mater. Sci., 19, 2749 (1984).
3. K. L. Choy, B. Derby, Colloq. C2, supply au J. de Phys. II, 1, September, 697 (1991).
4. K. G.Stjernberg, H. Gass and H. E.Hintermann, Thin Solid Films, 40, 81 (1977).
5. J. R.Peterson, J. Vac. Sci. Technol., 11, 715 (1974).
6. M. B. Palmer and T. J. Hirt, J. Phys. Chem., 67, 709 (1963).
7. K. L. Choy, B. Derby, to be presented at Metal Matrix Composites III Conference in London, UK, December (1991).
8. K. L. Choy, B. Derby, to be published.
9. D. G. Konitzer and M. H. Loretto, Mater. Sci and Technol., 5, 627 (1989).

CVD COATING OF CERAMIC MONOFILAMENTS

JASON R. GUTH
Applied Sciences, Inc., PO Box 579, Cedarville, OH 45314

ABSTRACT

In many composite systems it has become apparent that coatings on the reinforcements are necessary to achieve high toughness materials. In order to examine materials which may be used as coatings on ceramic monofilaments and remain stable in high temperature, oxidizing environments, the deposition of a number of refractory metals has been attempted. The results of coating experiments using silicon carbide fibers as substrates as well as general observations concerning the prospects of continuously coating long lengths of fibers will be discussed. The materials studied include carbon, cobalt, zirconium, molybdenum, tantalum, tungsten, and iridium. Carbon has been deposited from methane and propylene onto both SiC and sapphire fibers. Deposition of the metals has been achieved by direct chlorination of the metals followed by hydrogen reduction at the fiber. Iridium(III)2,4-pentanedionate has been used to deposit iridium metal. All metals were deposited at low pressure in a hot wall reactor with fibers continuously spooled through the reactor.

INTRODUCTION

Ceramic materials have found uses in many high temperature structural applications. However, due to the inherent brittleness of most ceramic materials the interest in monolithic ceramics is being replaced by composites incorporating fibers or filaments as reinforcements in ceramic matrices[1]. In recent years it has become evident that weak interfacial strengths are a necessary condition to obtain a tough ceramic composite material[2,3]. To date, the materials found to best generate such weak interfaces are carbon and boron nitride, usually applied as a coating to the reinforcement fibers[3-5]. Both carbon and boron nitride are easily oxidized at fairly low temperature. Therefore, it is desirable to find other materials or compounds which are stable in oxidizing environments to above 1000°C, easily applied to fibers or filaments, and lead to a sufficiently weak interface for composite toughening. The purpose of this work is to investigate these first two issues and feed experiments aimed at defining the third.

The systems chosen to investigate the issues mentioned above consist of SiC (Textron SCS0) and single crystal sapphire (Saphikon) ceramic monofilaments and glass or ceramic matrices. In general, the composites consist of unidirectionally aligned filaments in a matrix densified by hot pressing. To apply coatings to the fibers before consolidation into composites, chemical vapor deposition (CVD) was chosen. The advantages of the CVD method include the wide range of elements and compounds that can be deposited, uniform surface coating that is not line-of-sight, and the high purity of deposited materials[5]. The fibers of interest are received and coated as single, long filaments which can be wound on spools and laid-up into tapes for composite fabrication.

EXPERIMENTAL

The CVD system, built specifically for coating fibers[6], is a hot wall system consisting of a 20(ID)X25(OD) mm fused silica tube in a 18 in (46 cm) vertical platinum wire wound furnace. The tube is contained by vacuum fittings between two boxes which house the fiber supply and take-up spools. Gases are metered using electronic mass flow controllers and may be introduced in the tube either at the top or bottom. A rotary vane vacuum pump in combination with a capacitance manometer, solenoid valve, and feedback controller are used to evacuate the system and to operate at reduced pressures. Metal chlorides are generated by passing chlorine gas over the heated metal in a side arm of the fiber coating muffle tube. Alternatively, metal chlorides or organometallic compounds may be contained in this side arm, sublimed, and carried into the muffle tube with argon or hydrogen carrier gas. Exhaust from the pump is passed through a scrubber and through a hydrogen burner before being exhausted. The entire apparatus, except for the electronic controllers, is housed under a fume hood surrounded by a flexible vinyl curtain. The furnace with metal chlorination process tube in place, spool boxes, and gas flow equipment and plumbing are shown assembled in Figure 1.

The fibers are held and wound onto 8" (20.3cm) plastic spools contained in vacuum tight stainless steel boxes with polycarbonate windows. The fiber is pulled through the reactor by rotating the bottom spool with a variable speed DC motor coupled by a vacuum tight rotary feedthrough. While turning, the spool is made to

Figure 1.
Photograph of the assembled CVD fiber coating apparatus.

move laterally to avoid winding the fiber on itself and damaging the coating. The only tension applied to the fiber is by the action of two brushes acting on the supply spool to prevent free rotation and uncontrolled payout of the fiber.

In order to prevent coatings on the fiber outside of the furnace hot zone and the flux of corrosive gases into the spool boxes, shield tubes cover the fiber from the boxes to approximately 5 cm into the furnace. Fused silica tubes with 2-5 mm ID are held by teflon retaining rings within the vacuum flange joint between the fiber spool boxes and the process tube fitting. Argon gas flows into the boxes and out the shield tubes to remove coating gases from these areas.

Carbon coatings have been obtained at 900-1100°C with propylene flows of 5-20 sccm and Ar flows of 95-200 sccm. Carbon has also been obtained at 1100-1300°C with 1-4% methane in argon with total flow rates of 200-600 sccm. Metallic coatings have been obtained by passing chlorine over heated metal followed by subsequent hydrogen reduction at the fiber surface. Typical conditions are 5 sccm Cl_2 plus 5 sccm Ar flowing over the metal foil or wire at 350-500°C. Typical coating conditions are 20-30 torr (2.7-4 kPa), 800-900°C with 100 sccm H_2. Iridium coatings were obtained by passing 10 sccm Ar over iridium(III)2,4-pentanedionate held at 240°C. The metal deposited at 800°C and 20 torr (2.7kPa) with 150 sccm H_2. In most cases the fiber is pulled at 2-4 cm/min and 50 sccm Ar flows into each spool box.

Coatings were examined by SEM after cutting or breaking the fiber to reveal a step in the coating. Coatings removed from the inside of the process tube were analyzed by XPS and X-Ray diffraction. Fibers were also analyzed before and after running through the coating system under identical conditions but without the metal, organometallic compound, or carbon gas present.

RESULTS

Carbon coatings have been applied to both SiC and sapphire fibers. Generally it is difficult to obtain smooth carbon films on either fiber. In attempting to increase the deposition rate, most often a sooty deposit, probably caused by homogeneous gas phase nucleation, is obtained. This is shown in Figure 2 on both SiC and sapphire fibers. The sapphire fiber has been coated twice. Once with a thin smooth layer and subsequently with a thicker, rougher, less dense layer. The existence of the first thin coating does not influence the morphology of the second coating. Figure 3 shows smooth coatings on both types of fiber. Both coatings duplicate the surface morphology of the fiber. In Figure 3b the coating has been removed from the sapphire fiber. Typically, any dense carbon coating greater than 0.5-0.75 micron will easily spall off of the sapphire fiber. If the linear speed of the fiber is too great, the coating can be seen spalling off as the coated fiber exits the reactor and cools.

Metallic coatings have been obtained by direct chlorination of Co, Mo, Ta, W, and Zr. Although these metals may not have the desired high temperature oxidation resistance, this method was chosen to allow for coating a variety of metals with one simple piece of equipment and to avoid the problems associated with storing and disposing of a large number of metal compounds. One common problem experienced in the coating of SiC fibers is nucleation. Figure 4a shows Mo on SiC. The wide separation of

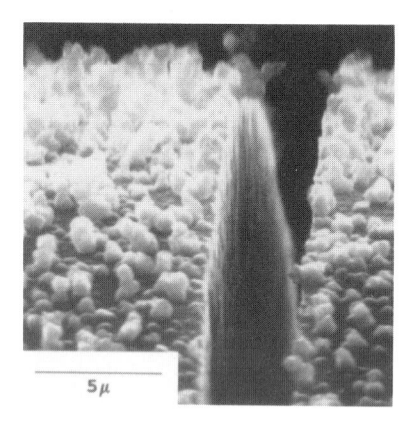

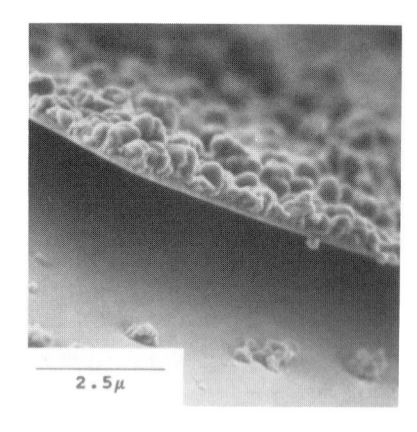

Figure 2. SEM micrographs of rough carbon coatings on a) SiC and b) sapphire fibers.

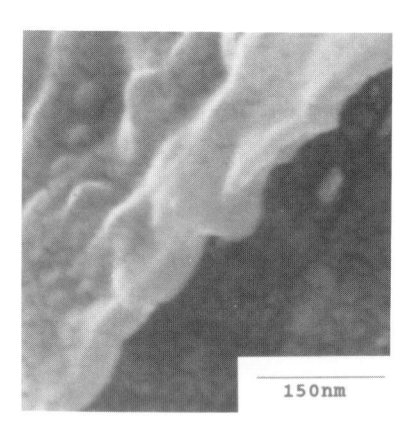

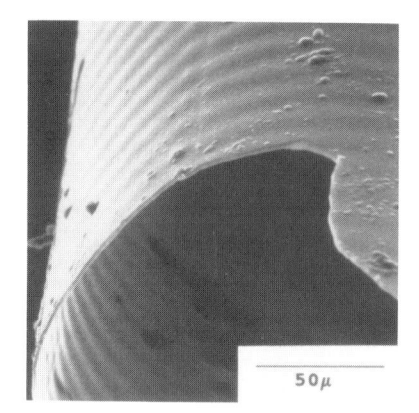

Figure 3. SEM micrographs of smooth carbon coatings on a) SiC and b) sapphire fibers. In b) the fiber has been removed.

the nucleation sites requires that a coating be relatively thick before it will be continuous. Figure 4b shows a thicker Mo based multilayer coating. Analysis of the coating reveals Mo closest to the fiber surface followed by molybdenum oxide. It is undetermined whether the oxide is a result of an air leak into the reactor or of a reaction of $MoCl_5$ which condensed on the fiber then oxidized after exposure to air. Figure 5 shows a uniform, thick coating of Zr on SiC fiber. Since iridium does not react or form a stable chloride, the acetylacetylate was used as a source. Figure 6 shows a multiphase coating from this reagent on SiC fiber. The phases found are Ir closest to the fiber followed by carbon. While carbon has been shown to contaminate Ir deposited from this source in previous work [7], this two phase deposit has

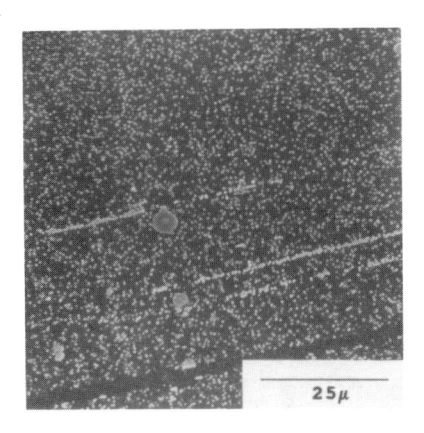

Figure 4. Mo coatings on SiC fiber.

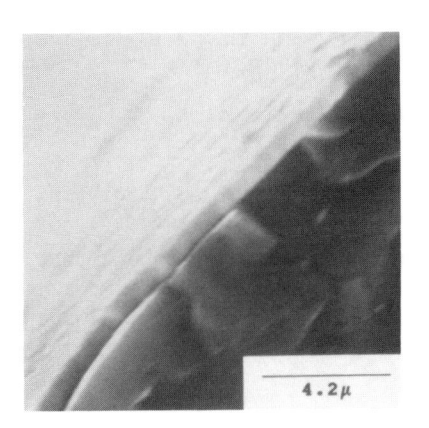

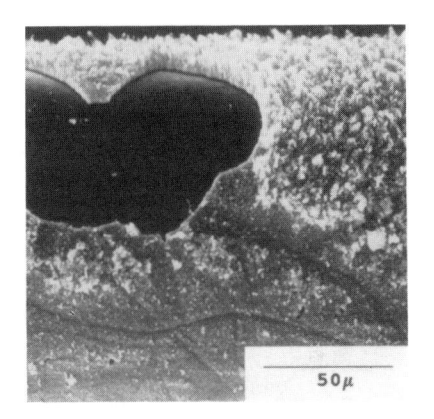

Fig. 5. Zr on SiC fiber.   Fig. 6. Ir on SiC fiber.

not been reported. In contrast to other CVD reactors, the substrate in fiber coating reactors experiences a wider variety of gas phase compositions. It is proposed that the organometallic decomposes to deposit the metal and, subsequently, downstream the ligand decomposes to deposit carbon.

DISCUSSION

The results of this research point out two major concerns for continuously coating filaments. These are deposition rate and deposit morphology. In addition, these two phenomenon are interrelated. In many cases attempting to increase the deposition rate adversely affects the deposit quality. The deposition on polycrystalline fibers, such as the SCS0 SiC fiber, appears to be partially controlled by surface nucleation. The density of nucleation sites may in some cases determine the structure of the coating and also the minimum thickness coating that will be continuous over the surface of the fiber. This is particularly evident in the early stages of growth of a Mo film as shown in Figure 4. Thicker films of the other metals studied generally tend to show a structure that would develop from nucleation such as observed in Figure 4. A similar structure has been observed [8] upon annealing sputtered Pt thin films on SiC where an initially uniform thin film is shown to aggregate into islands.

Several points may be made concerning the viability of coating continuous filaments by chemical vapor deposition. Surface chemistry in metal-SiC systems; such as nucleation, wetting, and surface diffusion; play a very important role in determining coating morphology. Attempting to increase the deposition rate in order to coat longer lengths of fiber quicker may lead to poor quality coatings as observed with carbon coatings. While longer furnace lengths may seem attractive to increase the overall coating rate, other problems may develop. These include unwanted reactions such as etching of the fiber by reaction byproducts (HCl, HF) and deposition of second phases (carbon from organic ligands, condensation of metal compounds). Other important points concerning continuous production include the density and strength of metal films, the adherence of coatings, and the difference in thermal expansion between coating and fiber.

REFERENCES

1. T. Mah, M.G. Mendiratta, A.P. Katz, and K.S. Mazdiyasni, Ceram. Bull. 66, 304 (1987).

2. R.J. Kerans, R.S. Hay, N.J. Pagano, and T.A. Parthasarathy, Ceram. Bull. 68, 429 (1989).

3. R.A. Lowden and K.L. More in Interfaces in Composites, edited by C.G. Pantano and E.J.H. Chen (Mater. Res. Soc. Proc. 170, Pittsburgh, PA 1990) pp.205-214.

4. N.D. Corbin, C.A. Rossetti, Jr., and S.D. Hartline, Ceram. Eng. Sci. Proc. 7, 958 (1986).

5. R.N. Singh and A.R. Gaddapati, J. Am. Chem. Soc. 71, C-100 (1988).

6. J.R. Guth, Wright Laboratories Technical Report, in press.

7. J.T. Harding, V. Fry, R.H. Tuffias, and R.B. Kaplan, AFRPL TR-86-099, 1986. (AD-A178337)

8. V.M. Bermudez and R. Kaplan, J. Mater. Res. 5, 2882 (1990).

This research was supported by WL/ML under contracts F33615-88-C-5402 and F33615-89-C-5604 at Wright-Patterson AFB, OH.

FACILITY FOR CONTINUOUS CVD COATING OF CERAMIC FIBERS

ARTHUR W. MOORE
Union Carbide Coatings Service Corporation, 12900 Snow Road, Parma, OH 44130

INTRODUCTION

The development of new and improved ceramic fibers has spurred the development and application of ceramic composites with improved strength, strength/weight ratio, toughness, and durability at increasingly high temperatures. For many systems, the ceramic fibers can be used without modification because their properties are adequate for the chosen application. However, in order to take maximum advantage of the fiber properties, it is often necessary to coat the ceramic fibers with materials of different composition and properties. Examples include (1) boron nitride coatings on a ceramic fiber, such as Nicalon silicon carbide, to prevent reaction with the ceramic matrix during fabrication and to enhance fiber pullout and increase toughness when the ceramic composite is subjected to stress[1]; (2) boron nitride coatings on ceramic yarns, such as Nicalon for use as thermal insulation panels in an aerodynamic environment, to reduce abrasion of the Nicalon and to inhibit the oxidation of free carbon contained within the Nicalon[2]; and (3) ceramic coatings on carbon yarns and carbon-carbon composites to permit use of these high-strength, high-temperature materials in oxidizing environments at very high temperatures[3,4].

This paper describes a pilot-plant-sized CVD facility for continuous coating of ceramic fibers and some of the results obtained so far with this equipment.

EXPERIMENTAL

A pilot-plant-sized inductively heated CVD furnace with a hot zone of 150mm diameter by 300 mm long was adapted for continuous coating of yarn as shown in Figure 1. The deposition chamber is formed by a graphite susceptor which is heated with a 10 kHz 30 kW induction power supply and thermally insulated with Grade WDF graphite felt manufactured by the National Electrical Carbon Company. The temperature is read and controlled by optical pyrometry on the wall of the susceptor. Reactant gases are introduced into the CVD chamber using a water-cooled stainless steel two-gas injector.

The yarn feed and collection apparatus was added to the existing CVD furnace as shown in Figure 1. Ceramic yarn was drawn from the feed spool (in the chamber beneath the furnace) over two Teflon^R pulleys (one mounted on a tension indicator) through the deposition chamber and wound onto a collection spool (in the chamber above the furnace) after passing over two more Teflon^R pulleys, one of which was mounted on a reversing actuator to provide uniform spreading of the yarn over the length of the collecting spool. The collecting spool and reversing actuator are driven with electric motors via rotary shaft vacuum seals. Once the yarn is in place, the transparent vacuum covers are installed to cover the yarn feed and collection apparatus; and the whole system is evacuated using a vacuum booster (blower) pump backed by a rotary piston vacuum pump. The system is capable of temperatures up to 2000°C, pressures as low as 0.05-0.1 Torr during deposition, and yarn speeds up to 3 meters/minute.

Figure 1. Arrangement for Continuous Yarn Coating in Pilot Plant CVD Furnace.

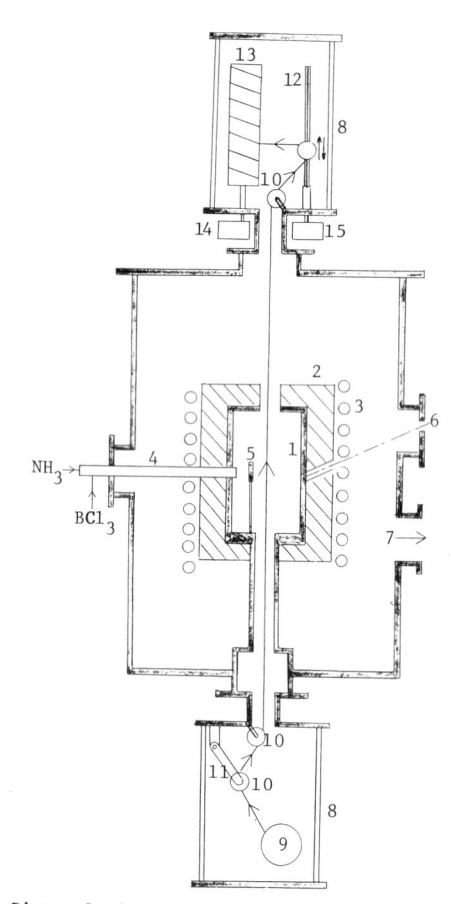

1. Graphite Susceptor Enclosing Deposition Chamber
2. Graphite felt Thermal Insulation
3. Induction Heating Coil
4. Two-Gas Injector
5. Gas Deflector
6. Sight Hole for Optical Pyrometry
7. Exhaust to Vacuum Pumps
8. Vacuum Covers for Yarn Feed and Collection Apparatus
9. Yarn Feed Spool
10. Teflon[R] Pulleys
11. Yarn Tension Indicator
12. Reversing Actuator
13. Yarn Collecting Spool
14. Motor Drive for Yarn Collecting Spool
15. Motor Drive for Reversing Actuator

A typical yarn coating experiment was carried out as follows. A 500 meter spool of Nicalon was wound onto the graphite feed spool and strung through the furnace and attached to the collection spool. A short length of yarn was motor driven through the furnace to make sure that the equipment was working properly. The plexiglass covers were then installed, the system pumped down, and then induction heated to the desired coating temperature. After running the yarn for a few minutes to obtain a vacuum-heat-treated reference sample, the reactant gases (NH_3 and BCl_3) were fed through the injector to start the PBN deposition; and the yarn speed was adjusted to obtain the desired coating thickness. Usually, near the end of the run, the gases were shut off so that a second vacuum-heat-treated sample could be obtained for comparison with the coated yarn.

The coated yarns were characterized by weight per unit length, tensile strength and modulus, scanning electron microscopy (SEM), and scanning Auger microscopy (SAM). The coating thickness was estimated from the yarn weight increase per unit length by assuming a density of 1.40 g/cc for PBN grown at 1100°C and assuming that all of the filaments in the yarn were coated equally.

RESULTS AND DISCUSSION

Examples of PBN coating conditions and some properties of PBN-coated Nicalon NL 202 yarn are given in Table I. Deposition temperature was varied from 1075-1200°C, pressures were 40-80 microns of HG, and yarn speeds were 13-61 cm/minute. A 60 cm/minute speed corresponds to about 30 seconds of residence time in the deposition chamber. Coating 9129 was made using additional pulleys which allowed three yarn passes through the deposition zone, thus increasing the residence time by a factor of three.

Table I

PBN Coating Conditions and Properties of Coated Nicalon NL 202 Yarn

Sample	Deposition Temp. °C	Deposition Pressure µ Hg	NH_3/BCl_3 Ratio	Yarn Speed cm/min	PBN Coating Thickness, Microns From Yarn Wt. Increase	From SAM	Approximate B/N Ratio in Coating	Yarn Tensile Strength MPa	Yarn Tensile Modulus GPa
9028	1075	45	3.6	61	0.15	0.1	1.0	2360	210
9110	1080	40	3.1	59	0.1	0.1	NM	2690	210
9129	1080	55	2.9	59*	0.4	0.3	1.3-2.4	2430	210
9035	1100	47	3.3	13	0.7	0.7	1.1-1.5	2380	~190
9032	1200	80	3.5	61	0.3	0.35	1.0-1.4	1290	~190

* Three yarn passes through furnace. All others one pass.
NM - Not Measured
Tensile strength and modulus of as-received Nicalon NL 202 were 2430-2770 MPa and 193-200 GPa, respectively.
Tensile strength of Nicalon NL 202 heated to 1000-1100°C for one minute at 0.5 Torr argon was ~2000 MPa.

The average PBN coating thicknesses as calculated from yarn weight gain were in good agreement with thicknesses determined by scanning Auger microscopy. Typical SAM composition depth profiles for a thin PBN coating (9028) and a thick coating (9035) are shown in Figures 2 and 3. The SAM results showed that the PBN coatings were either stoichiometric in boron and nitrogen or they were boron-rich, and the composition was affected by the NH_3/BCl_3 ratio in the reactant gases. Results suggest that a NH_3/BCl_3 ratio of at least 3.5 is needed to obtain a stoichiometric PBN coating. Significant quantities of oxygen and carbon were found in the coatings. Some of the oxygen and carbon may be diffusing out of the Nicalon during the coating process. Increased oxygen near the surface of the coating may be due to reaction of the PBN with atmospheric moisture[5].

SAM and SEM results showed that the PBN coating thickness was fairly uniform along the length of the Nicalon yarn and from filament to filament with the yarn bundle but more variable around the circumference of each filament. Filament-to-filament coating-thickness variation was much greater in PBN-coated Nicalon cloth. For example, in a sample of Nicalon cloth coated at 1080°C and 200 microns Hg for four minutes, the outermost filaments in the fiber bundle received a PBN coating which was 0.7 micron thick while the innermost filaments were coated with only about 0.1 micron of PBN.

The strand tensile strength of the Nicalon yarn which was PBN-coated at 1075-1100°C was within the range of values for the as-received Nicalon, showing that the PBN coating conditions caused little or no degradation of the Nicalon. In contrast, the tensile strength of Nicalon NL 202 which was heated to 1000-1100°C in low-pressure argon was 20% less than that of the as-received Nicalon. Coating the Nicalon at 1200°C reduced the strength by about 50%.

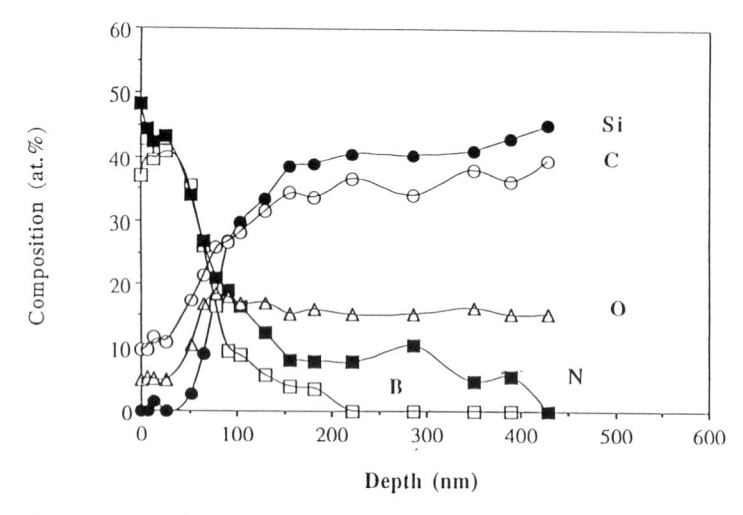

Figure 2. Scanning Auger Composition Depth Profile for PBN-Coated Nicalon NL 202 Sample 9028.

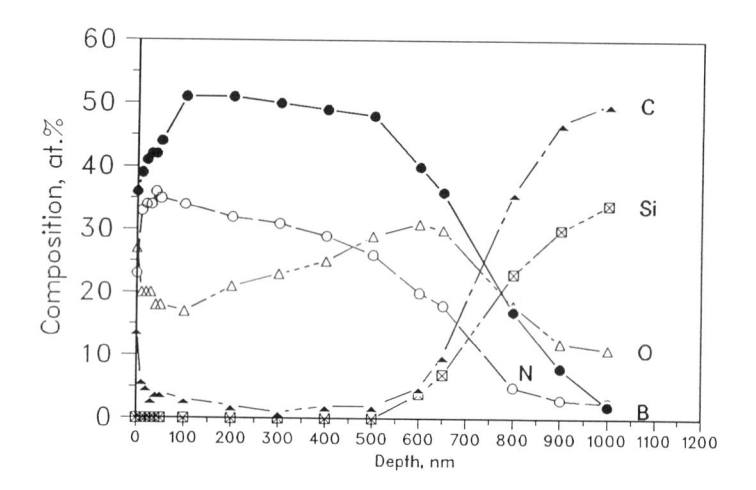

Figure 3. Scanning Auger Composition Depth Profile for PBN-Coated Nicalon NL 202 Sample 9035.

The Nicalon NL 202 yarn with 0.1-0.2 microns of PBN coating (such as samples 9028 and 9110) were woven into plain- and harness-weave cloth, and the cloth pieces were sewn into thermal insulation panels for evaluation at high temperatures in an aerodynamic environment[2]. Although the PBN-coated Nicalon was easier to handle than heat-cleaned (air oxidized to remove the polyvinyl acetate sizing) or vacuum-heat-treated Nicalon, the PBN coating was not entirely satisfactory as a high-temperature sizing. To control loose filaments in the sewing and weaving operations, it was necessary to cross- wrap the Nicalon with 30-denier rayon, which is later removed by oxidation after the panels are fabricated. The PBN-coated Nicalon was not affected by air oxidation at 800°C, but substantial degradation occurred after air oxidation at 1000°C.

A lithium aluminosilicate (LAS) matrix composite was made using the thickly coated fibers from Sample 9035, and it yielded a relatively good flexural strength of 320 MPa. Results with thinner PBN coatings were poorer so it appears that a coating thickness of at least 0.5 micron is needed to prevent matrix/coating/fiber interactions. A barium magnesium alumino-silicate (BMAS) matrix composite made using Nicalon coated with 0.7 micron of PBN yielded a flexural strength of 450 MPa at room temperature and 320 MPa at 1100°C.

A few trials were made in which carbon yarn was coated with PBN. In one set of experiments, Thornel T40R carbon yarn (3,000 filaments of average diameter about 6.5 microns) was coated with PBN at 1200°C, 1400°C, and 1680°C. Details of the coating conditions and results are given in Table II. The calculated average coating thicknesses, based on assumed densities for the PBN coatings, ranged from 0.15 microns for the 1200°C coating to 0.25 microns for the 1680°C coating. SEM showed that the PBN coatings on the Thornel T40R were less uniform than those on the 500-filament 15-micron diameter Nicalon NL 202 yarn.

Table II

PBN Coating Conditions of Oxidation Rate of
PBN-Coated Carbon Yarn Type Thornel T40R

Sample	Deposition Temp. °C	Deposition Pressure μ Hg	Coating Thickness Microns	Oxidation Rate in Dry Air at 600°C 0.5% Burn Off Min^{-1}	Oxidation Inhibitor Factor
9025	1200	50	0.15	1.49×10^{-4}	25
9026A	1400	100	0.2	1.34×10^{-4}	28
9026B	1680	200	0.25	1.32×10^{-4}	28

Oxidation rates for bare Thornel T40R = 3.75×10^{-3} Min^{-1}

Oxidation tests were performed on the PBN-coated T40R yarn at 600°C in dry air at 0-5% burn off. The PBN coatings reduced the oxidation rate to about 4% of that of the untreated carbon yarn. The inhibition factor for the 1200°C PBN coating is almost as large as that for the 1400°C and 1680°C coatings.

CONCLUSIONS

Modification of an existing CVD furnace has made available a pilot-plant-sized facility for continuous coating of ceramic yarns by CVD. Coatings at very low pressure are possible because all of the coating apparatus, including the yarn feeding and collecting equipment, is under vacuum. With a single-yarn pass through the furnace, Nicalon NL 202 yarn can be coated with 0.1-0.2 microns of BN at a yarn speed of 60 cm/minute so that a 500 meter spool of Nicalon can be coated in about 14 hours. Coating capacity was tripled by adding pulleys to permit three yarn passes through the furnace. Further capacity increases can be obtained by using three feed and collection systems of the type shown in Figure 1 and by altering gas flow patterns and deposition chamber design to increase the deposition rates.

The continuous CVD facility yields fairly uniform coatings of PBN on 500-filament, 15-micron diameter Nicalon NL 202 yarn, but coatings on 3,000-filament, 6.5 micron diameter Thornel T40R carbon yarn are less uniform than desired. Method for enhancing filament separation during CVD coating are being explored.

PBN coatings 0.1-0.7 microns thick can be deposited on Nicalon NL 202 yarn at 1000-1100°C with little or no loss of strength. Coatings 0.1-0.2 microns thick may enhance the performance of Nicalon in a high-temperature aerodynamic environment, but coatings 0.5 micron thick or more are needed for good performance of the Nicalon in ceramic composites.

Many different kinds of CVD coatings can be applied to ceramic yarns using the facility described here, and some will be explored depending on interest and time available. Wide ranges of deposition temperature and pressure are possible, limited by only the stability of the ceramic yarns and by the thermodynamics and kinetics of the CVD processes.

ACKNOWLEDGMENTS

Thanks are due to NASA Ames Research Center for support and permission to report some of information obtained under Contract Report NAS2-13109(JWS). The SAM data of Figure 2 was provided by Richard Jones of Dow Corning. John Brennan of United Technologies Research Center supplied the SAM data of Figure 3 and also the information on the strength of ceramic composites made with the PBN-coated Nicalon. The oxidation rate data of Table II were provided by Ed Stover of B. F. Goodrich.

REFERENCES

1. R. W. Rice, Ceram. Eng. Sci. Proc. 2 (7-8), 661 (1981).

2. A. W. Moore, Contract NAS2-13109(JWS), NASA Ames Research Center, to be published (1992).

3. J. R. Strife and J. E Sheehan, Amer. Ceram. Soc. Bull. 67, 369 (1988).

4. S. M. Gee and J. A. Little, J. Mats. Sci. 26, 1093 (1991).

5. T. Matsuda, J. Mats. Sci. 24 2353 (1989).

THE OXIDATION STABILITY OF BORON NITRIDE THIN FILMS ON MgO AND TiO2 SUBSTRATES

XIAOMEI QIU*, ABHAYA K. DATYE*, ROBERT T. PAINE**
and LAWRENCE. F. ALLARD***

Center for Microengineered Ceramics and Departments of *Chemical and Nuclear Engineering and **Chemistry, University of New Mexico, Albuquerque, NM 87131.
***High Temperature Materials Laboratory, Oak Ridge National Laboratory, Oak Ridge, TN 37831.

ABSTRACT

The stability of BN thin film coatings (2-5 nm thick) on MgO and TiO2 substrates was investigated using transmission electron microscopy (TEM). The samples were heated in air for at least 16 hours at temperatures ranging from 773 K - 1273 K. On MgO supports, the BN thin film coating was lost by 1073 K due to a solid state reaction with the substrate leading to formation of Mg2B2O5. No such reaction occurred with the TiO2 substrate and the BN was stable even at 1273 K. However, the coating appeared to ball up and phase segregate into islands of near-graphitic BN and clumps of TiO2 (rutile). The oxidizing treatment appears to promote the transformation from turbostratic BN to graphitic BN.

INTRODUCTION

Boron nitride thin coatings have attracted a great deal of interest for modifying the interface in fiber-reinforced composites to improve fiber pullout and prevent interfacial reaction [1]. The high temperature oxidation stability of boron nitride may provide distinct advantages over graphite despite the higher cost. However, there is no definitive data in the literature on the oxidation resistance of thin films of BN. Bulk BN powders have been reported to oxidize at 1073 K following parabolic oxidation kinetics [2]. The extent of oxidation was monitored in this particular study by following changes in weight as the sample was progressively oxidized. Lavrenko et al. [3] reported that the oxidation resistance of BN powders was greatly influenced by sample pretreatment, for example the extent of annealing in N_2. They found that oxidation of pyrolytic BN occurred at negligible rates at 1173 K. However, measurable weight losses were detected in TGA measurements at 1373 K, and quite rapid oxidation occurred at 1473 K. Significant oxidation of the graphitic BN did not occur at temperatures below 1223 K. Interestingly, they observed a weight loss with oxidation time for the pyrolytic BN but an increase in weight for graphitic (more ordered) BN. This they attributed to differences in porosity of the sample: pyrolytic powders, being more porous, led to loss of volatile boron oxide, which was retained in the graphitic samples that had a denser microstructure. Borek et al. [4] studied the effect of crystallinity and surface area of BN on its oxidation resistance. While they observed that the high surface area, poorly crystallized samples oxidized at rates that were an order of magnitude faster than the more crystalline samples, they concluded that if oxidation rates were normalized to surface area, no effect of crystallinity could be detected. All of these previous studies used TGA analysis to study oxidation behavior.

The implicit assumption was that oxidation of BN would lead to a change in sample weight as the N was replaced by O to form B_2O_3. This would cause an increase in sample weight until the volatilization temperature of B_2O_3 was reached, at which point the weight would start declining. This method is quite suitable for the study of bulk samples, but is not very sensitive when the oxidation behavior of thin films of BN is being investigated. We have therefore chosen to use transmission electron microscopy (TEM) to study transformations in the BN as it is progressively oxidized at higher temperatures.

EXPERIMENTAL

The samples were prepared by grinding together equal weights of a BN polymeric precursor with the oxide substrates: MgO or TiO_2 powder. The oxide powders we used had controlled morphology particles which facilitated detection of thin film coatings without any sample preparation, as described previously [5]. The preparation of BN polymeric precursors involves crosslinking of functionalized borazenes using N-containing molecules such as trimethyldisilazane. Further details are provided elsewhere [6]. The samples were first pyrolyzed in flowing N_2 at 1473 K to allow the BN to spread on the oxide surface and form a coating of turbostratic BN that was 2-4 nm thick. The BN-coated samples were then heated in air for at least 16 hours at temperatures ranging from 773 K - 1273 K.

RESULTS

In a previous study [7] we found that BN films only a few nm thick on MgO substrates were stable to air oxidation overnight at 973 K. Fig. 1 shows a micrograph of BN on MgO smoke particles after the initial pyrolysis in N_2 at 1473 K. When this sample was heated in air at 873 K (fig. 2a) there was no change in the BN coating. However, when heated at 1073 K the BN appeared to react with the MgO to form $Mg_2B_2O_5$. Fig. 2b shows that the substrate has now transformed to $Mg_2B_2O_5$ (as identified by its lattice spacings) and an amorphous surface layer is left behind. This solid state reaction occurred over the entire sample and caused the MgO cubic particles to change shape completely, forming elongated grains of the magnesium orthoborate phase. However, in some areas of this sample we observed a few patches of the BN film which were still intact after the oxidizing treatment. Therefore, it was not clear in those experiments whether the loss of BN was aided by the availability of a reactive substrate. We have now compared the oxidation behavior of BN on MgO substrates with that on TiO_2 substrates. We found that the TiO_2 proved to be an unreactive substrate, and there was no evidence for loss of BN coatings at temperatures where the $Mg_2B_2O_5$ had formed. Fig. 3 shows a micrograph of the BN/TiO_2 after pyrolysis in N_2 at 1473 K. When this sample was subsequently heated in air at 1273 K, we found that the BN coating had phase segregated into large grains of mesographitic BN and clumps of TiO_2 (rutile). Figs 4a and b show the morphology of the TiO_2 and BN respectively after the 1273 K calcination in air. The grains of BN are quite thin and tend to lie flat on the carbon film support. It appears that the oxidizing treatment actually promotes the crystallization of the BN, with the structure changing from turbostratic to graphitic.

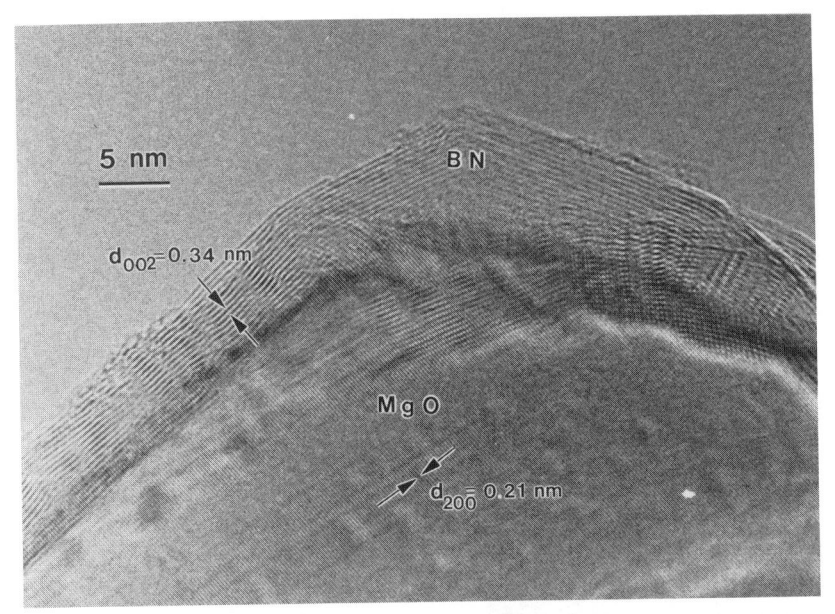

Figure 1. BN coating on MgO surface

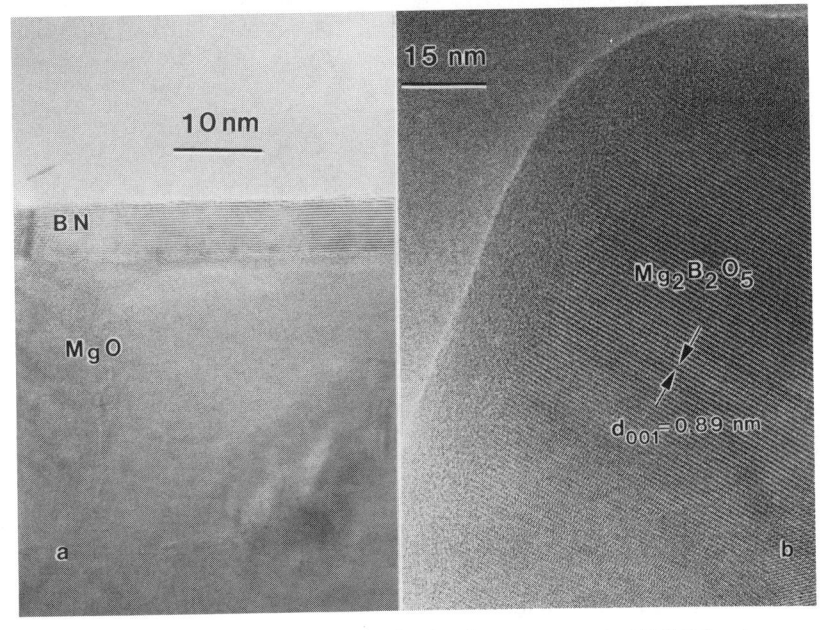

Figure 2. (a) BN coating on MgO after heat treatment at 873 K in air;
(b) BN coating on MgO after heat treatment at 1073 K in air

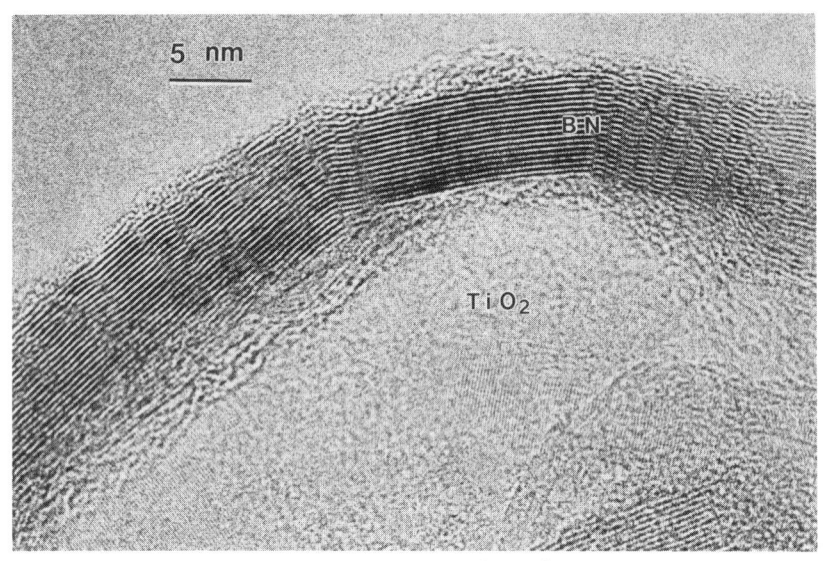

Figure 3. BN coating on TiO$_2$ surface

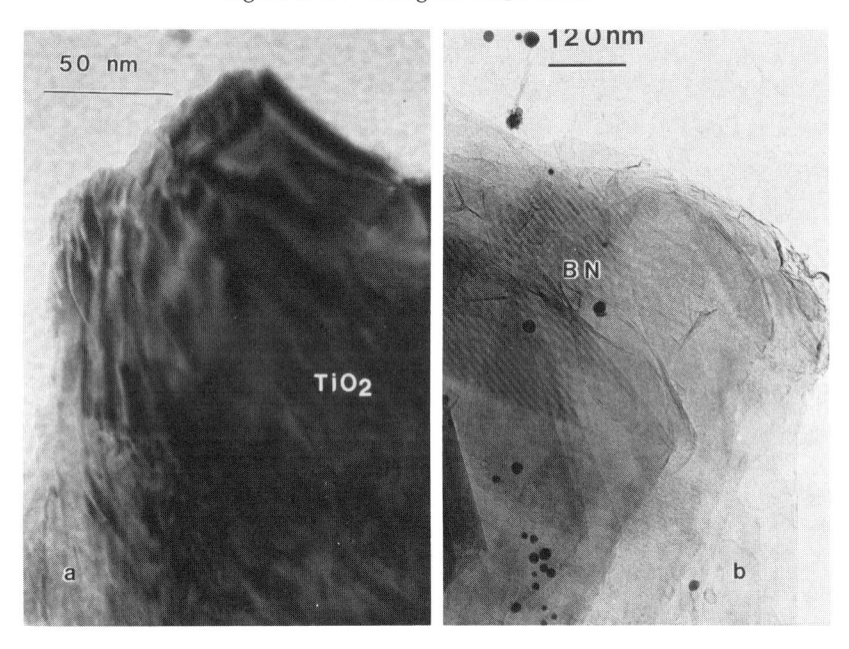

Figure 4. (a) BN coating on TiO$_2$ after heat treatment at 1273 K in air; (b) BN coating on TiO$_2$ after heat treatment at 1273 K in air

CONCLUSIONS

We have shown that the loss of BN thin films under oxidizing conditions is dependent very much on the substrate. On MgO, the BN coating is lost by 1073 K due to a solid state reaction with the substrate, which leads to formation of $Mg_2B_2O_5$. However, on TiO_2 substrates there is no such reaction, rather the BN and TiO_2 tend to phase segregate at 1273 K. The TEM results show that the BN is able to survive even an overnight calcination at 1273 K in air, and that the oxidizing treatment tends to favor sintering of the BN into larger grains. The degree of ordering of the turbostratic BN also improves with the oxidizing treatment leading to a structure that is closer to graphitic BN. The work presented here demonstrates that the interaction with specific substrates should be investigated in order to establish stability limits for BN thin films under oxidizing conditions. Furthermore, treatment under oxidizing conditions should be used if more ordered, graphitic BN is desired.

ACKNOWLEDGEMENTS

The work was supported by NSF grant CTS 8912366. Partial support from the UNM/NSF Center for Microengineered Ceramics is also gratefully acknowledged. A portion of the TEM work on this project was performed at the Microbeam Analysis Facility within the Department of Geology at the University of New Mexico. The high resolution TEM studies were conducted at the Oak Ridge National Laboratory, and were sponsored by the U.S. DOE, Asst. Secretary for Conservation and Renewable Energy, Office of Transportation Technologies, as part of the High Temperature Materials Laboratory User Program, under contract DE-AC05-84OR21400 with Martin Marietta Energy Systems Inc.

REFERENCES

1 R. N. Singh and M. K. Brun, Adv. Ceram. Mater. 3, 235 (1987).
2 N. G. Coles, D. R. Glasson and S. A. A. Jayaweera, J. Appl. Chem. 19, 178 (1968).
3 V. A. Lavrenko and A. F. Alexeev, Ceramics Intl. 12, 25 (1986).
4 T. T. Borek, D. A. Lindquist, G. P. Johnston, S. L. Heitala, D. M. Smith and R. T. Paine, submitted to J. Am. Ceramic Soc.
5 A. K. Datye, R. T. Paine, C. K. Narula and L. F. Allard, Mater. Res. Soc. Proc. 153, 97 (1989).
6 R. T. Paine, C. K. Narula, R. Schaefer and A. K. Datye, Chem. Mater., 1, 486 (1989).
7 A. K. Datye, X. Qiu, T. T. Borek, R. T. Paine and L. F. Allard, Mater. Res. Soc. Symp. Proc. 180, 807 (1990).

PART V

Organometallic CVD

CVD OF SiC AND AlN USING CYCLIC ORGANOMETALLIC PRECURSORS

[*]L. V. Interrante, [**]D. J. Larkin, and [*]C.Amato
[*]Department of Chemistry, Rensselaer Polytechnic Institute, Troy, New York 12180-3590
[**]National Aeronautics and Space Administration, Lewis Research Center, Cleveland, Ohio 44135

ABSTRACT

The use of cyclic organometallic molecules as single-source MOCVD precursors is illustrated by means of examples taken from our recent work on SiC and AlN deposition, with particular focus on SiC. Molecules containing $(SiC)_2$ and $(AlN)_3$ rings as the "core structure" were employed as the source materials for these studies. The organoaluminum amide, $[Me_2AlNH_2]_3$, was used as the AlN source and has been studied in a molecular beam sampling apparatus in order to determine the gas phase species present in a hot-wall CVD reactor environment. In the case of SiC CVD, a series of disilacyclobutanes, $[Si(XX')CH_2]_2$ (with X and X' = H, CH_3, and $CH_2SiH_2CH_3$), were examined in a cold-wall, hot-stage CVD reactor in order to compare their relative reactivities and prospective utility as single-source CVD precursors. The parent compound, disilacyclobutane, $[SiH_2CH_2]_2$, was found to exhibit the lowest deposition temperature (ca. 670 °C) and to yield the highest purity SiC films. This precursor gave a highly textured, polycrystalline film on the Si(100) substrates (70% with a SiC<111> orientation).

INTRODUCTION

The use of single-source precursors for MOCVD of refractory materials such as AlN and SiC affords potential advantages in terms of controlling the composition, deposition temperature and microstructure of the deposited product. The need to control these factors may be particularly critical in a CVI reactor environment, where the thermal/chemical stability of the fiber reinforcement, or of the fiber-matrix interface, may set limits on the processing temperature or where the use of mixed precursor sources may lead to local variations in composition and/or microstructure [1].
The choice of organometallic molecules for use as single-source precursors has been largely an empirical process, with few guidelines available relating to the relationship between molecular structure and such factors as decomposition temperature, resulting film composition, and microstructure. In the case of SiC CVD, precursors such as methyltrichlorosilane (MTS) and the methylsilanes have been employed for many years; however, temperatures well in excess of 1000 °C along with added H_2 are usually needed to obtain SiC free of elemental C or Si and significant variations in composition are often experienced as the deposition conditions are varied [1,2]. Moreover, the HCl produced as a byproduct of the decomposition of chlorosilanes can be corrosive to fiber reinforcements or Si substrates [1].

The Use of Cyclic Organometallics as Precursors to AlN and SiC

A key objective of our research efforts has been to develop precursors and procedures that will be effective in depositing refractory materials such as AlN and SiC in high purity at relatively low temperatures (<500 °C)

and pressures (< 10 torr) under thermal CVD conditions. Such procedures would be advantageous for many applications for CVD in electronics or in the processing of structural components where the substrate, the interface chemistry or the device structures built into the substrate may set effective limits on the processing temperature.

Another major objective of our program has been to obtain a detailed understanding of selected thermal CVD processes and, through this understanding, to develop precursor structure/function relationships that will be useful in the design of new precursor systems. In this context, we have selected two particular systems for detailed study, leading to the production of AlN and SiC thin films, respectively. In both cases the selected precursor systems are cyclic compounds which provide all of the required elements of the product thin film in the form of a volatile molecular entity of the type, $[A(XX')B(YY')]_n$ (where A, B = Al, N; or Si, C; X,X',Y,Y' = H, CH_3, $CH_2SiH_2CH_3$ etc. and n = 2 and 3). Our prior studies have provided ample evidence for the efficacy of such cyclic species as both AlN and SiC precursors, as well as a base of fundamental information relating to their synthesis, properties and pyrolysis chemistry [3]. As a class of molecular species, such cyclic compounds often afford higher volatilities compared to analogous acyclic compounds, perhaps due to their more compact molecular structures, as well as a core structural unit that can be used as a framework for attaching different substituent groups.

AlN Deposition Using Organoaluminum Amides

In the case of the cyclic organoaluminum amides of the type, $[R_2AlNR'_2]_n$, both dimeric (n = 2) and trimeric (n = 3) structures are known, depending on the nature of the R and R' groups; however, the most volatile species are those with relatively simple (R, R' = Me or H) substituents [4]. The compound $[Et_2AlN_3]_3$ has been studied by Gladfelter and coworkers and is reported to give films at quite low deposition temperatures; however, C-contamination appears to be a significant problem [5]. We have examined a series of N-Me and N-H amides and have found that, of these, only $[Me_2AlNH_2]_3$ gives satisfactory results as a MOCVD precursor for AlN [6].

Our initial studies of this precursor were carried out in a simple hot-wall CVD reactor and resulted in the deposition of high quality, polycrystalline thin films of AlN on Si and other substrates at considerably lower temperatures than had been required previously to obtain films of this type [7]. The apparent resemblance between the six-membered $(AlN)_3$ rings in this precursor and those that make up the wurtzite structure of AlN provided an additional stimulus for a more detailed study of the precursor-to-ceramic conversion process in this case. However, separate studies of this precursor system in solution, by NMR spectroscopy, revealed a rapid equilibrium involving the dimeric and, presumably, the monomeric form of this compound [8] and Generalized Valence Bond calculations were performed that indicated a planar structure and a significant stability for this monomer [9]. Recent mass spectral studies carried out in collaboration with J. Hudson have suggested that this trimer-dimer equilibrium extends also to the gas phase and that the active species in the CVD process may be one of these other forms of the $[Me_2AlNH_2]_3$ precursor [10]. We are currently engaged in a detailed study of the gas phase chemistry of the $[Me_2AlNH_2]_3$ compound by using a molecular beam mass spectrometry system to sample the gas phase species from a reactor held at different temperatures and pressures. Time-of-flight measurements are being carried out in this system in order to distinguish the parent molecular species from fragments produced in the mass spectrometer. The objective of these studies is to gain a better understanding of the gas phase chemistry occurring in a hot-wall CVD reactor using the [Me$_2$AlNH$_2$]$_3$ compound.

SiC Deposition with Disilacyclobutanes

Our study of SiC precursors has centered on a series of disilacyclobutanes which contain the (SiC)$_2$ ring but which differ in the nature of the substituents that are attached to this ring. The objective here is to examine a homologous series of compounds with a common structural core, so as to determine how the nature of the substituent influences the decomposition chemistry and composition/microstructure of the product film. The choice of the disilacyclobutanes as the core structure for this study was motivated by the results of prior decomposition studies of organosilicon compounds and the supposition of significant ring strain energy stored in the relatively small four-membered (SiC)$_2$ ring [11]. In comparison with the organoaluminum amides studied as AlN precursors, it should be noted that the nature of the bonding and the chemistry in these (SiC)$_2$ compounds differs considerably, with a greater prevalence of covalent bonding and radical decomposition processes in the case of the organosilicon species.

The specific compounds chosen for study in this initial examination of substituted 1,3-disilacylobutanes as SiC precursors are members of the structural type,

$$XX'Si\underset{CH_2}{\overset{CH_2}{\diamond}}SiXX''$$

where: X = Me, X' = H, X" = CH$_2$SiH$_2$CH$_3$ (I)
X = Me, X',X" = H (II)
and X, X', X" = H (III)

Preliminary results obtained for the initial compound in this series, [MeSi(H)μ(CH$_2$)$_2$Si(Me)CH$_2$SiH$_2$Me] (I), along with its synthesis, were described in earlier publications [12]. We have now prepared the 1,3-dimethyl-1,3-disilacyclobutane (II), as well as the parent 1,3-disilacyclobutane, [SiH$_2$CH$_2$]$_2$ (III), and obtained information regarding their decomposition chemistries and their utility as SiC precursors. The results of these studies are reported briefly herein.

EXPERIMENTAL

Preparation of Precursors

1,3-Dimethyl-1,3-disilacyclobutane was prepared as a ca. 50/50 cis/trans mixture by using a modification of the procedure described by Kriner [13]. This involved the Grignard coupling of CH$_3$Cl(OEt)SiCH$_2$Cl with Mg in THF, followed by reduction with LiAlH$_4$ in THF. The resultant mixture, consisting of mainly the four- and six-membered ring compounds [Si(Me)HCH$_2$]$_n$, was purified by atmospheric pressure fractional distillation (under N$_2$) giving an overall yield of ca. 25% for the 1,3-dimethyl-1,3-disilacyclobutane product mixture (b.p. 96-98 °C).

Disilacyclobutane was also prepared by using a method described in the literature [14], which involved the pyrolysis of 1,1-dichloro-1-silacyclobutane vapor by passing it through a 840 °C furnace tube packed with porcelain saddles. The resultant 1,1,3,3-tetrachloro-1,3-disilacyclo-butane was purified by sublimation and reduced in n-Bu$_2$O with LiAlH$_4$ at 0 °C. The overall yield was on the order of 30%.

Both compounds were characterized by ^{1}H, ^{13}C, and ^{29}Si NMR spectrosopy, as well as by mass spectrometry, and were >95% pure.

CVD Apparatus

A home-built, cold-wall, hot-stage CVD apparatus was used for these studies that was described briefly in a previous paper [12b]. The apparatus includes a stainless steel reactor chamber which contains the heated substrates, a bubbler loaded with the precursor, and a vacuum system consisting of a turbomolecular pump backed by a two-stage, rotary vane mechanical pump. The Si(100) substrates (resistivity = 1.1 Ω cm.; phosphorus doped) were cleaved into ca. 1.4 cm X 6.5 cm rectangular sections. Two or three of these substrates were then connected in a series circuit using tungsten clips with copper leads and resistively heated by using a current-limited DC power supply. The temperature is measured by means of a thermocouple attached to one of the wafers. The thermocouple was calibrated throughout the temperature range of interest prior to deposition under 1 Torr Ar pressure by using an optical pyrometer. The operating pressure in the reactor is monitored by a temperature-controlled capacitance manometer and controlled during the deposition by means of an electronically operated butterfly valve between the pumping system and the reactor chamber.

Decomposition Onset Determination

The CVD experiments using the LPCVD system described above were all carried out using a mass flow controlled argon carrier gas (10.0 sccm) while maintaining a constant reactor pressure of 1.0 ± 0.1 Torr. The precursor was loaded into a stainless steel bubbler while inside a N_2-dry box. Once the bubbler was attached to the CVD system, the precursor was subsequently freeze/thaw-degassed (backfilling with argon to remove the $N_{2(g)}$) before the start of the CVD experiment. During each LPCVD experiment, the partial pressure of the precursor was maintained at ca. 2.0 Torr by cooling the bubbler with an appropriate slush bath.

For the decomposition onset determination, a quadrupole mass spectrometer was connected to the reaction chamber downstream of the heated wafer sections. A series of liquid N_2-cooled traps were placed in the line leading to the mass spectrometer in order to trap out unreacted precursor and condensible byproducts. While flowing the precursor continuously through the reactor, the temperature of the Si(100) substrates was increased incrementally, as the intensity of the m/e = 15, 16 peaks for methane and m/e = 2 for hydrogen were monitored in the mass spectrometer. After each temperature increase, a reading was taken at each m/e setting until a relatively constant value was obtained. After reaching a maximum of ca. 900 °C, the substrate temperature was then decreased in intervals, taking readings again at each temperature setting. During the long period over which these measurements were performed (several hours), an appreciable thickness of SiC was deposited, insuring a fresh "SiC" surface throughout these decomposition studies.

RESULTS AND DISCUSSION

In our initial studies of SiC MOCVD we employed a substituted disilacyclobutane, 1,3-dimethyl-3-(methylsilyl)methyl-1,3-disilacyclobutane [MeSi(H)μ(CH$_2$)$_2$Si(Me)CH$_2$SiH$_2$Me] (I), that was obtained as the major volatile product from the Grignard coupling of MeSiCl$_2$CH$_2$Cl, after reduction [12,13].

Its ease of synthesis, compared to the simpler unsubstituted and methyl-substituted derivatives, and the observation that it gave near stoichiometric SiC films in a hot-wall CVD reactor environment, makes it an attractive choice as a CVD precursor. However, subsequent studies in our cold-wall reactor indicated that the composition of the SiC product varied considerably with the deposition conditions employed and that temperatures in excess of 750 °C were needed to achieve decomposition.

Figure 1. Structures of disilacyclobutanes used in this study.

The present study set out to compare this precursor with two related disilacyclobutanes, 1,3-dimethyl-1,3-disilacyclobutane [Si(Me)HCH$_2$]$_2$ (II), and 1,3-disilacyclobutane itself, [SiH$_2$CH$_2$]$_2$ (III). These precursors differ in the number of hydrogen atoms attached to the Si atoms of the disilacyclobutane ring, with 1, 2, and 4 H's present in (I), (II) and (III) respectively. They also differ in volatility, with the vapor pressure at 25 °C increasing from ca. 2 to 46 and 305 Torr from (I) to (III).

The results of the decomposition onset study for compound (I) are shown in Figure 2, and indicate a decomposition onset for this precursor of approximately 760 °C. Similar measurements were performed on both compounds (II) and (III), indicating decomposition onset temperatures of 740 and 670 °C, respectively. It is notable, however, that in the case of compound (III), no methane byproduct was observed, thus only the hydrogen m/e = 2 peak was employed to determine the decomposition onset temperature for this precursor.

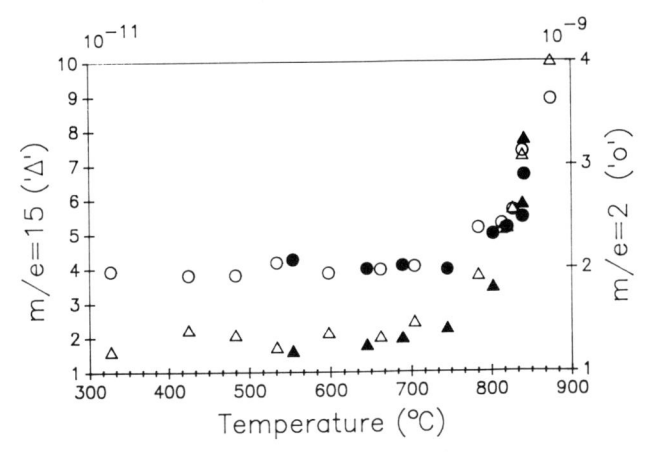

Figure 2. Deposition onset determination for compound I; closed symbols are for measurements performed on decreasing the substrate temperature.

Separate studies in which the substrates were held at the above onset temperatures, but where the temperature of the substrates varied locally, indicated that these decomposition onset temperatures also correspond approximately to the minimum temperatures for the production of a film.

The next series of experiments involved deposition for an extended period (ca. 1-2 h) at a particular temperature (ca. 700, 800, 900, 1000 and 1100 °C) in order to deposit at least 2 μm of the product film. The substrates were then removed from the reactor for compositional analysis by Auger spectroscopy. The deposition rates in these experiments varied from about 0.05 to 15 μm/h and the pressure in the reactor, as in the deposition onset studies, was maintained at 1 Torr total pressure.

The results indicate that the films obtained from precursors (I) and (II) under these conditions contain excess carbon, (as much as 16 atomic % for precursor (I) at 760 C) but that these films approach the stoichiometric SiC composition at substrate temperatures of 1100 and 1000 °C, respectively (Figure 3). On the other hand, the pyrolysis of precursor (III) produces Si-rich films (ca. 6%) at the lowest temperature studied (700 °C), but reaches the stoichiometric ratio by 800 °C.

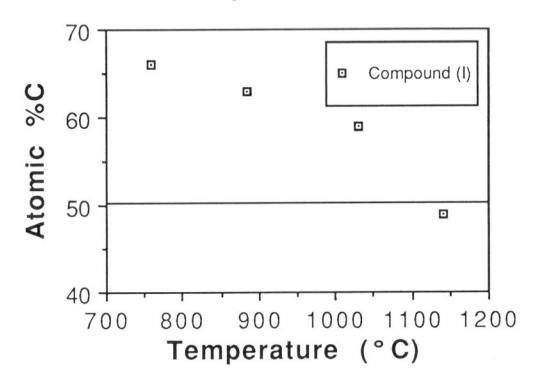

Figure 3. SiC composition as a function of temperature for films deposited by using compound I.

Powder X-ray diffraction studies were also carried out on these films using a pole figure diffractometer to analyze for possible preferred crystallite orientation. XRD patterns consistent with β-SiC were observed for films deposited by using all three precursors; however, in the case of precursors (I) and (II), only the films obtained at the higher temperatures (>900 °C) exhibited significant crystallinity. For precursor (III) on the other hand, even at the lowest deposition temperature studied (800 °C) the films were clearly crystalline. In this case, a pole figure resulting from analysis of the film deposited at 945 °C was obtained which indicated a strong (70%) fiber SiC(111) texture.

Thus, for these three precursors, the onset of crystallinity correlates with the composition of the product films and, as the composition approaches 1:1, increased crystallinity is evidenced. This is consistent with expectations regarding the inability of the crystalline SiC phases to tolerate significant variations from the 1:1 stoichiometry and the prior observations that excess C (or Si) tends to segregate to the SiC grain boundaries, thereby inhibiting crystallization [15]. The observation of preferred orientation of the SiC crystallites in the film obtained from precursor (III) at 945 °C can be also be understood on the basis of prior work, in which it was concluded that the SiC(111) plane would be preferred for crystal growth because it has a surface energy minimum resulting from a high atomic packing density [16].

SEM and FTIR transmission studies were also carried out on the SiC-coated Si wafer sections. The SEM studies indicated a wide variation of surface and cross-sectional film morphologies ranging from apparently fine grained, dense structures to rather porous deposits with elongated grains depending on the precursor employed and the deposition temperature. In general, the films were well adhered to the Si surface and had a rough, grainy, surface morphology. The FTIR studies showed only the expected 800 cm^{-1} band characteristic of SiC with no absorption bands in either the C-H or Si-H stretching regions.

After completion of a series of depositions at different temperatures for each precursor, the byproducts that had condensed in the main liquid N_2 cold trap in the CVD apparatus were analyzed by using both G.C./FTIR and mass spectrometry. In the case of (I), a mixture of hydrocarbons and methylsilanes were observed, including ethane, ethylene, acetylene, plus mono-, di-, tri- and tetramethylsilane. In the case of (II), the same C_2 hydrocarbons, along with methyl-, dimethyl- and trimethylsilane were detected, but no tetramethylsilane. Both precursors also gave hydrogen and methane as byproducts, as indicated from the decomposition onset studies.

In contrast, the byproduct distribution obtained from precursor (III) was relatively sparse, with only traces of the C_2 hydrocarbons along with methyl- and dimethylsilane detected, in addition to the hydrogen noted in the deposition onset determination.

Both the gas products observed and the compositional variations in the product films obtained using these three precursors can be understood on the basis of prior mechanistic studies of decomposition reactions of these and related 1,3-disilacyclobutanes. 1,3-Disilacyclobutane (III) and 1,3-dimethyl-1,3-disilacyclobutane (II), along with the tetramethyl derivative, have been studied in a stirred flow reactor by Auner and coworkers [11b]. The results of these and other studies [11a,14,17] have suggested that the decomposition onset temperature for these disilacyclobutanes varies inversely with the number of H-atoms substituted onto the Si atoms of the rings, with the parent disilacyclobutane decomposing at the lowest temperature. This presumably relates to the tendency of these H-substituted disilacylobutanes to form reactive silylenes (:SiRR' species) by 1,1-H_2 elimination or 1,2 Si-to-C H-transfer reactions that again form a silylene with resultant ring opening.

The product distribution from the stirred flow reactor study also parallels that observed herein for these precursors, where the distribution of methylsilanes observed reflects the bonding of Si in the initial precursor. Thus only compound I, which has up to four carbons around one Si, produces tetramethylsilane, compound II gives trimethylsilane as the highest methylated silane, and III, 1,3-disilacyclobutane, gives only up to dimethylsilane. This suggests that molecular rearrangements involving H transfer, as opposed to radical processes, are dominant in determining the silane reaction byproducts. Clearly more work needs to be done on these systems before a detailed reaction mechanism can be discussed with any confidence. Further deposition experiments along with surface decomposition studies are planned in order to help answer these remaining questions.

In comparing these three compounds as SiC CVD precursors, the parent disilacyclobutane is clearly the preferred compound for use in situations where the product purity and/or deposition temperature are critical factors. On the other hand, the preparation of this compound is still relatively difficult and it is certainly too costly to use at present for large scale CVD or CVI applications. Efforts to obtain this compound by more convenient methods and to examine alternative precursors of this type are also in progress.

REFERENCES

1. D.P. Stinton, A.J. Caputo, and R.A. Lowden, Ceram. Bull. 65(2) 347 (1986); S.M. Gupte, and J.A. Tsamopoulos, J. Electrochem. Soc. 136, 555 (1989); A.J. Caputo, R.A. Lowden, and D.P. Stinton, Oak Ridge Natl. Lab. Report No. ORNL/TM-9651, June 1985.

2. J. Chin, P.K. Gantzel, and R.G. Hudson, Thin Solid Films, 40, 57 (1977); P. Tsui and K.E. Spear, in "Emergent Process Methods for High-Technology Ceramics", R.F. Davis, H. Palmour III, and R.L. Porter, eds., Materials Science Research, Vol. 17, Plenum Press (1984), p. 371.

3. L.V. Interrante, C. Czekaj, and W. Lee, NATO ASI Proceedings, "Mechanisms of Reactions of Organometallic Compounds with Surfaces", D.J. Cole-Hamilton and J.O. Williams, eds., Plenum Publ. Corp. (1989) p. 205; L.V. Interrante, G.A. Sigel, C. Hejna and M. Garbauskas, Phosphorus, Sulfur and Silicon 41, 325 (1989); L.V. Interrante, B. Han, J.B. Hudson and C. Whitmarsh, Applied Surface Science, 46, 5 (1990).

4. L.V. Interrante, G.A. Sigel, C. Hejna and M. Garbauskas, Inorg. Chem. 28, 252 (1989).

5. D.C. Boyd, R.T. Haasch, D.R. Mantell, R.K. Schulze, J.F. Evans, and W.L. Gladfelter, Chem. Mater. 1, 119 (1989).

6. L.V. Interrante, L.E. Carpenter, C. Whitmarsh, W. Lee, G.A. Slack, and M. Garbauskas, Mats. Res. Soc. Symp. Proc. 73, 359 (1986).

7. L.V. Interrante, W. Lee, M. McConnell, N. Lewis, and E. Hall, J. Electrochem. Soc. 136, 472 (1989).

8. F.C. Sauls, C.L. Czekaj, and L.V. Interrante, Inorg. Chem. 29, 4688 (1990).

9. M.M. Lynam, L.V. Interrante, C.H. Patterson, and R.P. Messmer, Inorg. Chem., 30, 1918 (1991).

10. C. Amato, J. Hudson and L.V. Interrante, Mats. Res. Soc. Sympos. Proc. 168, 119 (1990); C. Amato, J. Hudson and L.V. Interrante, Mats. Res. Soc. Sympos. Proc. 204, 135 (1991).

11. (a) O'Neal, H. E.; Ring, M. A. J. Organometal. Chem. 213, 419 (1981); (b) N. Auner, I.M.T. Davidson, S. Ijadi-Mahhsoodi, F.T. Lawrence, Organomet. 5, 431 (1986).

12. (a) W. Lee, L.V. Interrante, C. Czekaj, J. Hudson, K. Lenz, and B-X. Sun, Mats. Res. Soc. Sympos. Proc. 131, 431 (1989); (b)D.J. Larkin, L.V. Interrante, J.B. Hudson and B. Han, Mats. Res. Soc. Sympos. Proc. 204, 141 (1991).

13. W.A. Kriner, J. Org. Chem. 29, 1601 (1964).

14. R.M. Irwin, J.M. Cooke, and J. Laane, J. Amer. Chem. Soc. 99 3273 (1977); N. Auner and J. Grobe, J. Organomet. Chem. 222, 33 (1981); N.S. Nametkin, V.M. Vdovin, V.I. Zav'yalov, and P.L. Grinberg, Izv. Akad. Nauk SSSR, Ser. Khim., 5, 929 (1965).

15. J.M. Blocher, Jr., in "Deposition Technologies for Films and Coatings", Noyes Publ., Park Ridge, NJ (1982), p. 335; M.H.J.M. de Croon and L.J. Giling, J. Electrochem. Soc. 137, 2867 (1990).

16. T. Takai, T. Halicioglu, and W.A. Tiller, Surf. Sci. 164, 341 (1985).

17. T.J. Barton, G. Marquardt, and J.A. Kilgour, J. Organometal. Chem., 85, 317 (1975).

CHEMICAL VAPOR DEPOSITION OF COPPER OXIDE THIN FILMS

YUNENG CHANG AND GLENN L. SCHRADER
Center for Interfacial Materials and Crystallization, Department of Chemical Engineering, Iowa State University, Ames, IA 50011

ABSTRACT

Copper oxide films were prepared by organometallic chemical vapor deposition of copper acetylacetonate in an oxygen-rich environment. The films were characterized by X-ray diffraction, Auger electron spectroscopy, X-ray photoelectron spectroscopy, and scanning electron microscopy. At 360°C, Cu_2O films were formed for an oxygen pressure of 150 torr and a copper acetylacetonate vapor pressure of 0.2 torr). The Cu_2O film was polycrystalline, but the orientation was primarily [111]. Differential scanning calorimetry indicated that O_2 assists decomposition of the organometallic precursor during pyrolysis.

INTRODUCTION

Transition metal oxide films have important optical, microelectronic, and energy-related applications [1]. Copper oxide (CuO and Cu_2O) films have been examined recently because of potential uses for solar cells, catalysts, and related superconductor materials.
Thermal oxidation [2], electrodeposition [3], and reactive sputtering [4] have been used to produce copper oxide films. Organometallic chemical vapor deposition (MOCVD) is potentially advantageous because of lower processing temperatures, reduced carbon contamination, and improved versatility.
Copper acetylacetonate ($Cu(CH_3COCHCOCH_3)_2$ or $Cu(acac)_2$) is volatile at moderate temperatures and is commercially available and less toxic than other β-diketonate derivatives. Previous research with metal acetylacetonate-MOCVD processes has established that oxidizing gases are necessary to control the stoichiometry of the films [5,6]. Ajayi found that hydrocarbons produced from $Cu(acac)_2$ decomposition were responsible for surface reduction of the copper oxide films [5]. Carbon contamination has also been investigated. Armitage reported that this contamination was diminished if the deposition was performed in air [6].
In this study, MOCVD processing conditions were determined for the production of Cu_2O films. Characterization of the films by X-ray diffraction (XRD), Auger electron spectroscopy (AES), X-ray photoelectron spectroscopy (XPS), and scanning electron microscopy (SEM) provided structural and compositional information. Differential scanning calorimetry (DSC) was employed to investigate the pyrolysis of $Cu(acac)_2$ under inert or oxidizing conditions.

EXPERIMENTAL

Deposition experiments were performed in a tubular quartz reactor. $Cu(acac)_2$ (Aldrich, 99%) was sublimed using He (Air Products) as the carrier gas. This stream was mixed with O_2 and He to obtain the desired partial pressure of oxygen. The depositions were performed on Si(100) (Unisil Corp., p-type). Depositions were performed for 10 minutes.

Instruments for characterization included: Siemens D-500 X-ray diffractometer, Perkin Elmer PHI 600 Scanning Auger, AEI 200B Spectrometer, and Cambridge S-200 scanning electron microscope.

Pyrolysis of Cu(acac)$_2$ was studied by a DuPont 2910 differential scanning calorimeter (DSC). Ten mg of Cu(acac)$_2$ was pyrolyzed in N$_2$ or dry air at a flow rate of 36 ml/min. The programmed heating rate was 5°C/min.

RESULTS AND DISCUSSION

Initially, depositions were performed under the following conditions: carrier gas flow rate (Q_T) = 100 sccm, temperature (T) = 360°C, total pressure (P_T) = 760 torr, oxygen partial pressure (P_{O_2}) = 150 torr, copper acetylacetonate pressure (P_{Cu}) = 0.20 torr. Particle-like aggregates can be observed by SEM (Figure 1A) with areas where no deposition has occurred. The irregular shape of the deposits suggests that gas-phase homogeneous nucleation dominated. The film morphology could be altered by increasing the carrier gas flow rate: continuous, smooth films were formed for carrier gas flow rates above 600 sccm (Figure 1B). The total pressure primarily affected the deposition rates.

Fig. 1 (A)
SEM of MOCVD film
(Q_T = 100 sccm)

Fig. 1 (B)
SEM of MOCVD film
(Q_T = 600 sccm)

With both P_T (760 torr) and Q_T (750 sccm) constant, depositions using a variety of reactant concentrations and substrate temperatures were performed to produce different film compositions. Crystalline phases were examined by XRD. In Table I, the processing conditions for producing films containing crystalline Cu$_2$O, CuO and Cu phases are listed. For a limited range of processing parameters, Cu$_2$O was the only crystalline phase present in the films: for P_{O_2} = 150 torr and P_{Cu} = 0.20 torr, Cu$_2$O films were deposited over a narrow temperature range (T = 360°C); at higher precursor concentrations (P_{Cu} = 0.25 torr), Cu$_2$O deposition occurred from T = 340 to 380°C.

The XRD pattern of a typical Cu$_2$O film is shown in Figure 2. A strong diffraction peak is observed at d = 2.464 Å and is assigned to the Cu$_2$O (111) plane. This (111) preferential orientation was found in most films dominated

Table I
Process condition for preparing Cu$_2$O films
(P_T = 760 torr and Q_T = 750 standard cm^3/m)

T°C	P_{O_2} (torr)	P_{Cu} (torr)	XRD (Å)*
340	150	0.2	2.18, 2.46
360	150	0.2	2.46
380	150	0.2	2.32, 2.46, 2.52
400	150	0.2	2.32, 2.52
340	150	0.25	2.46
360	150	0.25	2.46
380	150	0.25	2.46
400	150	0.25	2.52, 2.46, 2.32

*Reference compounds

	Plane	d-Spacing (A)	Intensity (%)
Cu	(111)	2.18	100
	(200)	1.82	78
Cu$_2$O	(110)	2.53	34
	(111)	2.46	100
	(200)	2.32	23
	(211)	2.14	17
CuO	(110)	2.75	12
	(002)	2.53	49
	(111)	2.52	100
	(111)	2.32	96
	(200)	2.31	30

Table II
Auger spectroscopy and XPS of MOCVD films

	MOCVD film	CuO	Cu$_2$O	Cu	CuCO$_3$
α_{Au}(eV):	1848.8	1851.6	1848.7	1851.2	1851.1
O$_{1s}$(eV):	530.5	529.6	530.4		
Cu 2p$_{3/2}$ peak position (eV):	932.7	933.8	932.7	932.8	934.6
Cu 2p$_{3/2}$ peak width (eV):	1.8	4.0	1.8	1.9	

Fig. 2
XRD of MOCVD film

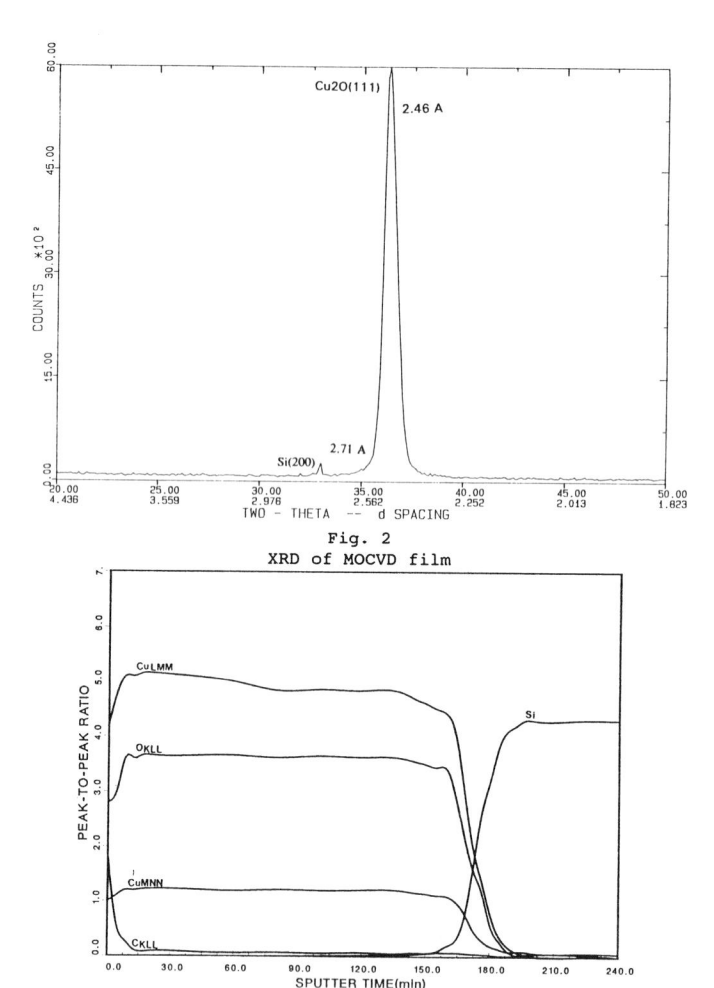

Fig. 3
Depth profile auger of MOCVD film

by the Cu$_2$O crystalline phase. A weak peak at d = 2.714 Å originates from the Si (200) plane of the substrate.

AES studies involved recording the peak-to-peak ratio (dN/dE) of the O(KLL) line, the C(KLL) line, the Cu(MNN) line, and the Cu(LMM) line. Two compositional zones are apparent in the film (Figure 3). In the outer surface layers (about 20 Å), copper, oxygen, and carbon were detected. The concentration of carbon dropped to below 1% under this surface layer: Cu and O were the primary components.

The possible existence of other phases was also investigated by depth profile XPS. After sputtering the surface layer, the XPS spectrum of the Cu 2p core electrons of this film (Figure 4) had two peaks at 932.7 eV (Cu 2p$_{3/2}$) and 953 eV (Cu 2p$_{1/2}$). The width (FWHM) of the Cu 2p$_{3/2}$ peak was 1.8 eV, and the distance between the Cu 2p$_{3/2}$ peak and the Cu 2p$_{1/2}$ peak was 20.3 eV. These results indicate the presence of Cu^0 or Cu^{+1}. As shown on Table II, copper carbonates (CuCO$_3$) or hydroxides (Cu(OH)$_2$) are not likely to be present. Also

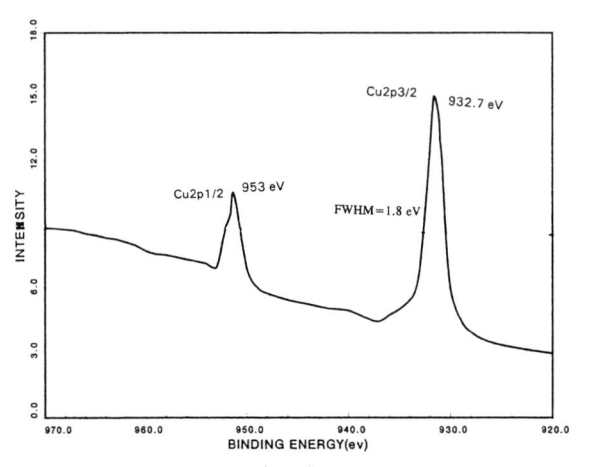

Fig. 4
XPS of MOCVD film

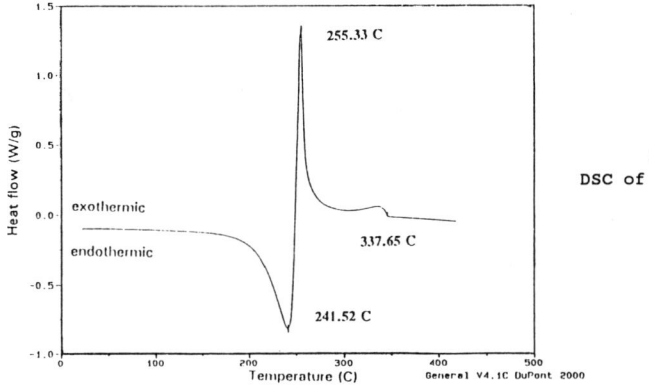

Fig. 5 (A)
DSC of Cu(acac)$_2$ in air

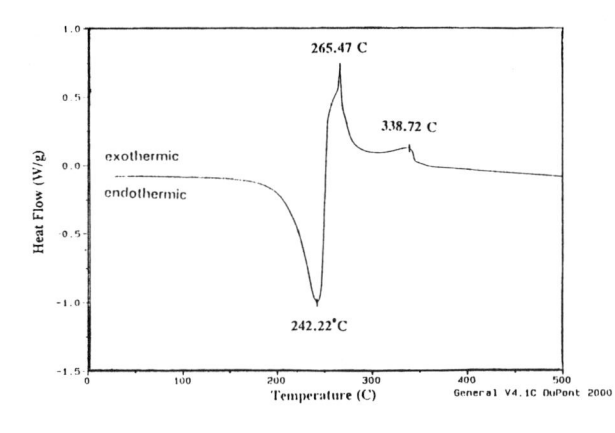

Fig. 5 (B)
DSC of Cu(acac)$_2$ in N$_2$

the absence of Cu 2p satellite peaks implies that Cu^{+2} compounds probably are not present (such as CuO). Comparison of the X-ray induced AES (XAES) peak position of Cu(LMM) in the spectrum (916.1 eV) with the XAES values for Cu (918.4 eV) and Cu_2O (916 eV) [9] indicates that the only detectable phase is Cu_2O.

DSC was employed to monitor the thermal energy during the decomposition of $Cu(acac)_2$. As shown in Figure 5A, there are three peaks for $Cu(acac)_2$ pyrolysis in air. The first endothermic reaction occured at 241°C; this was followed immediately by an exothermic peak located at 255°C. A third feature was a shoulder at 337°C. In the N_2 ambient (Figure 5B), the first endothermic peak was also found at about the same temperature. However, the second exothermic peak (265°C) was 10°C higher and also much broader than the second peak for the air studies. The position of the third peak was very similar to that for the air studies. Oxygen apparently assisted the second "step" of $Cu(acac)_2$ pyrolysis. A possible decomposition pathway for $Cu(acac)_2$ would involve the formation of reactive hydrocarbons indicated by the first endotherm (ligand dissociation of $Cu(acac)_2$). This reaction reached a maximum rate at 241°C. O_2 reacted with these hydrocarbons, as indicated by DSC: an exothermic step immediately follows the first reaction, with the highest reaction rate attained at 255°C. Clearly, deposition temperatures above a critical value were needed to initiate $Cu(acac)_2$ pyrolysis. O_2 is probably capable of oxidizing the hydrocarbons produced and thereby avoid reduction of the oxide film.

SUMMARY

Smooth Cu_2O films were deposited by MOCVD of $Cu(acac)_2$ under O_2-rich conditions. Cu_2O (111) oriented grains could be formed under specific processing parameters.

ACKNOWLEDGMENTS

The authors would like to thank Dr. J. Anderegg for his assistance in the XPS measurements, and Dr. A. Bevolo for his assistance in carrying out the AES measurements.

REFERENCES

1. W. M. Sears and E. Fortin, Thin Solid Films, 103, 303 (1983).
2. S. Nakahara, Thin Solid Films, 102, 345 (1983).
3. H. S. Potda and A. Mitra, Solar Energy Matl., 4, 291 (1981).
4. G. Beensh-Marchwicka and M. Slaby, Thin Solid Films, 88, 33 (1982).
5. O. B. Ajayi, M. S. Akanni, and J. N. Lambi, Thin Solid Films, 185, 123 (1990).
6. D. N. Armitage, N. I. Dunhill, R. H. West, and J. O. Williams, J. Cryst. Growth, 108, 683 (1991).
7. V. D. Castro, Appl. Surf. Sci., 28, 270 (1987).
9. M. M. Jones, Inorg. Chem., 1, 166 (1962).

MOCVD Growth of Copper and Copper Oxide Films From *Bis*-β-Diketonate Complexes of Copper. The Role of Carrier Gas on Deposit Composition.

WILLIAM S. REES, JR.* AND CELIA R. CABALLERO

Department of Chemistry and Materials Research and Technology Center, The Florida State University, Tallahassee, Florida 32306-3006, U.S.A.

ABSTRACT

An examination of thermal chemical vapor deposit elemental composition by EDAX has been completed for material films grown from Cu(acac)$_2$ and Cu(tmhd)$_2$ (acac = pentane-2,4-dionate; tmhd = 2,2,6,6-tetramethylheptane-3,5-dionate), using both hydrous and anhydrous carrier gas steams each of reducing (H$_2$), inert (N$_2$), and oxidizing (O$_2$) composition.

INTRODUCTION

The β-diketonate complexes of Cu(II) previously have been used as copper transport sources for chemical vapor deposition (CVD) both of elemental copper for electrical conduction,[1] and the copper oxide planes which form the basis of superconducting metal oxides.[2] Having an interest in each area,[3] we wondered what controlled the deposit composition which emanated from an identical source compound. There are numerous reports of utilizing Cu(acac)$_2$ or Cu(tmhd)$_2$ as the precursor for deposition of copper-containing thin films;[1] however, to our knowledge, no systematic examination has been reported for the role of carrier gas in determination of the composition of these deposits. We now report the results of a study exploring hydrous and anhydrous oxidizing, inert, and reducing carrier gas streams for Cu(acac)$_2$ and hydrous and anhydrous oxidizing and hydrous inert carrier gas streams for Cu(tmhd)$_2$.

EXPERIMENTAL

Substrate temperatures were determined by placement of a thermocouple abutting the substrate (**Figure 1**). Flow rates were controlled by manostat-type mass flow controllers. Deposit compositions were determined by employing an EDAX attachment on a JEOL scanning electron microscope. The substrate temperature and flow rate were adjusted to produce a growth rate of ~1μ/hour. The summary results presented for Cu(acac)$_2$ (**Table I**) and Cu(tmhd)$_2$ (**Table II**) each represent an average value (x̄) of data collected from six independent growth runs. The precursors were prepared by known methods, identified as pure by gas chromatography/ mass spectrometry

(GC/MS), and authentic by comparison of the observed melting points with those reported previously.[4] All manipulations for bubbler filling occurred in a Vacuum Atmospheres brand inert atmosphere glove box. SiO_2 substrates were cleaned by washing with 6 **M** HNO_3, three times with distilled water, and finally twice with electronic grade methanol, and oven dried at 130°C prior to use. The entire growth system was equilibrated at growth conditions for 15 - 20 minutes prior to introduction of the source material into the vapor phase. The optimum bubbler temperatures necessary for saturation of the carrier gas stream were ascertained by examination of GC/MS decomposition data for the two precursors.

Figure 1. Schematic diagram of the hot-walled CVD reactor used to deposit films from $Cu(acac)_2$ and $Cu(tmhd)_2$. Key: TC, thermocouple; Z1 - Z3, independent control temperature zones; S1 - S3, substrates to be coated; PH, pre-heater for deposition gasses; B_T, bubbler temperature; MFC, mass flow controller; G1 - G2, inlet gasses; note that both TC and G2 have variable positions due to the addition of bellows in-line. Flow through G2 was utilized to prevent premature decomposition of the source for those ambients with which it reacted rapidly.

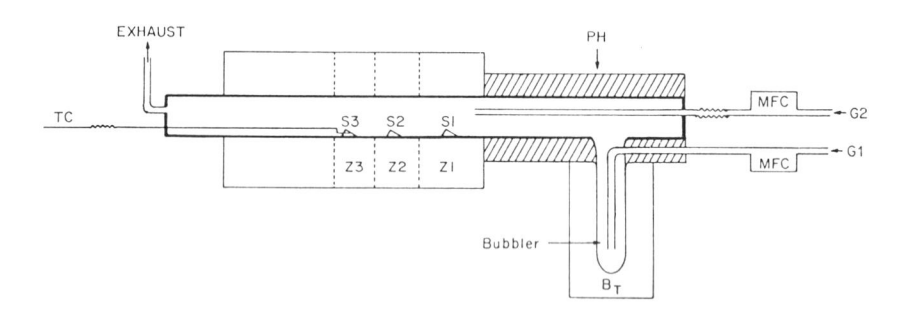

Table I. Effect of Thin Film Deposit Composition on Deposition Atmosphere for Cu(acac)$_2$ (1 μ/hour, x of 6 runs).

Atmosphere	Substrate temp. (°C)	Flow rate (sccm)	Deposit composition
Ar	387.1	186	Cu
O$_2$	360.7	135	Cu$_2$O
H$_2$	386.0	231	Cu
Ar + H$_2$O	371.5	231	Cu
O$_2$ + H$_2$O	367.4	135	Cu$_2$O
H$_2$ + H$_2$O	313.0	231	Cu

Table II. Effect of Thin Film Deposit Composition on Deposition Atmosphere for Cu(tmhd)$_2$ (1 μ/hour, x of 6 runs).

ATMOSPHERE	DEPOSIT COMPOSITION
O$_2$	Cu$_2$O
O$_2$ + H$_2$O	Cu$_2$O
Ar + H$_2$O	Cu

RESULTS

Both precursors produced only elemental copper under all conditions examined other than those utilizing oxidizing ambients. For all three sets of growth atmospheres explored (reducing, oxidizing, and inert) the inclusion of a water saturated carrier gas stream proved to be uncorrelated with deposit composition. No significant differences were found between Cu(acac)$_2$ and Cu(tmhd)$_2$ in the current study. Both yield Cu° under non-oxidizing atmospheres, and Cu$_2$O under oxidizing conditions. We presently explain these observations by invoking the known oxidation of elemental copper at growth conditions.[5] Thus, under all sets of deposition parameters investigated in the present study, *including oxidizing ones*, the initial deposit formed from Cu(acac)$_2$ or Cu(tmhd)$_2$ is Cu°. The film remains intact under reducing or inert conditions, whereas in an oxidizing ambient, it is converted *by a second reaction* from Cu° to copper oxide. This explanation is consistent with all data observed to date, and shows great potential for employment of sources containing copper-oxygen bonds in the deposition of high purity metal for ULSI electronic device use. For the case of superconducting metal oxides, clearly a two step mechanism is necessary for growth of copper-oxygen sheets from the *bis*-β-diketonate family of source compounds. However, this may not prove to be a detrimental feature, as that high quality superconducting metal oxides have been grown by chemical vapor deposition from *bis*-β-diketonate precursors.[2,3]

ACKNOWLEDGMENT

Defense Advanced Research Projects Agency contract number MDA 972-88-J-1006 is gratefully acknowledged for financial support of this work.

REFERENCES

1. a) S. Murarka and M. C. Peckeror, *Electronic Materials Science and Technology* Academic Press: New York; **1989**.
   b) H. H. Shin, K.-M. Chi, M. J. Hampden-Smith, T. T. Kodas, J. D. Farr, and M. F. Paffett, in *Chemical Perspectives of Microelectronic Materials II*, L. V. Interrante, K. F. Jensen, L. H. Dubois, and M. E. Gross, Eds., MRS Proceedings, Vol. 204, Materials Research Society: Pittsburgh, Pennsylvania; **1991**, pp. 421 - 426.
   c) G. Xue, J. Dong, and Q. Sheng, *J. Chem. Soc., Chem. Commun*, **1991**, 407.
   d) A. E. Kaloyeros, A. Feng, J. Garhart, K. C. Brooks, S. K. Ghosh, A. N. Saxena, and F. Luehrs, *J. Electronic Mater.*, **1990**, *19*, 271.
   e) A. E. Kaloyeros, A. N. Saxena, K. Brooks, S. Ghosh, and E. Eisenbraun, in *Advanced Metallizations In Microelectronics*, A. Katz, S. P. Murarka, and A. Appelbaum, Eds., MRS Proceedings Vol. 181, Materials Research Society Pittsburgh: **1990**, pp. 79-84.
   f) C. Trundle and C. J. Brierley, *Appl. Surf. Sci.*, **1989**, *36*, 102.
   g) S. Poston and A. Reisman, *J. Electronic Matls.*, **1989**, *18*, 79.
   h) D. Temple and A. Reisman, *J. Electrochem. Soc.*, **1989**, *136*, 3525.
   i) N. Awaya and Y. Arita, *Abst. Symp. VLSI Technol. #12-4*, Kyoto, **1989**, 103.
   j) C. Oehr and H. Suhr, *Appl. Phys. A*, **1988**, *45*, 151.
   k) C. R. Moylan, T. H. Baum, and C. R. Jones, *Appl. Phys. A.*, **1986**, *40*, 1.
   l) F. A. Houle, R. J. Wilson, and T. H. Baum, *J. Vac. Sci. Technol., A*, **1986**, *4*, 2452.
   m) D. Braichotte and H. van den Bergh, *Springer Ser. Opt. Sci.*, **1985**, *48*, 38.
   n) F. A. Houle, C. R. Jones, T. H. Baum, C. Pico, and C. A. Kovac, *Appl. Phys. Lett.*, **1985**, *46*, 204.
   o) C. R. Jones, F. A. Houle, C. A. Kovac, and T. H. Baum, *Appl. Phys. Lett.*, **1985**, *46*, 97.

- p) F. A. Houle, C. R. Jones, R. Wilson, and T. H. Baum, in *Laser Chemical Processing of Semiconductor Devices, Extended Abstracts*, F. A. Houle, T. F. Deutsch, and R. M. Osgood, Jr., Eds., MRS Proceedings Symposium B, 1984 Fall Meeting, Boston, Materials Research Society, Pittsburgh: **1984**, pp. 64 - 66.
- q) D. Braicholle and H. van den Bergh, *Springer Ser. Chem. Phys.,* **1984**, *39*, 183.
- r) R. L. Van Hemert, L. B. Spendlove, and R. E. Sievers, *J. Electrochem. Soc.*, **1965**, *112*, 1123 - 1126.

2. For an excellent review on the early work in this area, see: L. M. Tonge, D. S. Richeson, T. J. Marks, J. Zhao, J. Zhang, B. W. Wessels, H. O. Marcy, and C. R. Kannewurf, in: *Electron Transfer in Biology and the Solid State: Inorganic Compounds With Unusual Properties, Part III*; M. K. Johnson, R. B. King, D. M. Kurtz, Jr., C. Kutal, M. L. Norton, and R. A. Scott, Eds., American Chemical Society Advances in Chemistry Series, Number 226; American Chemical Society: Washington, D. C.; **1990**, pp. 351 - 368.

3. a) W. S. Rees, Jr. and C. R. Caballero, *Advanced Materials for Optics and Electronics*, **1992**, *1 (1)*, in the press.
   b) W. S. Rees, Jr., manuscript in preparation, to be submitted for publication.

4. R. C. Mehrotra, R. Bohra, and D. P. Gaur, *Metal β-Diketonates and Allied Derivatives*, Academic: New York; **1978**.

5. *Chemical Rubber Company Handbook of Chemistry and Physics, 58th Edition*, R. C. Weast, Ed., The Chemical Rubber Company: Cleveland, Ohio; **1977 - 1978**, pp. D-51 - D-60.

CHEMICAL VAPOR DEPOSITION OF TUNGSTEN AND MOLYBDENUM FILMS FROM M(η^3-C$_3$H$_5$)$_4$ (M=Mo, W)

Rein U. Kirss*, Jian Chen and Robert B. Hallock, Northeastern University, Department of Chemistry, Boston, MA 02115

Abstract

Chemical vapor deposition using tetra(allyl) tungsten and molybdenum precursors yielded amorphous tungsten and molybdenum carbide films on pyrex substrates. The films were characterized by Auger, ESCA, SEM, XRD and resistivity measurements. Volatile pyrolysis products consisted primarily of propene, C_3H_6.

Introduction

Microelectronic devices based on gallium arsenide technology require chemical and thermal compatibility of the metallization layers with the substrate. Chemical reactions between the gallium arsenide substrate and a metallic conductor which result in degradation of the interface and lead ultimately to the failure of the device. Such reactions are accelerated at elevated operating temperatures. The work of Williams and co-workers has led to the identification of the transition metals which are chemically inert to group III-V semiconductors. For gallium containing semiconductors, only tungsten is chemically stable toward reactions with either gallium or arsenic. [1] For indium containing semiconductors, molybdenum, osmium, rhenium, and tantalum can be added to the list of chemically unreactive conductors. Comparing the resistivities of these five metals, tungsten and molybdenum are most attractive candidates for metallization as they offer the lowest resistivities.
   Of the competing technologies available for the preparation of thin films (e. g. sputtering, molecular beam epitaxy, MBE, and chemical vapor deposition, CVD), CVD offers advantages in conformal coverage and scale-up of the process over the other deposition methods. [2] Current methods for CVD of Mo and W using volatile metal halides and a reducing gas such as hydrogen or silane suffer from the flammability of the reducing gas and the corrosive properties of the transition metal halides. The corrosivity of hydrogen halide products is of particular concern for deposition on III/V semiconductors. [3] Films prepared from tungsten carbonyl, $W(CO)_6$, were found to be contaminated with carbon and oxygen,

requiring a high temperature anneal (700°C) to reduce the residual contaminants to acceptable levels (e. g. < 0.1% Oxygen). [4]

Organometallic compounds with carefully selected ligand sets are attractive precursors for reducing growth temperatures without sacrificing purity by yielding hydrocarbon products, exclusively, upon pyrolysis. The search for alternate precursors is limited by at least two criteria: the source reagent must be sufficiently volatile to permit transport to the reactor and the reagent must decompose in the reactor to yield only the desired element. [5] Successful application of homoleptic allyl complexes of rhodium, iridium, and palladium for the CVD of high purity thin films [1b, 6] had been reported and prompted the present investigation of $W(\eta^3-C_3H_5)_4$ and $Mo(\eta^3-C_3H_5)_4$ as precursors for the CVD of tungsten and molybdenum films. $W(\eta^3-C_3H_5)_4$ was reported to react with hydrogen at 60°C producing tungsten metal, however, purity and application in CVD was not reported. [7]

Experimental

Tetra(allyl)tungsten, $W(\eta^3-C_3H_5)_4$, and tetra(allyl)molybdenum, $Mo(\eta^3-C_3H_5)_4$, were prepared using literature procedures. [8] The deposition apparatus consisted of a 1" O. D. pyrex tube placed in a Lindbergh tube furnace. The precursor (0.25-0.5 g) was sublimed directly into the hot zone under vacuum (10-20 mTorr). Substrates were supported on an alumina boat inside the reactor tube. Highly reflective amorphous films were deposited on pyrex glass and silicon substrates over the temperature range 300-400°C at 10-20 mTorr. Pyrex substrates were cleaned with hexane prior to deposition while the silicon substrates were rinsed sequentially with dilute HF and distilled water. During deposition studies performed under hydrogen, H_2 was leaked into the reactor through a needle valve. The reactor pressure was monitored with a Welch vacuum gauge (760 -10^{-3} Torr range).

A one piece glass trap was constructed with ground glass joints and two stopcocks to allow isolation and removal of the trap from the pyrolysis apparatus. After deposition of a metal film, the trap was removed and attached to a vacuum line. Solvent ($CDCl_3$ or C_6D_6) and the contents of the trap were vacuum transferred to an NMR tube cooled to -196°C. The tube was flame sealed, thawed and analyzed by ^1H NMR spectroscopy. Products were identified by comparison of the chemical shifts with authentic samples.

Results and Discussion

Dark, reflective films were deposited from $M(\eta^3-C_3H_5)_4$ on pyrex glass over the temperature range 300-400°C at 1-2 x 10^{-2} Torr. Below 250°C for tungsten and below 350°C for molybdenum, no deposition was observed. Some of the tungsten precursor was recovered from the walls of the reactor tube at the exit from the furnace at 250°C and identified by ^1H NMR. [8] Under the ultimate vacuum attained in the deposition reactor (10^{-2} Torr), the sublimation of the precursors was accompanied by decomposition to black residues in the sublimation vessel. The films were found to be smooth, featureless, and amorphous by scanning electron microscopy and x-ray diffraction, respectively. For Rh, Ir and Pd deposition from homoleptic allyl derivatives, reflective, amorphous films were formed on pyrex glass substrates. Adhesion of the films was judged to be good based on the Scotch tape test.

For tungsten-containing films grown under vacuum, growth rates ranged from 0.05-0.18 μ/h at 10-20 mTorr (Table 1). These growth rates were similar to those observed for deposition of Rh (0.15μ/h) and Ir (0.10 μ/h) from the corresponding tris(allyl) metal complexes at a pressure of 3-10x 10^{-2} Torr. [6b] Auger electron spectroscopy of the films revealed approximately 40 atom % tungsten, 44 atom % carbon, 10 atom % oxygen and 6 atom % silicon after sputtering through the surface layers. The level of oxygen contamination was similar to the 10% observed in WC films prepared from $(Me_3CCH_2)_3W\equiv CCMe_3$, despite base pressures of only 10^{-2} Torr and was consistent with observations with other CVD precursors. [9] The films appeared to be resistant to aqua regia, a property associated with metal carbides. [10] Binding energies from ESCA analysis were used to differentiate between metal, metal bound oxygen, and metal bound vs free carbon. For a tungsten film deposited at 400°C under vacuum, electron binding energies of 31.69 eV (width 1.67 eV), and 33.78 eV (width 1.44 eV) were observed in the W(4f) region after sputtering with Ar^+. Reported binding energies for tungsten metal are 31.2 and 33.4 eV, respectively for the $W(4f_{7/2})$ and $W(4f_{5/2})$ electrons. [11] Binding energies of 31.77 and 33.91 eV were reported for WC prepared by the CVD from $(Me_3CCH_2)_3W\equiv CSiMe_3$ and 31.7 and 33.9 eV for sputtered WC, suggesting that tungsten carbides were present in the films grown from $W(\eta^3-C_3H_5)_4$. [9] For a molybdenum film deposited at 350°C under vacuum, an electron binding energy of 228.7 eV (width 1.60 eV) was observed in the Mo(3d) region after sputtering with Ar^+. Electron binding energies for Mo metal (Mo $3d_{5/2}$) were reported as 227.6 eV. [12] A molybdenum evaporation boat coated with carbon paint was resistively heated to generate a thin molybdenum carbide film. An electron binding energy of 228.3 eV (width 1.60 eV) was observed in the Mo(3d) region for this sample, suggesting that the films prepared from $Mo(\eta^3-C_3H_5)_4$ were primarily molybdenum carbides. The observed C_{1s} binding energies of 283.9 (width 1.54. eV) and 283.6 eV (width 1.77. eV), for W and Mo films, respectively were consistent with the presence of carbidic carbon (283.7 eV from the CVD of WC, 283.6 eV for molybdenum carbide prepared as above). [9] Some graphitic carbon was observed at 285.0 eV (width 1.57 eV) in all samples. [13] Four-point sheet resistivity measurements of the tungsten-containing films indicated that the films were insulating ($\rho > 10^6$ $\mu\Omega$-cm, the limit of the apparatus). Slightly better success was achieved in the deposition of molybdenum containing films from Mo(η^3-C_3H_5)$_4$ on pyrex and silicon substrates, although higher pyrolysis temperatures were required. Sheet resistance measurements ($\sim 10^6$ $\mu\Omega$-cm) were approximately six orders of magnitude greater than for pure molybdenum (5.7 $\mu\Omega$-cm). Attempts to prepare pure metal films by deposition under hydrogen yielded metal oxide films. The latter result can be traced to a failure to scrub the hydrogen carrier gas.

Table I: Characteristics of Tungsten and Molybdenum Containing Films on Pyrex Glass Grown From M(η^3-C_3H_5)$_4$ (M=Mo, W)

Growth Temp. (°C)	M	P$_{H2}$ (mTorr)	Thickness μ	Growth [a] Rate (μ/h)	Resistivity [b] ρ ($\mu\Omega$-cm)
450	Mo	0	0.72	0.080	5.1×10^6
	Mo	30	0.14	0.028	8.4×10^3
400	W	0	1.44	0.18	insulating
	W	30	0.42	0.052	insulating
	Mo	0	0.73	0.080	3.2×10^6
	Mo	30	0.67	0.11	4.5×10^4
350	W	0	1.36	0.14	insulating
	Mo	0	0.46	0.058	insulating
300	W	0	0.57	0.057	insulating
	W	30	0.33	0.041	1.0×10^6

a. from SEM measurement of thickness as a function of time
b. four point measurement $\rho = \frac{\text{Resistance} \times \text{Area}}{\text{length}}$

The volatile products from the deposition reactions using M(η^3-C_3H_5)$_4$ precursors were trapped at -196°C and analyzed by ^{1}H NMR spectroscopy. In all cases, propene and propane were the major products, accounting for >95% of the products. Minor amounts of benzene, ethylene and ethane were observed. Hexadienes and other C_6

products were not detected. The ratio of C3 products was independent of temperature over the range 350-450°C. The absence of C6 hydrocarbon products argues against a bond homolysis pathway involving the intermediacy of allyl radicals. Pyrolysis of (η^3-C3H5)2Pd produced carbon-free films and a mixture of propene (67%) and hexadiene (33%). [6a] The formation of hexadiene in the latter case was proposed to result from a bond homolysis pathway involving generation of allyl radicals. The gas phase thermal decomposition of 1,1'-azo-2-propene produced 1, 5-hexadiene with less than 0.1% propene, while diene products, exclusively, were observed in the pyrolysis of three isomeric methylallyl azo derivatives (E, E)-4, 4'-azo-2-isobutene, (Z, Z)-4, 4'-azo-2-isobutene, and 3, 3'-azo-isobutane. [14] Generation of 2-methylallyl radicals by addition of H· to 1, 3-butadiene at low temperature demonstrated that allylic-allylic radical reactions occurred exclusively by combination and not by disproportionation. [15] High purity HgTe and CdTe films have been deposited using (η^1-C3H5)2Te. [16] Coupling of allyl radicals to hexadiene in the pyrolysis of both (η^1-C3H5)N2(η^1-C3H5) and (η^1-C3H5)2Te was the dominant reaction observed upon pyrolysis. [17] Carbon rich films have been reported in ZnSe growth using (η^1-C3H5)2Se as a precursor for Se. [18] In the latter case, a 2:1 ratio of propene to hexadiene was observed for the volatile products. [17] An intramolecular "ene" type pathway was proposed to explain the high carbon content of the pyrolysis residues. A similar pathway was used to rationalize the observation of propene and oligomers derived from CH2=CHC(S)H in the pyrolysis of di(allyl)sulfide. [19] Based on these data, a bond homolysis pathway appeared to correlate with primarily C6 products and carbon-free thin films.

Three alternative pathways, β-hydride elimination, α-hydride eliminations and C-H activation of vinylic C-H bonds, were considered to account for the observed distribution and nature of the volatile products as well as the elemental composition of the films. Thermal β-hydride elimination reactions of an η^3-allyl ligand was recently demonstrated in the formation of allene and propyne complexes upon heating a benzene solution of Cp*2Ta(η^3-C3H5) (reaction 1). [20] The

1

absence of allene and propyne in our pyrolysis and film growth studies using M(η^3-C3H5)4 argue against intramolecular β-hydride elimination pathways. Furthermore, β-hydride elimination pathways do not readily account for the formation of strong metal carbon bonds implied by the metal-bound carbon detected in the films. Pathways involving α-hydride elimination or vinylic C-H activation both increase the metal-carbon bond order are tentatively proposed to account for the observed analytical results on the Mo and W containing films (reactions 2 and 3). Stable tungsten alkylidene and propadienylidene complexes (e. g. (CO)5W=C=C=CR2 for R=^{i}Pr, ^{t}Bu, Ph) are known. [21] The ^{1}H NMR spectrum of M(η^3-C3H5)4 is consistent with unsymmetrically bound allyl ligands. [8] Decomposition pathways involving activation of the vinylic C-H bonds of the allyl group and hydrogen migration may generate a metallocyclobutene with a W=C bond (reaction 3). Tungsten metallocycles and alkylidene compounds are well documented. [22] Further experiments directed toward understanding the pyrolysis pathways for metal allyl complexes and their role in forming carbon impurities in metal films are in progress.

$$M(\eta^3\text{-}C_3H_5)_4 \xrightarrow{-C_3H_6} \left[(\eta^3\text{-}C_3H_5)_2 M\diagdown\!\!=\!\!\diagup\right] \xrightarrow{-RH} MC_x \quad \quad 2$$

M=Mo, W

(reaction 3 scheme)

M=Mo, W; R=C3H5

Acknowledgment

The authors are indebted to Professor Joseph Gardella and Mr. Gary Jones at the Center for Biosurfaces, funded by NSF as an Industry University Cooperative Research Center, located at SUNY Buffalo for the ESCA spectroscopy. This work has also been supported in part by the Air Force Office of Scientific Research, AFOSR 91-0207.

References

1. a. H. D. Kaesz, R. S. Williams, R. F. Hicks, J. I. Zink, Y-J. Chen, H-J. Muller, Z. Xue, D. Xu, D. K. Shuh, and Y. K. Kim, New J. Chem., 14, 527 1990. b. H. D. Kaesz, R. S. Williams, R. F. Hicks, Y-J. Chen, Z. Xue, D. Xu, D. K. Shuh, and H. Thridandam, in Chemical Perspectives of Microelectronic Materials, edited by M. E. Gross, J. Jasinski, and J. T. Yates (Mater. Res. Soc. Proc. 131, Pittsburgh, PA 1989) p. 395.
2. M. L. Green and R. L. Levy, R. A., J. Metals June, 83 (1985).
3. a. A. Sherman, "Chemical Vapor Deposition for Microelectronics", (Noyes Publications, Park Ridge, N. J., 1987). b. G. A. Prinz, Phys. Rev. Lett., 54, 1051 (1985). c. J. R. Creighton, J. Electrochem. Soc., 136, 271 (1989)., d. P. I. Lee, J. Cronin, and C. Kaanta, ibid., 136, 2108 (1989).
4. a. M. Diem, M. Fisk, and J. Goldman, Thin Solid Films, 107, 39 (1983). b. G. J. Vogt, J. Vac. Sci. Technol., 20, 1336 (1982). c. I. M. Watson, J. A. Connor, and R. Whyman, Polyhedron, 8, 1794 (1989). d. N. S. Gluck, G. J. Wolga, C. E. Bartosch, W. Ho, and Z. Ying, Appl. Phys. Lett., 61, 998 (1987).
5. For a recent review of organometallic precursors for CVD see G. S. Girolami, and J. E. Gozum, J. E. in Chemical Vapor Deposition of Refractory Metals and Ceramics, edited by T. M. Besmann and B. M. Gallois (Mater. Res. Soc. Proc. 168, Pittsburgh, PA, 1990) p. 319.
6. a. J. E. Gozum, D. A. Pollina, J. A. Jensen, and G. S. Girolami, J. Am. Chem. Soc., 110, 2688 (1988). b. D. C. Smith, S. G. Pattillo, N. E. Elliott, T. G. Zocco, C. J. Burns, J. R. Laia, and A. P. Sattelberger, in Chemical Vapor Deposition of Refractory Metals and Ceramics, edited by T. M. Besmann and B. M. Gallois (Mater. Res. Soc. Proc. 168, Pittsburgh, PA, 1990) p. 369.
7. W. Oberkirch, PhD thesis, Techn. Hochschule, Aachen., 1963.
8. a. R. Benn, T. H. Brock, M. C. F. B. Dias, P. W. Jolly, A. Rufinska, G. Schroth, K. Seevogel, and B. Wassmuth, Polyhedron, 9, 11 (1990). b. F. A. Cotton, and J. R. Pipal, J. Am. Chem. Soc., 93, 5441 (1971).
9. Z. Xue, K. G. Caulton, and M. H. Chisholm, Chem. Mater., 3, 384 (1991).
10. P. Schwarzkopf, and R. Kieffer, "Refractory Hard Metals" (Macmillan Co., New York, 1953.)
11. R. L. Colton, and J. W. Rabalais, Inorg. Chem., 15, 236 (1976).
12. D. R. Wheeler, and W. A. Brainard, J. Vac. Sci. Technol., 15, 24 (1978).
13. L. Ramquist, B. Ekstig, E. Kallne, E. Norelund, and R. Manne, J. Phys. Chem. Solids 30, 1849 (1969).
14. R. J. Crawford, J. Hamelin, and B. Strehlke, J. Am. Chem. Soc., 93, 3810 (1971).
15. R. Klein, and R. D. Kelley, J. Phys. Chem., 79, 1780 (1975).
16. R. Korenstein, W. E. Hoke, P. J. Lemonias, K. T. Higa, and D. C. Harris, J. Appl. Phys. 62, 4929 (1987).
17. R. U. Kirss, D. W. Brown, K. T. Higa, and R. W. Gedridge, Jr.,

Organometallics, <u>10</u> 3589 (1991).
18. S. Patnaik, K. F. Jensen, and K. P. Giapis, J. Cryst. Growth, <u>107</u>, 390 (1991).
19. G. Martin, M. Ropero, and R. Avila, Phosphorous and Sulfur, <u>13</u>, 213 (1982).
20. V. C. Gibson, G. Parkin, and J. E. Bercaw, Organometallics, <u>10</u>, 220 (1991).
21. M. I. Bruce, Chem. Rev., <u>91</u>, 197 (1991), and references therein.
22. R. Davis, and L. A. P. Kane-Maguire, in <u>Comprehrensive Organometallic Chemistry Volume 3</u>, edited by G. Wilkinson, F.G.A. Stone, and E.W. Abel, (Pergamon Press, NY 1982), Chapt. 29.2 and references therein.

THE CHEMICAL VAPOR DEPOSITION OF PURE NICKEL AND NICKEL BORIDE THIN FILMS FROM BORANE CLUSTER COMPOUNDS

SHREYAS KHER AND JAMES T. SPENCER*

Department of Chemistry and the Center for Molecular Electronics, Syracuse University, Syracuse, New York 13244-4100

ABSTRACT

Several borane cluster compounds, such as pentaborane(9) and their corresponding metal complexes have been investigated in our laboratory for their utility as unique source materials for synthesizing metal/metal boride thin films by MOCVD. In this paper we report the preparation of thin films of nickel boride from the thermal decomposition of *nido*-pentaborane(9) in the presence of anhydrous nickel chloride [$NiCl_2$] in the vapor phase. Crystalline nickel boride thin films of controlled composition ranging from 0.1 to several microns have been readily prepared by controlling the temperature and the flow rate of the pentaborane(9) into the reaction chamber. The nickel boride films on GaAs were thermally annealed to form the Ni_7B_3 phase as hexagonal crystals in a Ni_3B matrix. These films have been characterized by AA, AES, EDXA, SEM, XRD and electron diffraction. The phases were determined primarily by X-ray and electron diffraction experiments.

INTRODUCTION

The formation of high purity metal boride thin films, such as nickel boride, in a controlled stoichiometric fashion has received considerable interest primarily due to the variety and importance of the applications of these materials [1]. The fundamental interest in the preparation, theoretical modelling and solid state characteristics of these metal borides arises both from their unique physical properties among solid state materials and to their wide structural diversity. Metal borides are typically extremely refractory materials with melting points frequently far in excess of pure metal or other metal binary systems. These metal borides are also exceptionally hard materials which are not significantly affected even in the most rigorous of chemical environments. The transition and lanthanide metal borides are typically rather good electrical conductors with some displaying superconducting properties at low temperatures. Finally, the incorporation of boron in metal films, especially nickel, has been shown to greatly enhance the strength and hardness of the alloy [13]. Thus, metal borides have found increased use in not only traditional applications such as hard coatings for cutting tools [3-6] but also in thermally and chemically taxed aerospace components, high-energy optical systems [7] and as new magnetic materials [8-12]. Relatively strain-free nickel boride thin films have been prepared previously from the pyrolytic chemical vapor deposition of mixtures of nickel tetracarbonyl, diborane, and carbon monoxide in an inert carrier [6,13]. The intrinsic difficulties in this process arise from the use of both nickel tetracarbonyl and diborane which are relatively expensive and extremely toxic and flammable reagents. Alternative methods for the facile preparation of metal boride thin films of controlled stoichiometry which do not rely upon these precursors is therefore of importance.

Numerous techniques have been studied for the preparation of these materials including molecular beam epitaxy (MBE), sputtering and chemical vapor deposition (CVD). The control of the stoichiometry in multicomponent films prepared by chemical vapor deposition techniques has relied on varying the ratio of individual source compounds in the vapor phase. These source materials typically deposit at significantly different rates on the substrate at a given temperature, often making the formation of a homogeneous film difficult.

The formation of metallic thin-films with varying boron content has recently been explored using the borane polyhedral cluster-assisted deposition process (CAD) [14,15]. Through the use of this borane cluster chemical transport and deposition process, films ranging in thickness from 0.1 micron to several microns have been readily prepared with controlled composition. This technique has been shown to be generally applicable to the formation of a variety of metal boride materials including members of both the transition and lanthanide metal series.

In this paper we report the preparation of nickel boride thin films using the cluster assisted deposition technique from the thermal decomposition of *nido*-pentaborane(9) in the presence of anhydrous nickel chloride [NiCl$_2$] in the vapor phase. These films have been characterized by AA, AES, EDXA, SEM, XRD and electron diffraction. The initially formed nickel boride films were annealed to form the Ni$_7$B$_3$ phase as hexagonal crystals in a Ni$_3$B matrix. Finally, we report the identification the phases present in the nickel boride films in both the as-deposited and annealed films by X-ray and electron diffraction experiments.

EXPERIMENTAL

Materials. Pentaborane(9) was from our laboratory stock. Decaborane(14) was obtained from the Callery Chemical Company and sublimed under vacuum prior to use. The anhydrous nickel(II)chloride (99%) and the standard solutions for atomic absorption spectrophotometry were used as received from the Strem Chemical Company and the Aldrich Chemical Company, respectively.

Thin Film Formation. The nickel boride thin films were prepared by the application of a vapor phase borane cluster-assisted deposition technique [14,15]. In the experiment, using inert atmosphere techniques [16], 1.0 g (17 mmol) of anhydrous (99%) nickel(II)chloride was placed in a quartz boat in the quartz reaction chamber shown schematically in Figure 1. A flask containing several grams of neat pentaborane(9) liquid at room temperature, B$_5$H$_9$, was connected

Figure 1. Apparatus for the Preparation of Nickel Boride Thin Films [15].

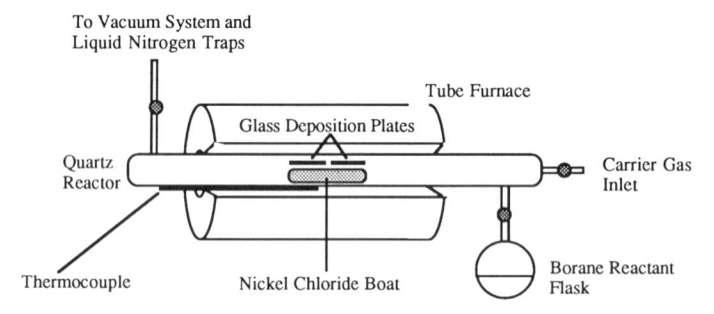

to the reactor (pentaborane vapor pressure at 25°C = 209 Torr [16]). The pentaborane reservoir flask was maintained at -78° C during the entire experiment by use of an external constant temperature bath jacketing the flask (acetone/dry ice). The entire apparatus, including the flask containing the pentaborane(9), was evacuated for 15 min., during which time a vacuum of 1 x 10^{-5} Torr was achieved. The reactor was slowly heated to 555°C under dynamic vacuum. After obtaining a stable temperature, the stopcock to the pentaborane(9) flask was opened slightly to allow a vapor of the borane to pass over the hot NiCl$_2$ while under dynamic vacuum conditions. The unreacted pentaborane(9) and other reaction byproducts were trapped downstream in a liquid nitrogen-cooled trap. The deposition was continued for 1 to 3 hours, during which time the formation of a mirror-like thin film was observed to coat the walls of the reactor and the deposition substrates held above the $NiCl_2$ boat (Figure 1). The stopcock to the borane flask was then closed and the reactor was allowed to cool slowly to room temperature. The reactor was filled with dry nitrogen and the film was removed from the reactor for further study. Free standing (substrate-free) films were obtained from the walls of the reactor and was found to be very reflective, flexible and strong materials. Film compositions ranging from 54% to 99% nickel have been prepared from these experiments. EDXA analysis showed that the film contained no chloride or other heavy element contaminations. Auger electron spectroscopy further showed that the film contained only nickel and boron. A scanning electron micrograph of a typical nickel boride film is shown in Figure 2. Similar results were obtained using decaborane(14) [15a,d].

Pentaborane(9) has the advantage of being a highly volatile liquid and can therefore be more easily controlled in a flow system than the previously used decaborane(14) cluster. Decaborane(14), however, has the advantage of being an air-stable material at room temperature while pentaborane(9) reacts violently with air.

RESULTS AND DISCUSSION

The nickel boride thin films formed from the thermal reaction of $NiCl_2$ with boron hydrides using the cluster-assisted deposition process produces uniform (AES depth profiling [15]), conformal thin film materials. The films prepared in this work from the pentaborane(9) cluster and $NiCl_2$ were formed either adhered to various substrates, such as copper, SiO_2, GaAs or pyrex, or as a free-standing (substrate-free) material. The as-deposited films were found to be primarily Ni_3B by a comparison of the X-ray diffraction data for the films and the previously reported data for the known nickel boride phases (Table I). Typical scanning electron micrographs, Figure 2A, showed that the films were relatively uniform materials which displayed relatively small surface voids. The reported structure for both Ni_3B (and Ni_7B_3), shown in Figure 2B, consists of isolated boron atoms at the center of tricapped

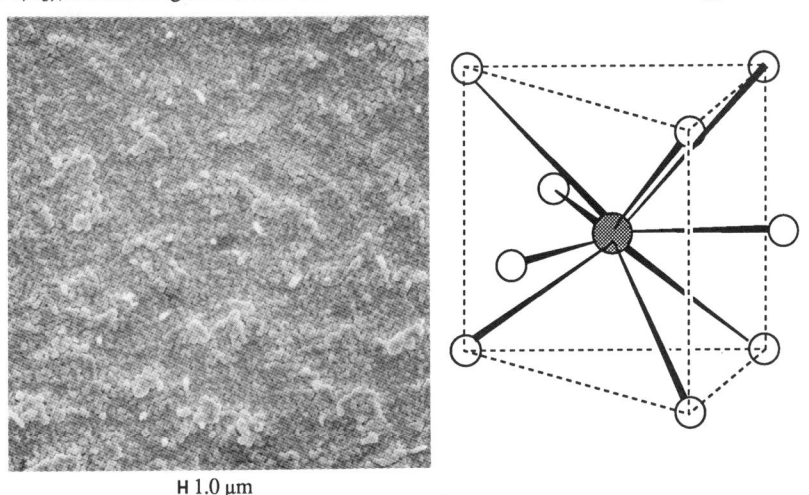

Figure 2. A. Scanning electron micrograph of a nickel boride film (Ni_3B) deposited at 555°C on a quartz substrate from anhydrous $NiCl_2$ and Pentaborane(9) in vacuum. B. Structure of nickel boride [Ni_3B and Ni_7B_3]. The shaded sphere is a boron atom and the open spheres are nickel atoms (trigonal prismatic arrangement).

trigonal prisms of metal atoms [17,18]. Upon annealing, significant crystalline modifications occur. After the annealing process, perfect hexagonal crystallites were observed embedded in a rather uniform, conformal matrix (Figure 3). X-ray diffraction experiments show that the bulk of the sample consisted of the Ni_3B and Ni_7B_3 phases with experimental data identical to the literature data (Table I). In order to establish the phase identity of these hexagonal crystallites, however, electron diffraction experiments of the crystals were undertaken. A "close-up" scanning electron micrograph view and the corres-ponding electron diffraction pattern for a typical hexagonal crystallite are shown in Figure 4. A comparison of the electron diffraction data with the known data for the nickel boride phases clearly identified the hexagonal crystalline modification as the Ni_7B_3 phase. This phase apparently grew from the Ni_3B matrix during the high temperature annealing process.

Table I. Crystallographic Data for Selected Nickel-Boron Phases.

Phase/ Unit Cell	Space Group/ Struct. Type	a, Å	b, Å	c, Å	other	Intense Reflections	Ref.
Ni Cubic	Fm3m fcc	3.5238				d=2.03, 2θ=44.64, I/I$_1$=100 d=1.76, 2θ=51.95, I/I$_1$=42 d=1.25, 2θ=76.15, I/I$_1$=21	19
Ni_3B Ortho-rhombic	Pbnm Fe_3C	5.211	6.619	4.389		d=2.35, 2θ=38.30, I/I$_1$=100 d=2.12, 2θ=42.65, I/I$_1$=80 d=1.97, 2θ=46.07, I/I$_1$=80 d=3.31, 2θ=26.94, I/I$_1$=20	17,18
Ni_2B Tetragonal	I4/mcm $CuAl_2$	4.993		4.249		d=1.98, 2θ=45.83, I/I$_1$=100 d=2.50, 2θ=35.92, I/I$_1$=24 d=2.13, 2θ=42.44, I/I$_1$=22 d=3.54, 2θ=25.16, I/I$_1$=9	2-6, 14,
Ni_4B_3 Ortho-rhombic	Pbnm o-Ni_4B_3	11.953	2.981	6.569		d=2.07, 2θ=43.73, I/I$_1$=100 d=2.47, 2θ=36.37, I/I$_1$=80 d=2.25, 2θ=40.07, I/I$_1$=80 d=5.97, 2θ=14.84, I/I$_1$=20	2-5
Ni_4B_3 Monoclinic	C2/c m-Ni_4B_3	6.430	4.882	7.818	β = 103.18	d=2.18, 2θ=41.42, I/I$_1$=100 d=2.05, 2θ=44.18, I/I$_1$=100 d=1.92, 2θ=47.34, I/I$_1$=100 d=3.85, 2θ=23.10, I/I$_1$=40	2-5
Ni_7B_3 Hexagonal	P6$_3$/mmc	6.96		4.38	c:a = 0.629	d=6.036, hkl=100 d=3.550, hkl=101 d=3.485, hkl=110 d=3.018, hkl=200	20
NiB Rhombic	Pbnm FeB	2.925	7.396	2.966		d=2.31, 2θ=33.99, I/I$_1$=100 d=2.72, 2θ=32.93, I/I$_1$=74 d=1.86, 2θ=48.97, I/I$_1$=68 d=3.68, 2θ=24.18, I/I$_1$=16	7

ACKNOWLEDGEMENTS

We wish to thank the National Science Foundation (Grant No. MSS-89-09793), the Donors of the Petroleum Research Fund administered by the American Chemical Society, the Wright-Patterson Laboratory (Award No. F33615-90-C-5291), and the Industrial Affiliates

Program of the Center for Molecular Electronics for support of this work. We would also like to thank Prof. F. Machizaud (Univ. of Nancy, France) and Prof. Stephen D. Hersee (Univ. of New Mexico) for their assistance in this work.

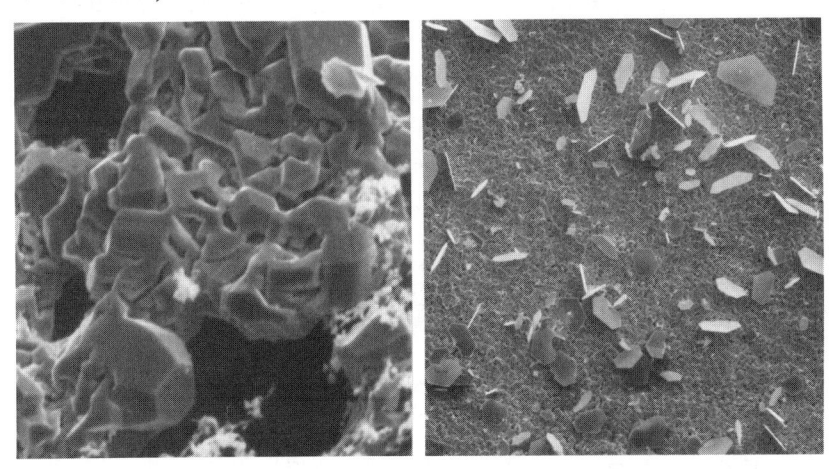

A. ⊢ 1.0 μm   B. ⊢ 10.0 μm

Figure 3. Scanning electron micrograph of annealed nickel boride films deposited from anhydrous NiCl$_2$ and Pentaborane(9) in vacuum at 555°C and annealed at 830°C for 38 h. A. Film deposited on gallium arsenide showing the large fused crystallites if Ni$_7$B$_3$. B Film deposited on quartz showing the hexagonal crystals of Ni$_7$B$_3$ embedded in the Ni$_3$B matrix.

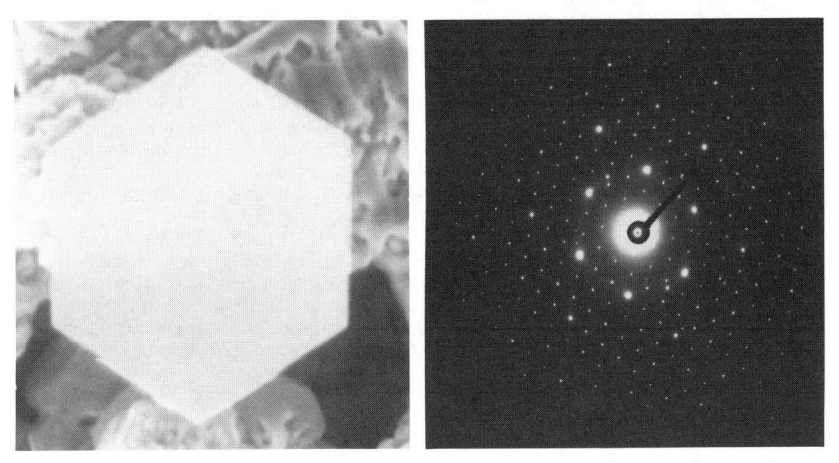

A. ⊢ 1.0 μm   B.

Figure 4. A. Scanning electron micrograph of a hexagonal crystal of nickel boride (Ni$_7$B$_3$) deposited from anhydrous NiCl$_2$ and pentaborane(9) in vacuum at 555°C and annealed at 830°C for 38 h on a quartz substrate. B. Electron diffraction pattern from the crystal shown in the SEM micrograph. The planes for the diffraction pattern are in the (001) zone.

REFERENCES

1. P. A. Dowben, J. T. Spencer and G. T. Stauf, *Mat. Sci. Eng. B* B2, 297 (1989).
2. A. N. Campbell, A. W. Mullendore, C. R. Hills and J. B. Vandersande, *J. Matl. Sci.* 23, 4049 (1988).
3. R. J. Patterson, *U.S. Patent* No. 3,499,799 (1970).
4. R. S. Lewandowski, *U.S. Patent* No. 4,522,849 (1985).
5. L. E. Branovich, W. B. P. Fitzpatrick, M. L. Long, Jr. *U.S. Patent* No. 3,692,566 (1972).
6. A. W. Mullendore and L. E. Pope, *Thin Solid Films* 267 (1987).
7. Z. Knittl, *Optics of Thin Films*, (Wiley and Sons, Publ., New York, 1976).
8. I. Bakonyi, *J. Magn. Magn. Mat.* 73, 171 (1988).
9. R. H. Mutlu and A. Aydinvraz, *J. Magn. Magn. Mat.* 68, 328 (1987).
10. N. Lundquist, H. P. Myers and R. Westin, *Phil. Mag.* 7, 1187 (1962).
11. S. N. Kaul and M. Rosenberg, *Phys. Rev.* B25, 5863 (1982).
12. I. Bakonyi, P. Panissod, J. Durand and R. Hasegawa, *J. Non-Cryst. Solids* 61/62, 1189 (1984).
13. M. Skibo and F. A. Greulich, *Thin Solid Films* 225 (1984).
14. Z. Zhang, Y.-G. Kim, P. A. Dowben and J. T. Spencer, *Mat. Res. Soc. Symp. Proc.* 131, 407 (1989).
15. (a) S. Kher and J. T. Spencer, *Chem. Mater.* in press. (b) J. A. Glass, Jr., S. Kher and J. T. Spencer, *Chem. Mater.* in press. (c) J. A. Glass, Jr., S. Kher, S. D. Hersee, G. O. Ramseyer and J. T. Spencer, *Mat. Res. Soc. Symp. Proc.* 204, 397 (1991). (d) J. A. Glass, Jr., S. Kher, Y.-G. Kim, P. A. Dowben and J. T. Spencer, *Mat. Res. Soc. Symp. Proc.* 204, 439 (1991).
16. D. F. Shriver and M. A. Drezdzon, *The Manipulation of Air-Sensitive Compounds* (Wiley-Interscience, New York, 1986).
17. (a) R. M. Minyaev and R. Hoffmann, *Chem. Mater.* 3, 547 (1991). (b) J. K. Burdett and E. Canadell, *Inorg. Chem.* 27, 4437 (1988). (c) P. Mohn and D. G. Pettifor *J. Phys. C: Solid State Phys.* 21, 2829 (1988). (d) P. Mohn, *J. Phys. C: Solid State Phys.* 21, 2841 (1988). (e) J. K. Burdett, E. Canadell and G. J. Miller, *J. Am. Chem. Soc.* 108, 6561 (1986). (f) D. G. Pettifor and R. Podloucky, *J. Phys. C: Solid State Phys.* 19, 315 (1986). (g) D. R. Armstrong, *Theor. Chim. Acta* 64, 137 (1983). (h) D. R. Armstrong, R. G. Perkins and V. E. Centina, *Theor. Chim. Acta* 64, 41 (1983). (i) H. Ihara, M. Hirabayashi and H. Nakagawa, *Phys. Rev. B: Solid State Phys.* B16, 726 (1977). (j) P. G. Perkins and A. V. J. Sweeney, *J. Less.-Common Met.* 47, 165 (1976). (k) S. H. Liv, L. Koop, W. B. England and H. W. Myron, *Phys. Rev. B: Solid State Phys.* B11, 3463 (1975).
18. G. V. Samsonov, *Handbook of High-Temperature Materials, No. 2 Properties Index*, (Plenum Press, New York, 1964).
19. (a) R. Thompson, *Prog. Boron Chem.* 2, 173 (1970). (b) P. Schwarzkopf, R. Kieffer, W. Leszynski and K. Benesovsky, *Refractory Hard Metals, Borides, Carbides, Nitrides, and Silicides*; (MacMillan, New York, 1953). (c) B. Aronsson, T. Lundstrom, S. Rundqvist, *Borides, Silicides and Phosphides*, (Wiley, New York, 1965). (d) V. I. Matkovich, *Boron and Refractory Borides*, (Springer-Verlag, New York, 1977). (e) G. V. Samsonov, Y. M. Goryachev and B. A. Kovenskaya, *J. Less.-Common Met.* 47, 147 (1976). (f) W. N. Lipscomb, *J. Less.-Common Met.* 82, 1 (1981). (g) J. Etourneau and P. Hagenmuller, *Philos, Mag.* 52, 589 (1985).
20. (a) B. Punge-Witteler and U. Koster *Mat. Sci. Eng.* 97, 343 (1988). (b) A. Y. Petrov, V.V. Kovalev and M.M. Markus, *Dokl. Akad. Nauk SSSR* 198, 389 (1971). (c) F. Machizaud, F.-A. Kuhnast and J. Flechon *J. Physique* 97 (1981). (d) F. A. Kuhnast, F. Machizaud, R. Vangelisti and J. Flechon *J. Microsc. Spectrosc. Electron* 4, 55 (1979). (e) F. Machizaud, Personal Communication.

DEPOSITION OF TUNGSTEN NITRIDE THIN FILMS FROM (ᵗBuN)$_2$W(NHᵗBu)$_2$

Hsin-Tien Chiu and Shiow-Huey Chuang
Department of Applied Chemistry, National Chiao Tung University, Hsinchu, Taiwan, 30050, R. O. C.

ABSTRACT

The possibility of growing tungsten nitride thin films from (ᵗBuN)$_2$W(NHᵗBu)$_2$, a single-source molecular precursor with two nitrogen to tungsten double bonds, by low pressure chemical vapor deposition (LPCVD) was investigated. Deposition of thin films on silicon and glass substrates was carried out at temperatures 500 - 650 °C in a cold-wall reactor while the precursor was vaporized at 60 - 100 °C. Elemental composition of the thin films, studied by wavelength dispersive spectroscopy (WDS), is best described as WN$_x$ (x = 0.8 - 1.8). Elemental distribution within the films, studied by Auger depth profiling, is uniform. X-ray diffraction (XRD) studies show that the films have a cubic structure with a lattice parameter a = 4.14 - 4.18 Å. A stoichiometric WN thin film has a lattice parameter a equal to 4.154 Å. Volatile products, trapped at -196°C, were analyzed by nuclear magnetic resonance (NMR) and gas chromatography-mass spectrometry (GC-MS). Isobutylene, acetonitrile, hydrogen cyanide and ammonia were detected in the condensable mixtures.

INTRODUCTION

There are many transition metal nitrides possessing interesting mechanical and electrical properties [1]. Many of them are potentially useful interconnect materials for diffusion barriers and gates in vary large scale integrated circuits (VLSI) [2]. Growing the thin films of these materials by conventional chemical vapor deposition (CVD) usually requires high temperatures (> 1000 °C) when metal halides and ammonia / hydrogen or nitrogen / hydrogen mixtures are employed as the precursors. Thus, their usefulness is severely limited. On the other hand, there are many interests in the deposition of mixed-element thin films from single-source precursors, molecules having all of the desirable elements. Usually, metallo-organic compounds are employed here as the precursors to grow thin films at temperatures much lower than the ones recorded previously. For example, metal carbides and oxides, such as TiC, WC, SiO$_2$, and TiO$_2$, have been deposited successfully into thin films by this route using Ti(CH$_2$ᵗBu)$_4$, (ᵗBuCH$_2$)$_3$W≡CᵗBu, Si(OEt)$_4$, and Ti(OⁱPr)$_4$, respectively [3 - 6]. Previously, our laboratory has shown that an ethylimido tantalum complex, (Et$_2$N)$_3$Ta=NEt, can be used to deposit polycrystalline cubic TaN thin films at relatively low temperatures 500 - 650 °C, with minor carbon incorporation [7]. In order to test whether it is a general route to metal nitrides by using metal imido complexes as the precursors, we performed the following metallo-organic chemical vapor deposition (MOCVD) experiments employing another metal-imido complex, (tBuN)$_2$W(NHtBu)$_2$, **1**, as the precursor. Our observations are reported below.

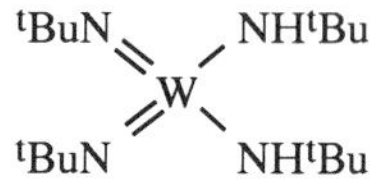

$$\text{tBuN} \underset{\text{tBuN}}{\overset{\text{NHtBu}}{\diagdown \!\!\!\! \diagup}} W \underset{\text{NHtBu}}{\overset{}{\diagup \!\!\!\! \diagdown}}$$

1

EXPERIMENTAL

(tBuN)$_2$W(NHtBu)$_2$ was synthesized and purified according to a procedure described by Nugent [8]. Deposition of thin films on silicon and glass substrates was performed in a low-pressure cold-wall reactor at temperatures 500 - 650 °C while the precursor was vaporized into the reactor at 60 - 100 °C. Argon or hydrogen flowing at 10 sccm was used as the carrier gas to assist the CVD process while the pressure was around 0.5 torr. The films were analyzed by scanning electron microscopy (SEM), wavelength dispersive spectroscopy (WDS), X-ray diffraction (XRD), and Auger electron spectroscopy (AES). An U-trap was placed between the reactor and the pump to trap condensable gas phase products at -196 °C. These products were analyzed by nuclear magnetic resonance (NMR) and gas chromatography-mass spectrometry (GC-MS).

RESULTS AND DISSCUSSION

(tBuN)$_2$W(NHtBu)$_2$ is a solid compound at room temperature and has sufficient volatility when sublimed under vacuum. Thus, it is suitable for low pressure MOCVD studies. Uniform gray and metallic shining thin films were grown on Si(100), Si(111) and glass substrates at temperatures exceeding 500 °C. The scanning electron micrograph of a typical thin film is shown in Fig. 1. The growth rates of the thin films, estimated from the thickness, were 20 - 100 Å/min. Some films on silicon substrates did not adhere well. This problem was probably due to a large mismatch of lattice

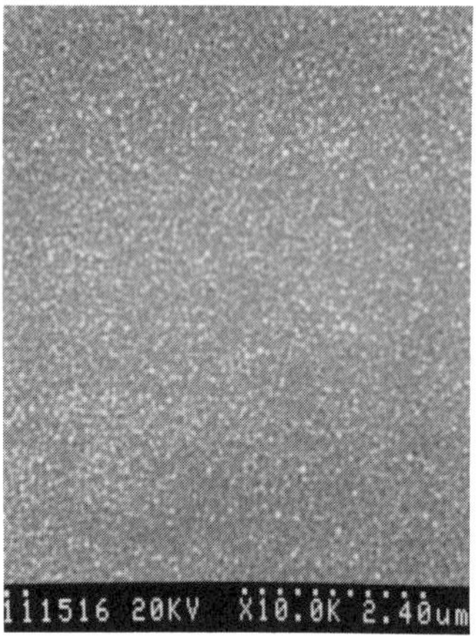

Figure 1 Scanning electron micrograph of a thin film

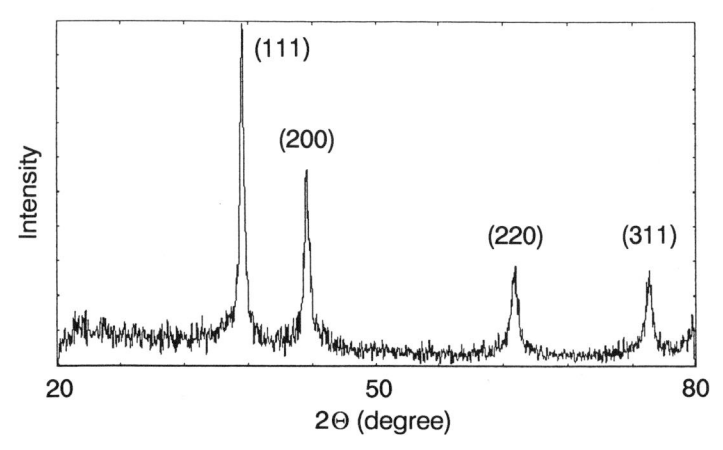

Figure 2   X-ray diffraction pattern of a thin film

parameters between the substrates and the thin films (a_{Si} = 5.43 Å, $a_{thin\ film}$ = 4.14 - 4.18 Å. See below for the thin film characterization).

The XRD pattern of a thin film, deposited at 600 °C using hydrogen as the carrier gas and annealed at the same temperature for 1 h, is shown in Fig. 2. It shows major Cu Kα peaks at angles 2Θ equal to 37.42°, 43.54°, 63.32°, and 75.75°. These peaks are characteristic of a cubic structure, a = 4.154 Å, and are assignable to (111), (200), (220), and (311) diffractions, respectively. Although the pattern does not match that of any known tungsten nitride or tungsten carbide phase exactly, it is close to the diffraction pattern of β-W$_2$N, a = 4.126 Å, which shows (111), (200), (220), and (311) reflections at 37.73°, 43.85°, 63.73°, and 76.51°, respectively [9]. Lattice parameter of the thin films versus the temperature of deposition is shown in Fig. 3. In Fig. 4, N/W and C/W ratios of

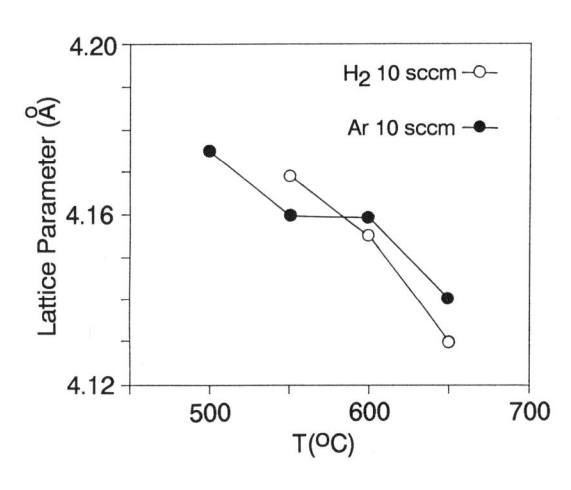

Figure 3
Effect of temperature of deposition on lattice parameter

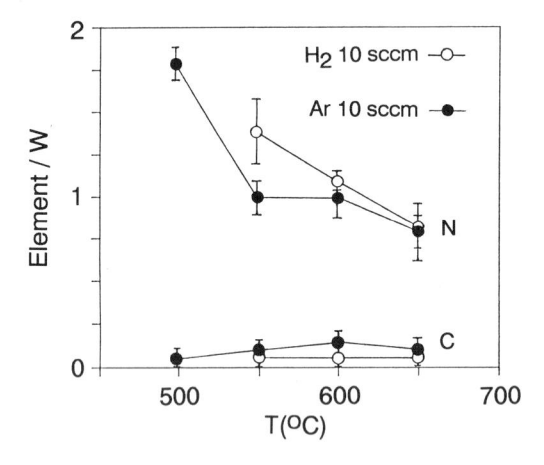

Figure 4
Effect of temperature of deposition on N/W and C/W ratios

the thin films, derived from WDS studies, are plotted against the temperature of deposition. Figs. 3 and 4 indicate that with increasing temperature of deposition, both the N/W ratio and the lattice parameter of the films decreased. Also, it is clear that the lattice parameter decreased with decreasing N/W ratio. As shown in Fig. 4, it appears that the films deposited using hydrogen as the carrier gas contain less carbon. Reduction of carbon concentration in tungsten carbide by hydrogen has been reported [10]. Auger depth profile of a thin film, shown in Fig. 5, indicates that the concentrations of tungsten, nitrogen, carbon, and oxygen on the surface were 35%, 27%, 6%, and 26%, respectively. After several hundred angstroms were removed by sputtering, uniform distribution of the elements within the film is observed. The elemental concentrations in the bulk obtained here does not agree with those found by WDS. This is attributed to the preferential sputtering and has been observed for other early transition metal nitride thin films [11].

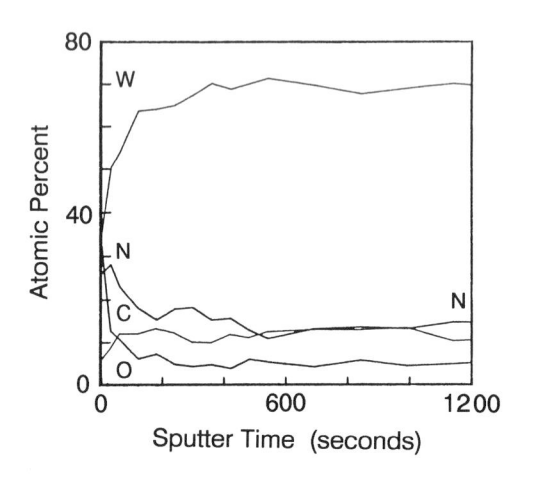

Figure 5
Auger depth profile of a thin film

Volatile products, condensed at -196 °C, were analyzed by NMR and GC-MS. The major constituent being identified is isobutylene. Other gases including acetonitrile, hydrogen cyanide, and ammonia were identified also. Detection of noncondensable gases, such as hydrogen, nitrogen, and methane, was attempted by installing a pressure gauge between the U-trap and the pump. In the deposition experiments conducted without passing any carrier gas, a pressure increase was observed by this gauge, indicating that the evolution of hydrogen, nitrogen or methane is possible. Based on these observations, we speculate that in the thermal decomposition of $(^tBuN)_2W(NH^tBu)_2$, tertbutyl groups were removed as isobutylene. The mechanism might be a γ-hydrogen elimination process followed by carbon-nitrogen bond cleavage. Acetonitrile, hydrogen cyanide, and possibly methane might be generated through a β-methyl activation process which is also responsible for the incorporation of carbon into the films.

In conclusion, this preliminary study has shown that $(^tBuN)_2W(NH^tBu)_2$, a bisimido tungsten complex, can be utilized as a single-source precursor to grow cubic tungsten nitride thin films, having the composition WN_x (x = 0.8 - 1.8), by low pressure MOCVD. The N/W ratio and the lattice parameter are affected by the temperature of deposition. Using hydrogen as the carrier gas can reduce the incorporation of carbon significantly.

ACKNOWLEDGMENTS

We appreciate the support of the National Science Council and the Ministry of Education of the Republic of China. We also thank the Instrument Center of NSC, the Institute of Materials Science and Engineering, NCTU, and the Materials Research Laboratory, ITRI, for assisting the analyses.

REFERENCES

1. L. E. Toth, Transition Metal Carbides and Nitrides, (Academic Press, New York, 1971).
2. Y. Pauleau, in Microelectronic Materials and Processes, edited by R. A. Levy (Kluwer Academic Publishers, Dordrecht, The Netherlands, 1989) pp.646 - 647; pp. 658 -660.
3. G. S. Girolami, J. A. Jensen, D. M. Pollina, W. S. Williams, A. E. Kaloyeros, and C. M. Allocca, J. Am. Chem. Soc. 109, 1579 (1987).
4. Z. Xue, K. G. Caulton, and M. H. Chisholm, Chem. Mater. 3, 384 (1991).
5. T. Jin, T. Okuhara, and J. M. White, J. Chem. Soc., Chem. Commun. 1987, 1248.
6. J.-P. Lu and R. Raj, J. Mater. Res. 6, 1913 (1991).

7. H.-T. Chiu and W.-P. Chang, J. Mater. Sci. Lett., in press.
8. W. A. Nugent and R. L. Harlow, Inorg. Chem. 19, 777 (1980).
9. "Powder Diffraction File", JCPDS International Center for Diffraction Data, (1982) File No. 25-1257.
10. F. H. Ribeiro, R. A. Dalla Betta, G. J. Guskey, and M. Boudart, Chem. Mater. 3, 805 (1991).
11. S. Ingrey, M. B. Johnson, and R. W. Streater, J. Vac. Sci. Technol. 20, 968 (1982).

DEPOSITION AND CHARACTERIZATION OF METALORGANIC CHEMICAL VAPOR DEPOSITION ZrO$_2$ THIN FILMS USING Zr(thd)$_4$

Jie Si, Chien H. Peng, and Seshu B. Desu
Department of Materials Engineering, Virginia Polytechnic Institute and State University, Blacksburg, VA 24061.

ABSTRACT

Excellent quality ZrO$_2$ thin films were successfully deposited on single crystal silicon wafers by metalorganic chemical vapor deposition (MOCVD) at reduced pressures using tetrakis(2,2,6,6–tetramethyl–3,5–heptanedionato) zirconium, [Zr(thd)$_4$]. For substrate temperatures below 530°C, the film deposition rates were very small ($\leq$ 1 nm/min). The film deposition rates were significantly affected by: (1) source temperature, (2) substrate temperature, and (3) total pressure. As–deposited films are stoichiometric (Zr/O = 1/2) and carbon free. Furthermore, only the tetragonal ZrO$_2$ phase was identified in as–deposited films. The tetragonal phase transformed progressively into the monoclinic phase as the films were subjected to high temperature post–deposition annealing. The optical properties of the ZrO$_2$ thin films as a function of wavelength, in the range of 200 nm to 2000 nm, are reported. The measured value of the dielectric constant of the as–deposited ZrO$_2$ films is around 19 in the frequency range of 5 kHz to 1000 kHz.

INTRODUCTION

ZrO$_2$ thin films, due to their variety of applications, are of considerable interest. For example, ZrO$_2$ films were utilized as thermal barrier coatings, buffer layers for the growth of high T$_c$ superconducting thin films, and in high–temperature optical filters. Many different methods, such as reactive electron beam evaporation [1], dc and rf magnetron sputtering [2], and chemical vapor deposition (CVD) [3], have been employed for the fabrication of ZrO$_2$ thin films. Among these techniques, CVD seems to be a very promising and compatible method for electronic and optical device applications. The primary issue of CVD ZrO$_2$ thin films is the identification of a good precursor. ZrCl$_4$, which was first used as a precursor for CVD ZrO$_2$, requires very high deposition temperatures (> 800°C) and could result in the formation of fine powders [4]. The organometallic alkoxides such as zirconium i–propoxide, Zr(OPr^i)$_4$, and zirconium t–butoxide, Zr(OBu^t)$_4$, are also not suitable as precursors due to their instabilities [4]. However, zirconium acetylacetonates appear to be most suitable precursors for ZrO$_2$ deposition. Use of zirconium triflurocetylacetonate, Zr(C$_5$H$_4$F$_3$O$_2$)$_4$, for MOCVD of ZrO$_2$ films was reported earlier by our research group [4]. Although good quality ZrO$_2$ films were obtained using zirconium trifluroacetylacetonate, the process parameter space is limited due to the possible fluorine contamination. Here we report on the use of an acetylacetonate without fluorine, namely, [Zr(thd)$_4$] for the deposition of excellent quality ZrO$_2$ films.

EXPERIMENTAL PROCEDURE

ZrO$_2$ thin films were deposited in a reactor which was described in a previous paper [4]. Depositions were carried out at reduced pressures (4 to 10 torr). Zr(thd)$_4$ was used as the precursor and the precursor temperature was varied from 230 to 240°C. The substrate temperature was also varied from 540 to 575°C. The precursor was carried by oxygen carrier gas into the reaction chamber. The film thicknesses were measured at the center of the silicon wafer by an ellipsometer at a wavelength of 632.8

nm. The films were also characterized by scanning electron microscopy (SEM) and electron spectroscopy for chemical analysis (ESCA) for surface morphology and stoichiometry, respectively. Both X-ray diffraction and Fourier transform infrared (FTIR) spectrometer were used for phase identification. Optical transmission spectra of the films in the range of 200 nm to 2000 nm were obtained by an UV-VIS-NIR spectrophotometer. For the dielectric property measurement, ZrO_2 thin film was deposited on the platinum-coated Si substrate and contacted with 2.14 x 10^{-4} cm^2 palladium electrodes. The dielectric constant was calculated from the film thickness and capacitance.

RESULTS AND DISCUSSION

Deposition Behavior

In order to determine the precursor volatilization temperature, the precursor, $Zr(thd)_4$, was characterized by thermogravimetric analysis (TGA). Fig. 1 shows the TGA spectrum of the precursor at heating rate of 10°C/min in flowing nitrogen. As can be seen from Fig. 1, the precursor starts to lose weight around 200°C. In order to have desirable vapor pressure, source temperatures in the range of 230 to 240°C were used for this study. The amount of source materials transported to the reactor can be controlled by adjusting the source temperature and the carrier gas flow rate. In the range of experimental parameters investigated, source temperature, substrate temperature, and total pressure, were found to have significant effects on the film deposition rate. Fig. 2 shows the variation of the deposition rate with the substrate temperature at source temperatures of 230°C and 240°C, at a pressure of 6 torr. The deposition rate increased with increasing substrate temperature and source temperature (Fig. 2). It was also found that the deposition rate increased with increasing total pressure, as can be seen in Fig. 3. Therefore, the deposition rate of ZrO_2 thin films using $Zr(thd)_4$ can be easily controlled (from 6 to 18 nm/min) by adjusting the above parameters.

Surface Morphology and Composition

As-deposited MOCVD ZrO_2 thin films were specular, crack-free, and adhered well to the substrates. The as-deposited thin films were featureless under SEM examination. Fig. 4 shows the surface morphology of the MOCVD ZrO_2 film which was annealed at 1000°C for one hour. The average grain size was estimated to be around 100 nm in this annealed sample.

The ESCA spectra of the film before and after five minutes of argon sputtering are illustrated in Fig. 5. The carbon peak in Fig. 5(a) was due to carbon adsorption on the sample surface. After five minutes of ion sputtering, only Zr and O peaks were observed (Fig. 5(b)). This indicates that the carbon content of the film is below the ESCA detectable limit. Fig. 6. shows the narrow scans of Zr_{3d} and O_{1s} peaks for an as-deposited specimen. From the presence of Zr-3d peak at 181.7 eV, we have concluded that the oxidation state of Zr is 4+. Quantitative analysis of ESCA data (Fig. 6) indicated that the film is stoichiometric ZrO_2 (Zr/O ratio = 1/2).

Phase Identification of ZrO_2 Thin Films

Fig. 7 displays the X-ray diffraction patterns for the ZrO_2 thin film as a function of post-deposition annealing temperature. Only the tetragonal phase was identified in the as-deposited thin film, while the formation of the monoclinic phase was noted in the sample which was annealed at 600°C. After the sample was subjected to a high temperature (1000°C) heat treatment, it was found that the monoclinic phase was the major phase in the thin film.

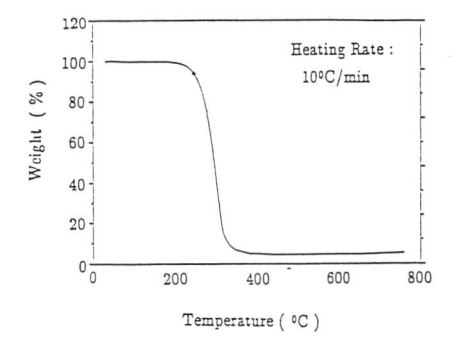

Fig. 1. Thermo-gravimetric Analysis of Zr(thd)$_4$

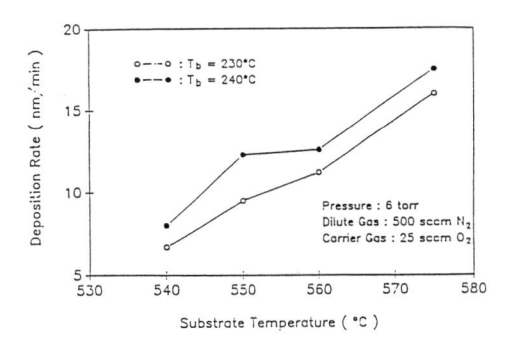

Fig. 2. Deposition Rate as Functions of Substrate and Source Temperature

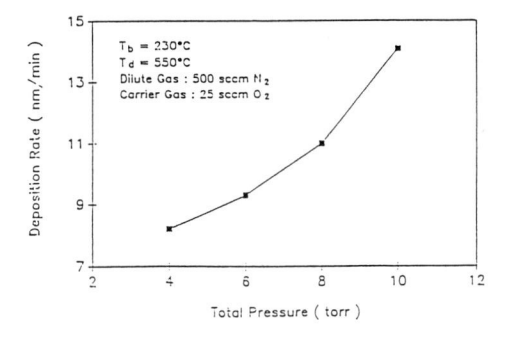

Fig. 3 Deposition Rate as a Function of Total Pressure

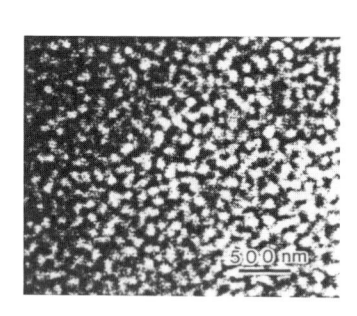

Fig. 4. SEM of ZrO_2 Film Annealed at 1000°C

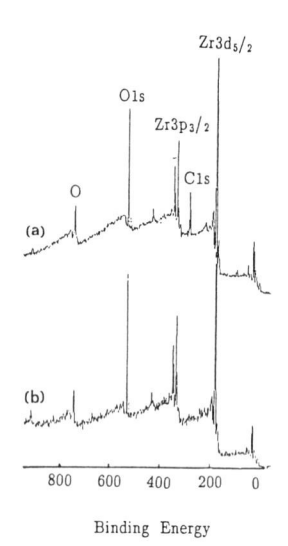

Fig. 5. ESCA Spectra of ZrO_2 Film (a) Before Sputtering (b) After Sputtering

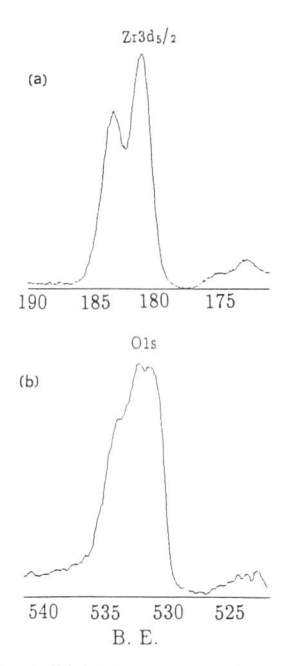

Fig. 6. ESCA Narrow Scan (a) $Zr3d_{5/2}$ (b) O1s

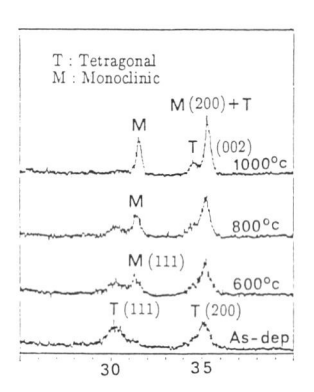

Fig. 7. X-ray Diffraction Pattern of MOCVD ZrO_2 Film (350 nm)

The FTIR absorption spectra of the same samples as a function of annealing temperature are shown in Fig. 8. The six prominent absorption peaks of the post–deposition annealed samples agreed very well with that of the CVD monoclinic ZrO_2 thin films obtained by Tauber [3]. Since the as–deposited film consisted of only tetragonal phase according to X–ray diffraction result, we attribute the two absorption peaks in the FTIR spectrum (Fig. 8) of as–deposited sample to the lattice vibration modes of the tetragonal phase. However, the absorption peaks of tetragonal phase could not be identified in the spectra of the annealed samples, which may be due to the overlap with the major peaks of the monoclinic phase.

There is a large scatter in the reported data about the phases existing in ZrO_2 thin film [1, 3, 5]. However, there is a general agreement that the number and nature of the phases in the film are a sensitive function of the deposition process, deposition parameters, grain size, and thickness [6]. In our study, the formation of the metastable tetragonal phase at the deposition temperature (e.g. 550ºC) may be related to the high deposition rate, film stress, and/or to the fine grain size. When the films were subjected to high–temperature heat treatments, the thermodynamically favored monoclinic phase formed gradually.

Optical and Electrical Properties

The UV–VIS–NIR transmission spectra of the as–deposited and annealed MOCVD ZrO_2 thin films on fused quartz substrates are shown in Fig. 9. As can be seen from Fig. 9, the transmittance drops down to 0% at $\lambda = 207$ nm (the absorption edge), while its value is around 90% in the visible and near infrared range. The envelope method [7] was used to calculate the refractive index (n) and the thickness of the film. The thickness of the film was calculated to be 370 nm. Fig. 10 shows the refractive index of the as–deposited and annealed ZrO_2 films as function of wavelength. The indices of refraction increased with the increasing annealing temperature. The change of the refractive index upon annealing can be attributed to the phase transformations of the films.

Fig. 11 shows the plot of dielectric constant, k, as function of frequency. An average value of dielectric constant is 19 and its value slightly decreased with increasing frequency. This value of the dielectric constant is larger than that of the CVD ZrO_2 film obtained from $ZrCl_4$ (k = 17) [3], but smaller than that of single–crystal monoclinic ZrO_2 (k =22) [8].

SUMMARY

Due to its stability and acceptable vapor pressure, $Zr(thd)_4$ is a preferred precursor for depositing excellent quality MOCVD ZrO_2 thin films. The as–deposited ZrO_2 thin films were fine–grained and stoichiometric. In addition, the deposition rate can be easily controlled from 6 to 18 nm/min by adjusting the deposition parameters in the range of experimental parameters investigated. Only the tetragonal phase was observed in the as–deposited films. The tetragonal phase transformed progressively into the monoclinic phase as the films were subjected to high–temperature post–deposition annealing. The FTIR spectra of both the tetragonal and monoclinic ZrO_2 thin film were obtained. The optical properties of the ZrO_2 thin films as a function of wavelength, in the range of 200 nm to 2000 nm, were investigated. The measured value of the dielectric constant of the as–deposited ZrO_2 film is around 19 in the frequency range of 5 kHz to 1000 kHz.

Fig. 8. FTIR Spectra of MOCVD ZrO$_2$ Film (350 nm)

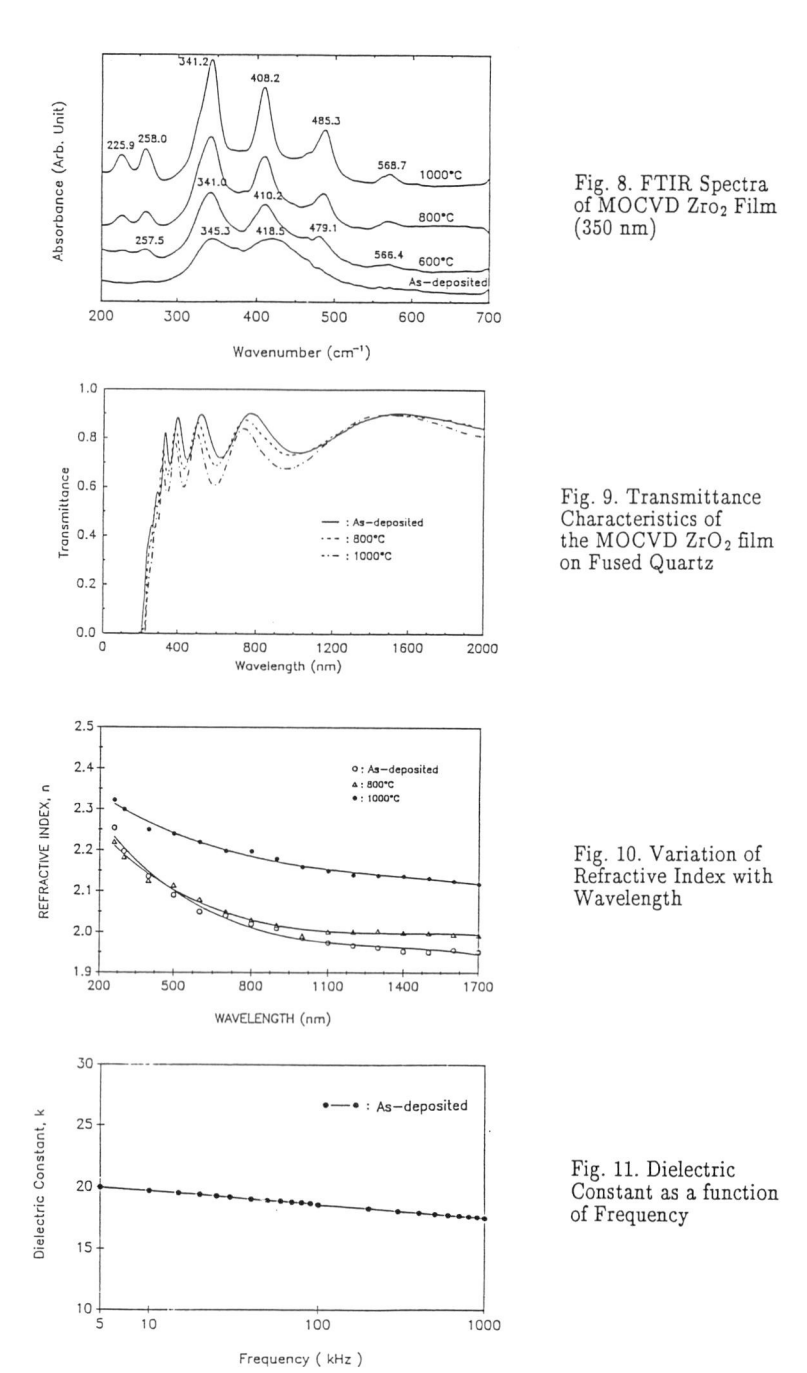

Fig. 9. Transmittance Characteristics of the MOCVD ZrO$_2$ film on Fused Quartz

Fig. 10. Variation of Refractive Index with Wavelength

Fig. 11. Dielectric Constant as a function of Frequency

REFERENCES

1. M. Ghanashyam Krishna, K. Narasimha Rao, and S. Mohan, Appl. Phys. Lett. 57(6), 557(1990).
2. C. Deshpandey and L. Holland in Proceedings of the 7th International Conference on Vacuum metallurgy, Iron Steel Institute of Japan, Tokyo, 1, 276(1982).
3. R. N. Tauber, A. C. Dumbri, and R.E. Caffrey, J. Electrochem. Soc., 118, 747(1971).
4. S. B. Desu, S. Tian, and C. K. Kwok, Mat. Res. Soc. Symp. Proc., 168, 349(1990)
5. C. K. Kwok and C. R. Aita, J. Appl. Phys. 66(6), 2756(1989).
6. E. N. Farabaugh, A. Feldman, J. Sun, and Y. N. Sun, J. Vac. Sci. Technol. A 5(4), 1671 (1987).
7. P. J. Harrop and J. N. Wankiyn, Brit. J. Appl. Phys., 18, 739 (1967).
8. J. C. Manifacier, J. Gasiot, and J. P. Fillard, J. Phys. E: Sci. Instrum., 9, 1002 (1976).

Advanced Dielectrics Deposited by LPCVD

Andrew P. Lane, SPC Texas Instruments, Dallas, TX 75265,
Arthur Chen, Neal P. Sandler, Dean W. Freeman, and Barry S. Page,
Lam Research Corporation, Fremont, CA 94538

ABSTRACT

Ta_2O_5 and Nb_2O_5 films were deposited using the Lam Research Integrity™ reactor. The chemical precursors used were pentaethoxides of Ta and Nb. Typical films were deposited at a rate of 4 nm/min with uniformities of <1.5% 1σ in the presence of O_2 at 470°C. Annealing the Ta_2O_5 films did not change the O/Ta ratio. Annealing the Nb_2O_5 films increased the O/Nb ratio to 2.5/1 at 850°C. Interfacial SiO_2 grew to 4 nm after annealing at 850°C for both Ta_2O_5 and Nb_2O_5. The as-deposited films were amorphous, but became crystalline above 600°C and 700°C for the Nb_2O_5 and Ta_2O_5 films respectively. The TEM observations on crystallization is supported by x-ray diffraction data.

INTRODUCTION

Decreasing device geometries require thinner dielectric films for storage capacitors. The need to reduce dielectric film thickness presents major processing problems. A near term solution for 64 Mbit and 256 Mbit devices is the use of alternative dielectrics such as oxides of Ta, Nb, Zr, Hf and Y. These oxides have dielectric constants, K, between 16 and 40, and may be deposited by several techniques, including sputtering and LPCVD.

The formation of Ta_2O_5 films has been studied extensively over the years utilizing anodic oxidation [1] and thermal oxidation [2] of tantalum films, reactive sputtering [3-7], plasma deposition [8], and chemical vapor deposition [9-13]. The formation of Ta_2O_5 (and other high dielectric constant films) from metal halides and organometallic reactants by CVD has been reviewed recently [14]. Previous CVD studies based on the organometallic compound tantalum pentaethoxide, $Ta(OEt)_5$, as the precursor molecule have led to encouraging results [9-13]. However, as-deposited films require annealing at 800°C - 850°C in O_2 to attain acceptable electrical properties, such as low leakage current density and high dielectric breakdown field. In addition, film deposition parameters were not optimized for coating the large-area substrate wafers (150 mm to 200 mm in diameter) required in ULSI fabrication.

In this paper, we describe the deposition of Ta_2O_5 and Nb_2O_5 films by thermal LPCVD of $Ta(OEt)_5$ and $Nb(OEt)_5$, and O_2 in a single-pass, laminar flow reactor. Precise control of residence time, temperature, pressure, and optimized O_2-to-$Ta(OEt)_5$ ratio are key to obtaining conformal and uniform films of excellent quality over large-area substrates.

EXPERIMENTAL

The Ta(OEt)$_5$ source had a purity of 99.999% and was used without further purification. The vaporized source was carried by ultra dry N$_2$ and mixed with O$_2$ prior to introduction to the reaction chamber. Typical deposition and annealing conditions are listed in Table 1. Planar and trench-etched 150-mm diameter silicon wafers were used as test substrates.

Deposition

Ta(OEt)$_5$ and Nb(OEt)$_5$ source temperature	40 °C
Source vaporizer temperature	200 °C
Delivery temperature	120 °C - 140 °C
Total gas flow rate	3-6 slm
Mole ratio O$_2$/Ta(OEt)$_5$ and O$_2$/Nb(OEt)$_5$	120:1
System pressure	400 mTorr - 600 mTorr
Deposition temperature	430 °C - 490 °C
Deposition time	Variable (typical 10 min)

Annealing
O$_2$ Anneal

Anneal temperature	600 °C - 850 °C
Time	10-30 min

Table 1. Film Deposition and Annnealing Conditions

The LPCVD reactor is a fully automated, hot wall, radial plate system from Lam Research Corporation (Lam Integrity™). It is designed for precise control of gas flow dynamics, reactant concentration, process temperature, and system pressure. The gas flow is laminar, single-pass and radial towards the center [15]. Generation of particulates by gas-phase nucleation is minimized since no recirculation occurs.

Films were characterized by Rutherford Backscattering Spectroscopy (RBS), Auger Electron Spectroscopy (AES), Secondary Ion Mass Spectroscopy (SIMS), X-ray Diffraction analysis (XRD), Scanning Electron Microscopy (SEM) and Transmission Electron Microscopy (TEM). Film thickness and refractive index were measured simultaneously by ellipsometry. Particle densities were determined with a laser reflectance instrument.

RESULTS AND DISCUSSION

The rate of Ta$_2$O$_5$ and Nb$_2$O$_5$ film deposition was 1.5 nm - 8 nm over the deposition temperature range of 430 °C - 490 °C. Typical film thickness uniformity of films 40 nm in thickness, deposited on 150 mm diameter wafers at 470 °C is ≤1.5% 1σ within wafer, wafer-to-wafer, and run-to-run. Average added particle densities do not exceed 0.2/cm^2 (>0.3 µm size; 6 mm edge exclusion). Figure 1 shows the Arrhenius behavior over this narrow, optimal temperature interval for the Ta$_2$O$_5$ deposition process indicating an activation energy of 25.3 Kcal/mole. Likewise, the Nb$_2$O$_5$ deposition process has an activation energy of 21.1 Kcal/mole. These data indicate a surface reaction rate limited process for both films.

The SEM cross-section micrograph in Figure 2 shows the excellent step coverage of >95% for these films on Si, stemming from the unique flow conditions and the surface-controlled reaction in this LPCVD reactor.

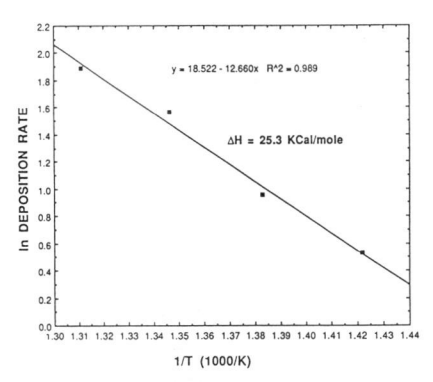

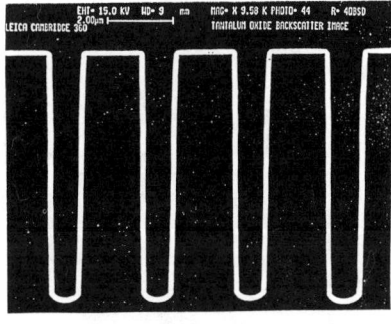

Fig. 1. Arrhenius plot for Ta_2O_5 deposition process from 430°C - 490°C

Fig. 2. Back scatter image of test sample cross-section showing uniform and conformal coverage of 7.2 μm deep and 1.2 μm wide trenches in silicon with 200 nm thick Ta_2O_5 film.

RBS analysis confirms the as-deposited Ta_2O_5 films have the correct stoichiometry. Annealing these films did not change either the stoichiometry or the oxide/silicon interface. The stoichiometry of the as-deposited Nb_2O_5 films was 2.2/1 for O/Nb which reached 2.5/1 at an annealing temperature of 850°C.

Figure 3 shows TEM cross-sections of Ta_2O_5 through a film of 60 nm thickness (as-deposited). The films are amorphous (see also XRD studies) and show no interactions of the Ta_2O_5 with the underlying substrate due to the buffering property of the approximately 1.2 nm thick native SiO_2 underneath. Films annealed at 850°C in O_2 (Figure 3) show the SiO_2 increased to 4 nm. Similar SiO_2-growth under the Nb_2O_5 films upon annealing is shown in Figure 4. The TEM studies also provided information concerning the crystallization of these films during annealing. The as-deposited films were amorphous but became crystalline above 600°C and 700°C for the Nb_2O_5 and Ta_2O_5 films respectively. The TEM work on crystallization is supported by x-ray diffraction data which appeared in Figures 5 and 6. The crystallization takes place above 700°C (700°C - 750°C) and the crystal structure corresponds to orthorhombic β-Ta$_2$O$_5$. Increasing the temperature to 900°C does not change the crystallinity or the Ta-O phases, but changes the lateral growth of the crystallites.

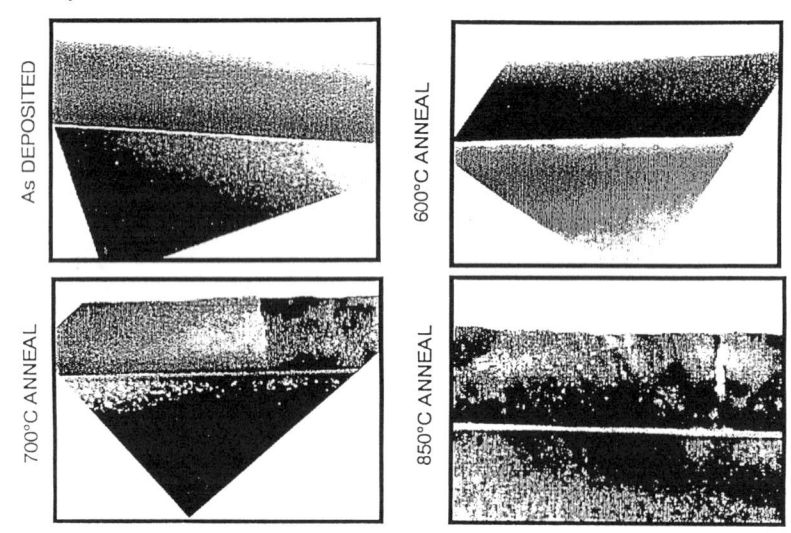

Fig. 3　TEM cross-sections of as-deposited and annealed Ta$_2$O$_5$ films on Si.

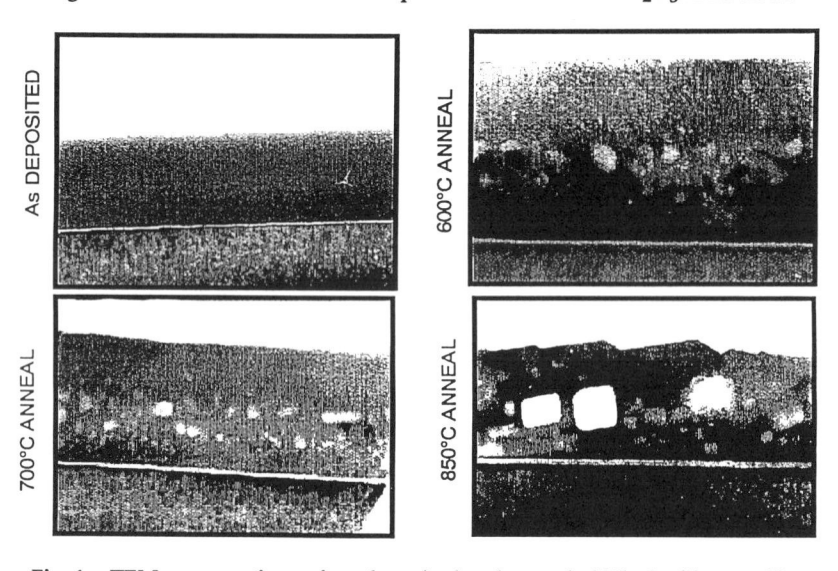

Fig. 4　TEM cross-sections of as-deposited and annealed Nb$_2$O$_5$ films on Si.

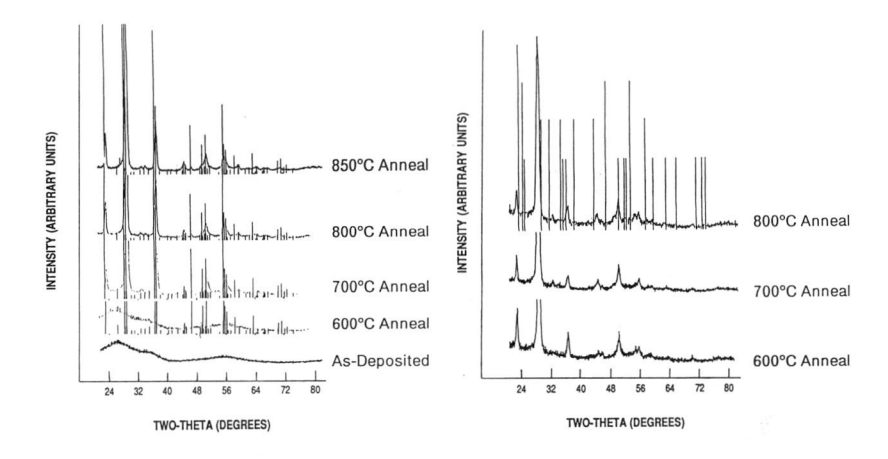

Fig. 5   XRD traces for Ta$_2$O$_5$ films annealed at different temperatures. The crystallization temperature is 700°C.

Fig. 6   XRD traces for Nb$_2$O$_5$ films annealed at different temperatures. The crystallization temperature is 600°C.

CONCLUSIONS

An LPCVD process has been presented for producing uniform, conformal and stoichiometric Ta$_2$O$_5$ films over large-area substrates 150 mm to 200 mm diameter wafers. Annealing crystallizes the Ta$_2$O$_5$ and Nb$_2$O$_5$ films above 700°C and 600°C, respectively. Annealing increase the interfacial oxide thickness from 1.2 nm to 4.0 nm. Preliminary results indicate Nb$_2$O$_5$ films are potentially applicable as storage capacitor dielectrics for advanced 256 Mbit DRAM devices.

ACKNOWLEDGMENTS

The authors wish to thank the following people for their valuable contributions to this work: B. Anderson, D. Mytton, and J. Whitlam for CVD, and E. O'Brien for electrical measurements.

REFERENCES

1. K. Ohta, K. Yamada, R. Shimizu, and Y. Tarui, IEEE Trans. Electron Devices ED-29 368 (1982)
2. T. Kato, T. Itoh, M. Taguchi, T. Nakumura, and H. Ishikawa, Symposium on VLSI Technology, Digest of Technical Papers, IEEE, Piscataway, NJ, (1983), p. 86.
3. G. S. Oehrlein, J. Appl. Phys., 59 1587 (1986).
4. H. Shinriki, Y. Nishioka, and K. Mukai, Extended Abstr. of the 19th International Conf. on Solid State Devices and Materials, Business Center for Academic Societies Japan, Tokyo (1987), p. 215.
5. C. Hashimoto, H. Oikawa, and N. Honma, IEEE Transact. Electron Devices 36 14 (1989).
6. H. Shinriki, Y. Nishioka, Y. Ohji, and K. Mukai, IEEE Transact. Electron Devices 36 328 (1989).
7. S. G. Byeon and Y. Tzeng, IEEE Transact. Electron Devices 37 972 (1990).
8. Y. Numasawa, S. Kamiyama, M. Zenke, and M. Sakamoto, IEDM 89 43 (1989).
9. M. Saitoh, T. Mori, and H. Tamura, International Electron Devices Meeting, Technical Digest, No. 32, p. 680, IEEE, Los Angeles (1986).
10. M. Matsui, S. Oka, K. Yamagishi, K. Kuroiwa, and Y. Tarui, Jpn. J. Appl. Phys., 27 506 (1988).
11. S. Zaima, T. Furuta, Y. Yasuda, and M. Iida, J. Electrochem. Soc., 137, 1297 (1990).
12. S. Zaima, T. Furuta, Y. Koide, and Y. Yasuda, J. Electrochem Soc. 137 2876 (1990).
13. H. Shinriki, IEEE Transact. Electron Devices 38 55 (1991).
14. K. F. Jensen and W. Kern, Chpt. III-1, p. 283 in THIN FILM PROCESSES II, J. L. Vossen and W. Kern, Editors; Academic Press, New York (1991).
15. D. Rafinjad, A. Runchel, J. Monkowski, and L. Wright, 1st Int. Conf., Transport Phenomena in Processing, to be published, Honolulu, Hawaii (1992).

PART VI

Diamond Films

ROUTES TO DIAMOND HETEROEPITAXY

ANDRZEJ BADZIAN AND TERESA BADZIAN
Materials Research Laboratory, The Pennsylvania State University, University Park, PA 16802

ABSTRACT

This paper reviews the status of diamond heteroepitaxy approached by chemical vapor deposition and by physical methods. Reported are experiments with cubic boron nitride and nickel conducted with the help of microwave plasma chemical vapor deposition. X-ray diffraction data confirm diamond heteroepitaxy on the (111) faces of cubic boron nitride crystals. Heteroepitaxy on nickel was not demonstrated yet nevertheless suppression of graphite nucleation was achieved by formation of nickel hydride.

HINTS FOR HETEROEPITAXY

Heteroepitaxial growth is a common practice for the family of $A^{III}B^V$ and $A^{II}B^{VI}$ semiconductor compounds and elements like Si and Ge. This has been achieved by several growth methods, such as molecular beam epitaxy, metallorganic CVD, growth from solutions and others.

Demonstration that homoepitaxial growth of diamond by CVD is feasible allows us to expect that diamond heteroepitaxy should be attainable because of the analogy of diamond to these other semiconductors. Till now this expectation did not materialize with one exception - heteropitaxial growth of diamond on micrometer size faces of cBN crystals.

Heteroepitaxial growth means that on the surface of single crystal A (over macroscopic area) another crystal B starts to nucleate and grow with the specific crystallographic orientation relationship between A and B crystals, which corresponds to some kind of lattice matching. The lack of perfect matching at the interface introduces a strain between these semicrystals. The strain is relaxed in specific ways. For example, when the crystal A is silicon and the crystal B is silicon doped with boron, misfit dislocations are formed just above the interface. This formulation of heteroepitaxy does not consider any transition interlayer between A and B, but instead direct chemical bonds are formed between atoms of the crystals A and B.

In the sense formulated above heteroepitaxy of diamond was not documented despite years of trials using a whole spectrum of substrates. However, there is some evidence that diamond was grown heteroepitaxially on the (111) face of cubic BN crystal over an area of ~10μm by direct current plasma CVD [1]. Cubic BN is the first candidate to study heteroepitaxy because of its good lattice matching (1.4% mismatch) and its solubility with diamond. The formation (under high pressure) of substitutional solid solutions between diamond and cubic BN indicates the formation of chemical bonds between B, C and N [2].

The first report on diamond heteroepitaxy on micrometer size cBN grit was presented by W. Yarbrough, A. Kumar and R. Roy [3,4]. Scanning Electron Microscope observations on diamond growth on cBN crystals support epitaxial growth [5,6].

Some information on heteroepitaxy can be extracted from studies of inclusions in diamond crystals. X-ray diffraction techniques have revealed for some mineral inclusions in natural diamond crystals that heteroepitaxial relationships exists between the inclusions orientation and diamond matrix [7]. Similar measurements of high pressure/high temperature diamond crystals grown from Ni and Co liquids indicate *in situ* formation on the Ni and Co nonstoichometric carbides with the NaCl-type structure. Inclusions of these carbides are aligned to the diamond matrix, according to Weissenberg photographs [8,9]. Diamond crystals grown by microwave plasma CVD, at specific CH_4 concentrations, have nano-size graphite inclusions aligned on the {111} planes of diamond, according to the electron diffraction data [10].

CVD DIAMOND NUCLEATION

Observations on diamond nucleation on non-diamond substrates can be summarized as follows:

a) Diamond nucleates on foreign substrates on nano-size areas from which it grows radially on the surface and forms a crystal grain growing perpendicular to the surface.
b) Nucleation takes place on an intermediary phase formed *in situ*, during the incubation phase of the process. There are several such phases which have been reported:
   * hydrogenated disordered carbons
   * hydrogenated carbides like SiC with disordered sphalerite structure
   * other carbides like Ti and Mo

Our understanding is that just after an intermediary phase is formed as a nano-size nucleation site, diamond starts to grow. This tendency of diamond nucleation is an obstacle in achieving the oriented growth. It means that routine CVD procedures should be modified. Several such modifications have been explored.

a) Y. Sato and M. Kamo at NIRIM achieved columnar growth with high crystal perfection in the <001> direction by controlling process paramters [11].
b) The group of researchers at Osaka University explored artificial epitaxy. A silicon wafer was lithographically prepared, etched and passivated to form a square pattern. Although the CVD diamond grains were only partly preferential [12].
c) Another procedure came from the following observation. Micrometer size powder grains of natural diamond spread over a silicon wafer produces films with preferred orientation perpendicular to the surface. This results from the fact that diamond cleaves along the {111} planes and the grains sit on the surface sticking to the {111} faces. These films had random orientation of CVD crystals in reference to the directions lying in the surface of the wafer. This experiment indicates that organization of seeding in three dimensions requires a new concept how to orient the seeds on the surface.
d) Frustration with lack of success on conventional heteropitaxy has revived a homoepitaxy concept along the lines described above. Recently M. Geis and his collaborators have achieved spectacular success with mosaic diamond films. They were able to etch a relief structure with half octahedron holes 100 x 100µm in silicon wafers and locate diamond octahedral grains (100µm) in the resulting pits such that all of them have the <001> orientation. Continuous quasi monocrystalline diamond films over the area of the 2" diameter were then prepared with further deposition by hot filament atop this three-dimensionally oriented crystal array. Over 95% of the seeds are oriented and lead to a mosaic crystal approaching a large area quasi monocrystalline film [13].

We can draw the following conclusions on CVD diamond nucleation:

* It is trivial to say that diamond homoepitaxy works because at the initial stage of the process interaction of diamond surface with activated species is similar to any arbitrary moment of the growth.
* However, non-diamond substrates interact in different ways than diamond surfaces. The surfaces undergo chemical changes and diamond nucleation starts on a nano-size area where a specific crystal phase is formed. This is a "point" at which diamond nucleates.
* The intermediate phases make heteroepitaxy difficult. Disordered structures of these phases contributed to this.
* Diamond layer growth immediately atop a lattice matched heteropitaxial substrate (in analogy to Frank-van der Merve mechanism) is the most desired growth. It is also advised to undertake efforts on finding such intermediary phases which would be susceptible to form an array of oriented (in three dimensions) nucleation sites. Diamond would be nucleated on them forming islands followed by formation of an oriented continuous film. This type of nucleation would be analogous to a Volmer-Weber growth mechanism.

APPROACHES WITH PHYSICAL METHODS

In order to avoid chemical problems with surface reactions inherently connected to the CVD processes efforts on developing physical methods have been pursued. Listed below are various reported methods which use the energy and momentum of particles and the interaction of laser beams with matter in order to aid in nucleation and heteropitaxial growth.

Carbon Ion $^{12}C^+$ Beam

Freeman et al. in 1978 bombarded diamond substrate heated at 700°C with selected $^{12}C^+$ ions of 900eV. They claim that several μm-thick single crystal diamond film (with non-diamond inclusion) was grown on diamond substrate [14]. This claim is not confirmed. Recent reports by Harvell Laboratory does not contain any crystallographic data [33].

Ion Implantation

Growth of diamond on Cu by implantation of 120eV $^{12}C^+$ into (111) single crystal Cu at ~900°C was reported by J.F. Prins and H.L. Gaigher [15]. The claim was challenged by S. Tong Lee and collaborators whose experiments made at 200eV showed that films were invariably highly oriented crystalline graphite [16]. Professor J. Prins reaffirmed the claim that his implantation-out diffusion method produced diamond films.

Laser Assistance

Before powerful lasers were available periodic pulses from a xenon-gas-discharge tube were used to create a supersaturation of carbonaceous species above the diamond substrate. In this way thermal CVD was improved with respect to the ratio of diamond/graphite nuclei (B.V. Derjaguin and D.V. Fedoseev, 1967) [17].

The periodic heating/cooling method was also successful in the case of lasers with wavelengths of 1.06 and 10.6µm. Polycrystalline films 5µm thick were grown on diamond substrates [18]. These experiments have not been confirmed.

In relation to these experiments phase transformation in graphite powder was induced by IR laser with very low yield, nano-size grains of diamond formed. In addition to diamond carbyne phase was formed from hydrocarbon droplets. The results on phase transformation of graphite was confirmed at Penn State [19].

Implantation-Pulsed Laser Process

A new two step process was proposed by J. Narayan, V.P. Godbole and C.V. White. The first step is implantation of carbon into copper at 60-120keV. Next the implanted layer was treated with nanosecond excimer laser (308nm) energy with energy density up to 5J-cm-². During the second step, laser melting of copper, diamond films are produced as flakes. The thickness of the flakes is about 500Å and the area is about 100µm² [20]. This method waits for confirmation.

Conclusions on Physical Methods

Films obtained by physical methods are usually rather thin (below 1000Å) and to prove that they were grown heteroepitaxially requires:

* identification of crystal structure of the film
* determination of relationship between atomic lattices of the substrate and the film
* determination if the film has correct chemical composition

It should be excluded at the beginning that the deposited film is not graphite or a disordered carbon phase. The diffraction data are essential to prove heteroepitaxy but at the same time the interpretation is not trivial. The electron diffraction patterns of thin films of diamond and graphite have some similarities. The evaluation of these patterns requires evaluating them for several crystallographic directions, accurate measurements of as many reflections as possible, and careful indexing of the planes. Misinterpretation of diffraction patterns of carbon films are well documented in the diamond literature.

X-ray diffraction requires the separation of undesired effects like double diffraction, parasitic scattering and other artifacts. It should be required that the diffractometer measurements be complimented by photographic techniques. It is also documented in the diamond literature that unfortunate combination of diffraction effects in the particular diffractometer produced diffraction peaks originated in silicon substrate. These diffraction results were attributed to heteroepitaxial diamond film on silicon. Revision of complementary characterization data indicated that a disordered carbon film was deposited on Si instead [21].

In summary it is the opinion of the authors of this assessment that claims of diamond heteroepitaxy by physical methods have not been satisfactorily proven.

EXPERIMENTS ON HETEROEPITAXY OF DIAMOND ON cBN

cBN and Ni are the most frequently considered candidates for diamond heteroepitaxy. The other candidates include: Cu, Si, MgO, LiF and TiB_2. In this paper we will report on experiments using microwave plasma CVD of diamond with cBN and Ni substrates.

Goal

Examination of the crystallographic relationship between the diamond film and cBN crystal substrate by x-ray diffraction.

Substrates

Progress in cBN synthesis by CVD and physical methods is very slow. Only recently [22-24] micrometer size crystals of cBN were prepared.

Single crystals grown by HP/HT method are not easily available. Large single crystals of cBN up to 3mm in size have been seen by the authors in High Pressure Laboratory of NIRIM, Tsukubashi, Japan [25] but there have been no reports, as yet, that they were used for heteroepitaxy experiments. In general, the quality of cBN crystals is inferior to the diamond crystals grown by HP/HT.

In this study we have used cBN single crystals synthesized by authors using HP/HT process with Mg_3N_2 solvent-catalyst and chamber described previously [26]. The crystals have a mostly tetrahedral habit. The maximum size of these crystals was 250μm. We have also used crystals of non-regular octahedral habit (80-120μm) kindly supplied by Dr. R. DeVries of General Electric.

Deposition Process

The cBN crystals were placed on natural diamond crystal to avoid contamination from the holder and they were cleaned before deposition by annealing in hydrogen plasma. 5μm thick diamond film was grown for the x-ray diffraction study. A 1%CH_4 and 99%H_2 mixture at 80Torr of pressure was used over a substrate with a temperature of 900°C. More details on deposition is given in references 27 and 28. For the nucleation study, films of 0.7-3μm were prepared.

X-ray Diffraction

X-ray oscillation method was used to determine crystallinity and orientation of diamond film in relation to the 250μm cBN substrate. X-ray photographs were taken with CuKα radiation and 57.3mm diameter camera. Clear resolution of diamond (a=3.567Å) and cBN (a=3.615Å) reflections was observed. The reflections appear as doublets. The diamond film is a single crystal (with a distorted structure because of the elongation of the reflections) that has the same orientation as the crystal of cBN. Table I lists distances between diamond and cBN reflections.

TABLE I

	Δ2θ	Distance calculated	observed
111	0.62°	0.31mm	0.3±0.1mm
220	1.17°	0.58	0.6±0.1
311	1.60°	0.80	0.8±0.1

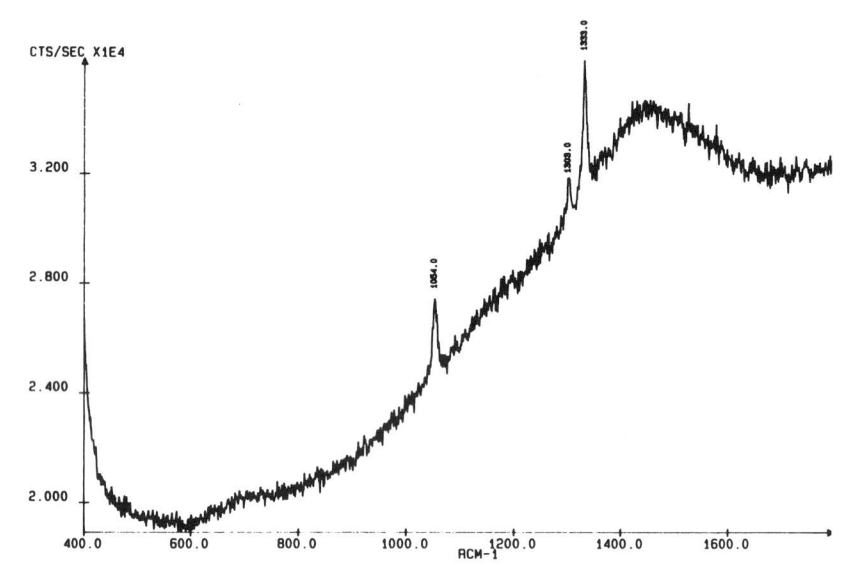

Fig. 1   Raman and luminescence spectra of 5μm diamond film on the (111) face of cBN tetrahedral crystal. Laser excitation wave-length 514.5nm.

Raman Spectra

The Raman spectrum of the crystal studied by x-ray diffraction is shown in Fig. 1. The spectrum contain cBN peaks at 1054 and 1304cm^{-1} and diamond signature at 1333m^{-1}. Strong luminescence with maximum at 1440m^{-1} correspond to disordered structure which is characteristic for (111) diamond homoepitaxy [28]. This defected form combines the effects of stacking faults and twinning on the {111} planes. These disorder effects became smaller when diamond film on cBN was grown with addition of B_2H_6 to the gas mixture. The strong luminescence measured along with Raman lines disappeared in this case, and the width of 1333m^{-1} diamond peak at half maximum was 3.0cm^{-1}.

Nucleation

A perfect cBN octahedral crystal which possess sphalerite type structure should have {111} faces terminated by B or N atoms alternatively. A tetrahedral single crystal should be terminated by B or by N atoms exclusively. The crystals used for substrates were too small to identify which atoms terminated which face. Deposition on the tetrahedra faces produced smooth surfaces with tarraces, an indication that growth proceed according to Frank-Van der Merve mechanism. Deposition on octahedral GE crystals, which were grown at unknown conditions, shows different behavior of the {111} faces (Fig. 2). Some of these faces have full coverage at early stages of growth.

Rows of microcrystals appeared along the <110> ledges. These crystals indicate a Volmer-Weber growth mechanism. We can only hypothesize about polarity of the {111} faces, which (111) face is terminated by B atoms and with ($\bar{1}\bar{1}\bar{1}$) is terminated by N atoms. They should show different behavior in respect of etching and growth processes as it was observed for all $A^{III}B^V$

Fig. 2   Nucleation of diamond on GE cBN crystals. Process conditions: 2%CH$_4$, 80Torr, 930°C. Bar corresponds to 100μm.

compounds. Indeed different etching behaviour for cBN was observed [25]. We think that in our nucleation experiments, boron terminated face is preferred for diamond nucleation because doping with B proceeds easily in contrast to N. So, the partial covering of the ($\bar{1}\bar{1}\bar{1}$) faces would seem to indicate N termination.

Conclusions

1. Diamond heteroepitaxy on cBN over area of 250μm was confirmed by x-ray diffraction.
2. Differences in nucleation density were observed on the {111} faces of cBN crystals.

EXPERIMENTS WITH NICKEL HYDRIDES

Goal

Formation of Ni hydrides in CVD diamond growth process.

Heteroepitaxy on Ni

Although true heteroepitaxy on Ni has not yet been achieved, diamond growth on Ni is still intriguing. The distribution of atoms on the {100} planes is close for Ni (a=3.524Å) and diamond (a=3.567Å). But, the crystallization of diamond in a CVD process does not take place on elemental

Ni. When Ni metal is in the contact with CH_4/H_2 plasma, both C and H dissolve into Ni immediately. So Ni is transformed from fcc structure to an interstitial alloy with C and H atoms located at random in the octahedral holes of fcc structure.

A similar phenomenon occurs with HP/HT synthesis of diamond, where Ni or Co form NiC_x or CoC_x, x~0.5 non-stoichometric carbides of NaCl type structure incorporated into diamond crystal matrix [8,9]. In the author's opinion, this HP/HT phenomenon is not important to CVD deposition and we concluded that Ni is not suitable for diamond epitaxy [27]. But Ni does take a different active role during diamond nucleation and growth processes for both HP/HT and CVD synthesis. For this reason, we undertook a study on participation of nickel hydrides in CVD process and preliminary results will be reported.

When the decomposition reactions of hydrocarbons on Ni surface are taken into account, the whole CVD process will become quite complicated. This is reflected in the fact that, depending on the process parameters, a variety of deposits on Ni were obtained. These deposits range in structure from pure graphitic deposits in the form of fibers to high quality single crystals of graphite or as diamond crystals which nucleate at Ni grain boundaries. Also reported was agglomerate of diamond crystals oriented in the <111> direction [29].

Nickel Hydrides

Formation of Ni hydrides was studied from the point of view of thermodynamic equilibria, which requires reproducible synthesis conditions [30]. Because of this requirement, high pressure method could be extensively explored. However, electric discharges in H_2, which are difficult to control but which are effective in the hydride formation were not extensively explored. The diatomic gaseous hydrides that form on the transition metals are known. Solid Ni hydride have have H/Ni ratio of 0.7-0.8 and its lattice constant expend 5.5% in comparison to the metal. This hydride is unstable and decomposes at room temperature after 40 hr [31].

Below we report on experiments conducted with the help of microwave plasma.

Thin Film of Ni on Diamond

Experiments were performed with a thin film of Ni sputtered on a part of surface of natural type Ia diamond cut along the (001) plane. The thickness of Ni film was less than 1μm. This substrate was used in diamond deposition with $1\%CH_4$, $99\%H_2$, pressure 80Torr and substrate temperature 900°C. On the part of the (001) surface not covered by Ni diamond homoepitaxy took place. Growth steps indicating the <110> growth direction are shown in Fig. 3a. On Ni coated part the morphology varies significantly. There are squares with the <110> sides and large flat areas that are single crystal areas (Fig. 3b).

Contamination with Ni was detected by electron microprobe and cathodoluminescence spectra show sharp lines which were assigned to Ni [32].

These experiments seem to indicate that growth of diamond proceed in the presence of Ni. We think that carbon coming from diamond crystal substrate and from CH_4 decomposition is dissolved in the Ni layer simultaneously with the process of hydrogenation which produces volatile NiH. No graphite formation was observed in this process.

Thin Film of Ni on Cu

Cu as substrate helps to eliminate the dissolution of diamond substrate in the Ni film which takes place in temperature above 600°C as described in the preceding section.

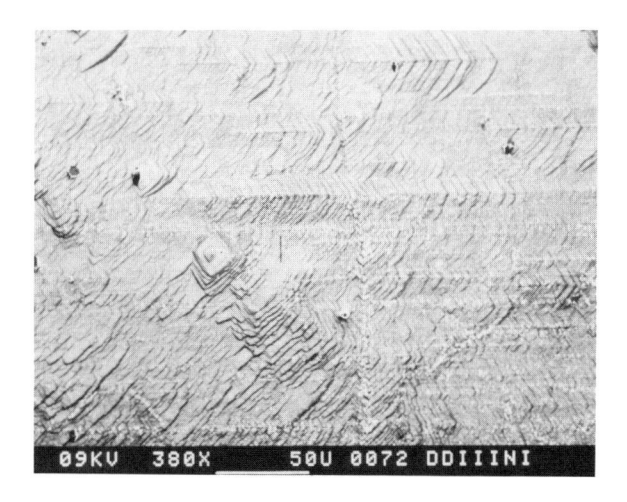

Fig. 3a Morphology of homoepitaxial diamond film on the (001). Process conditions: 1%CH$_4$, 70Torr, 875°C film thickness 17μm.

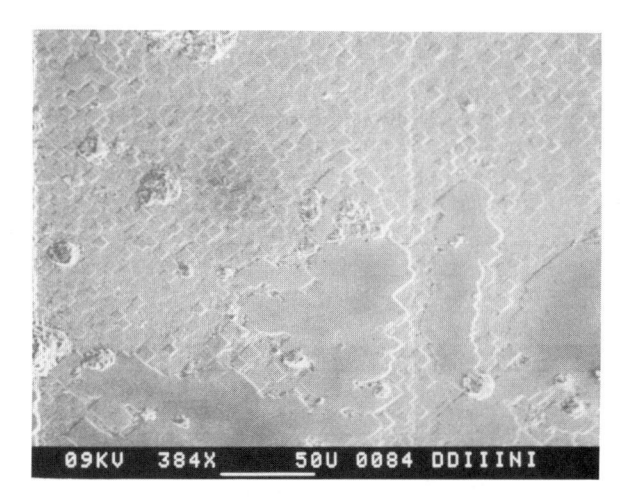

Fig. 3b Morphology of diamond film grown on Ni coated diamond surface. Bar corresponds to 50μm.

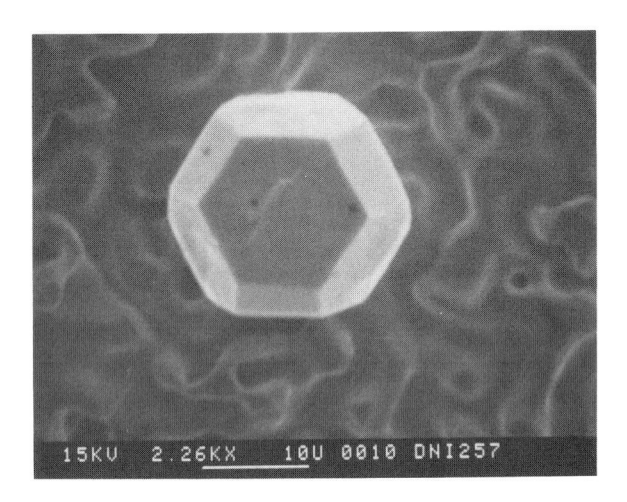

Fig. 4 Morphology of Ni surface and diamond crystal upon it. Bar corresponds to 10μm.

When bulk Ni is annealed in hydrogen plasma, it starts to melt at 1150°C, much below melting point of Ni 1450°C.

The first step of nucleation experiments with thin films of Ni (less than 1μm) on Cu was to anneal the films in a hydrogen plasma at 1000°C, below melting point of Cu (1080°C). Recrystallization of Ni and formation of Ni hydride takes place in this step.

In the next step, the temperature was decreased to 900°C and diamond nucleation was conducted with 1%CH_4 at 80Torr. Separate crystals of diamond were nucleated. Graphite was not formed at these conditions. Fig. 4 shows morphology of Ni layer recrystallized with some triangular features and truncated cubo-octahedral diamond crystal with the <111> orientation.

Conclusion

It was demonstrated that suppressing of graphite nucleation can be achieved when Ni metal is transformed into Ni hydride.

ACKNOWLEDGEMENTS

This work was supported in part by the Office of Naval Research (with funding from the Strategic Defense Initiative Organization's Office of Innovative Technology under the contract N0. 00014-86-K-0443 and the Diamond and Related Materials Consortium at The Pennsylvania State University. This material is based in part upon work supported by the National Science Foundation under Grant No. DMR-9104072. The Government has certain rights in this material.

We extend our thanks to Professors Rustum Roy and Russell Messier for discussions. We also express our gratitude to Diane Knight for her help in Raman Spectroscopy measurements, William Drawl for technical assistance and Ron Weimer for critical reading of the manuscript.

References

1. S. Koizumi, T. Murakami, K. Suzuki and T. Inuzuka, Appl.Phys. Lett. 57, 563 (1990).
2. A. Badzian, Mat. Res. Bull. 16, 1385 (1981).
3. W.Yarbrough, A.Kumar and R.Roy, presented at the 1987 MRS Fall Meeting Boston, MA,1987 (unpublished).
4. W. Yarbrough, J. Vac. Sci. Technol. A 9(3), May/Jun (1991).
5. R. Haubner, Refractory Metals and Hard Materials 9 (2), 70, (1990).
6. H. Saitoh (private communication).
7. J.W. Harris, in The Properties of Diamond, edited by J. E. Field (Academic Press, London, 1979), p.555.
8. A. Badzian and A. Klokocki, J.Crystal Growth 52, 543 (1981).
9. A. Badzian, in Advances in X-Ray Analysis, edited by C.S. Barret et al (Plenum Press, New York and London, 1987), vol.31,113.
10. W. Zhu, C. A. Randall, A.R.Badzian, J.Vac.Sci.Technol. A7(3),2315 (1989).
11. Y. Sato and M.Kamo, Surface and Coatings Technology, 39/40, 183 (1989).
12. J.S. Ma, H. J. Kawarada, T Yonehara, J. Suzuki, J. Wei, Y. Yokota, Y. Mori and A. Hiraki, in Technology Update on Diamond Films, edited by R.P.H. Chang, D. Nelsson, A. Hiraki (Mater. Res. Soc. Ext. Abs. EA-19,San Diego, CA 1989) pp.67-70.
13. M.W. Geis, H.I. Smith, A. Argoitia, J. Angus, G.H.M. Ma, J.T. Glass, J. Butler, C.J. Robinson, R. Pryor, Appl. Phys. Lett. 58(22), 2485 (1991).
14. H. Freeman, W. Temple and G.A. Gard, Nature 275, 634(1978).
15. J.F. Prins and H.L. Gaigher in New Diamond Science and Technology, edited by R. Messier, J.T. Glass, J.E. Butler, R.Roy (Pittsburg, PA 1990) pp.561-566.
16. S.T. Lee, S. Chen and G. Braunstein, X. Feng, I. Bello and W.M. Lau, Appl. Phys. Lett. 59, 785 (1991).
17. B.V. Derjaguin and Fedoseev, Surface Coatings and Technology 38 234 (1989).
18. I.G. Varsavskaja and Lavvrantiev, Archiwum Nauki o Materialach 7,127 (1986).
19. M. Alam, T. DebRoy, R. Roy and E. Breval, Carbon 27 (2), 289 (1989).
20. J. Narajan, V.P. Godbole and C.W. White, Science 252, 416 (1991).
21. J.L. Robertson, X.G. Jiang, P.C. Chow, S.C. Moss, Y. Lifshitz, S.R. Kasi, J.W. Rabalais and F. Adar in Technology Update on Diamond Films, edited by R.P.H. Chang, D. Nelson and A. Hiraki (Mater. Res. Soc. Ext. Abs. EA-19, Pittsburg, (1989) p. 9.
22. S. Komatsu, Y. Moriyoshi, M. Kasamatsu and K.Yamada, J. Phys. D: Appl. Phys. 24 1687 (1991).
23. H. Saitoh and W.A. Yarbrough, Appl. Phys. Lett. 58 (22),2482 (1991).
24. H. Saitoh and W.A. Yarbrough, Appl. Phys. Lett. 58 (20),2223 (1991).
25. O. Mishima, in Applications of Diamond Films and Related Materials, edited by Y. Tzeng, M. Yoshikawa, M. Murakawa, A. Feldman (Materials Science Monografs, 73, Elsevier Science Publishers B.V., 1991) pp.647-651.
26. A.Badzian and T. Kieniewicz-Badzian in High Pressure Science and Technology edited by B. Vodar and Ph. Marteau (Pergamon Press vol.2 , 1979) pp.1087- 1091.
27. A. Badzian, T. Badzian, R. Roy, R. Messier and K.E. Spear, Mat. Res. Bull. 23 (4), 531 (1988).
28. A. Badzian, T. Badzian, X.H. Wang and T. Hartnett in New Diamond Science and Technology edited by R. Messier, J.F. Glass, J.E. Butler and R. Roy (Mater. Res. Soc., Pittsburgh, PA 1990) pp.549-556.
29. Y. Sato, I. Yashima, H. Fujita, T. Ando and M. Kamo in New Diamond Science and Technology edited by R. Messier, J.F. Glass, J.E. Butler and R. Roy (Mater. Res. Soc., Pittsburgh, PA 1990) pp.371-376.
31. Metal Hydrides edited by W.M. Mueller, J.P. Blackledge and G.G. Libowitz (Academic Press, New York and London, 1968) p.629.
32. J. Lebens, B. Jacobi, K. Valaha, A. Badzian and T. Badzian (to be published).
33. M. Buckley-Golder, R. Bullough, M.H. Hayns, J.R. Willis, R.C. Piller, N.G. Blamires, G. Gard and J. Stephen, Diamond and Related Materials 1, 43 (1991).

DIAMOND DEPOSITION BY A NONEQUILIBRIUM PLASMA JET

D.G. KEIL, H.F. CALCOTE, AND W. FELDER
AeroChem Research Laboratories, Inc., P.O. Box 12, Princeton, NJ 08542

ABSTRACT

A nonequilibrium plasma jet has been used to deposit diamond films on a number of substrates, including silicon, silicon nitride, alumina, and molybdenum. Hydrogen is passed through a glow discharge and expanded through a supersonic nozzle to produce a highly nonequilibrium jet. Methane is added downstream of the nozzle, where it mixes and reacts with the nonequilibrium concentration of hydrogen atoms. The resulting supersonic jet strikes the substrate surface producing a high quality (determined by laser Raman spectrometry) adherent diamond film. Because of the low jet temperature, substrate cooling is unnecessary. Diamond deposition rates have exceeded 2 mg/kWh and 1 μm/h averaged over 16 cm^2 area; good quality films have been prepared at substrate temperatures below 600 K.

INTRODUCTION

The ability to produce CVD diamond films at practical rates and thicknesses blossomed with the use of "activated" hydrogen along with a carbon source gas [1,2]. This early work spawned the development of a wide variety of techniques to activate hydrogen (and carbon species) to deposit diamond films. Filament-assisted CVD (FACVD), microwave plasma-assisted CVD (MWCVD), thermal arc jet CVD (arc jet), and combustion gas CVD (torch) techniques have been successfully used to produce quality polycrystalline diamond films. Differences in the characteristic hydrogen activation processes of each technique define an operating "envelope" of deposition rate and diamond film quality (area, uniformity, morphology, adhesion, etc). Encouraging attempts have been made to connect them within a global description of diamond deposition[3]. However, even if similar chemistry is occuring in these CVD processes, there are practical criteria which differentiate them. The operating envelopes can be translated into advantages and disadvantages for the processes. Based on the current understanding of the CVD diamond mechanism and the characteristics of other CVD techniques, we have developed a new technique designed to optimize the diamond deposition environment. It utilizes efficient, low temperature activation of hydrogen in a nonequilibrium plasma, and expansion of the plasma into a supersonic jet which, when admixed with a carbon species, rapidly transports diamond precursors to the growing diamond film surface.

AEROCHEM NONEQUILIBRIUM PLASMA JET

The nonequilibrium plasma jet (NEPJ) was originally developed at AeroChem over 30 years ago[4]. Its utility as an efficient source of atomic hydrogen for synthesis was demonstrated in a successful program to deposit photovoltaic grade amorphous silicon. We have since demonstrated and patented[5] its application to diamond film technology, and we anticipate its utility in many CVD processes.

Apparatus

The basic apparatus, as illustrated in Figure 1, is divided into two sections separated by a nozzle across which supersonic flow develops. The high pressure side contains a glow discharge which activates a high velocity hydrogen stream. Glow, or "silent" discharges are generally characterized by moderately high voltages and low currents, as opposed to thermal arcs which operate at low voltages and very high currents. The glow discharge is supported between an axial water-cooled electrode and the nozzle. Thus all the hydrogen passes through the discharge and all the resultant plasma gases must flow through the nozzle, expanding into the deposition chamber. This chamber is maintained, by a vacuum pump, at a much lower pressure than that in the upper, discharge chamber. Under these conditions, the plasma gases accelerate at the expense of random gas molecule motions to form a directed supersonic jet of cooled active gas. Downstream of the nozzle methane is mixed into the jet and the jet impinges on a substrate surface where a diamond film is deposited. As practiced, this method does not directly expose the carbon-containing gas to the discharge.

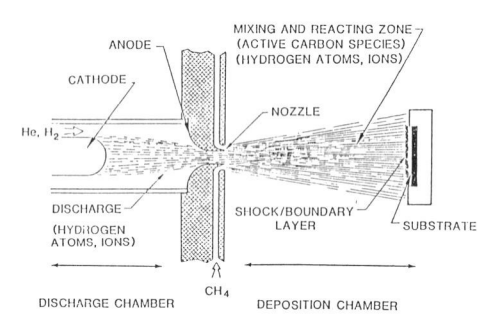

Figure 1 AeroChem Nonequilibrium Plasma Jet (NEPJ) apparatus for diamond depositions.

Process Characteristics

Glow Discharge. In a glow discharge ions, electronically excited species, and molecular dissociation products are formed at concentrations much in excess of that predicted based on the bulk gas temperature (which is much lower than the effective electron temperature). Hence it is a nonequilibrium plasma. Positive ions play a role both in maintaining the discharge through high energy collisions with the cold electrode surface and in transporting the current. These characteristics differentiate the glow discharge from an electric or thermal arc, in which electron "boil-off" from a hot electrode and electron transport of charge dominate in maintaining the discharge. In such arcs, the gas in the plasma is largely at thermal equilibrium, and active species concentrations result from the extremely high associated temperatures (typically >5000 K). Contamination by vaporized impurites can be problematic in both arc and FACVD processes. A glow discharge provides high hydrogen atom fluxes at low power levels, since energy isn't used to heat the hydrogen to the equivilent thermal dissociation temperature, as in arc or FACVD processes. The ultimate heating of the substrate by the nonequilibrium plasma jet is less than from an arc jet due to the lower bulk gas temperature. Atmospheric torch methods, where both the thermal and active species fluxes are high, also require substrate cooling.

Nozzle Expansion. The supersonic expansion of the plasma gases is another key element of the NEPJ for diamond deposition. When the pressure ratio across the nozzle is much greater than unity, random thermal motions are directed to form an expanded, supersonic jet of cooled gases. The supersonic expansion prevents feedback from the downstream chamber into the discharge chamber, effectively isolating the discharge processes from perturbations or manipulations of the downstream jet. This provides separate optimization of hydrogen activation and transport processes. Isolation also provides stability of operation in the presence of substrate motion, which is not possible in conventional MWCVD.

The supersonic jet rapidly transports active plasma species (hydrogen atoms, ions, etc.) toward the substrate. Three-body hydrogen atom recombination is low due to the short available reaction time and to the low pressure in the jet. Bimolecular reactions with positive activation energies are slower at the lowered jet temperature.

At the substrate surface, a shock wave forms, and the gas recovers the upstream temperature. The active species are transported across the resultant boundary layer to form the diamond film. The high gas velocities impinging on the substrate makes the boundary layer thin, facilitating transport to the surface.

Diamond Depositions

Experimental. Typical gas flow rates for results described here are 0.5-1 sL/s hydrogen, 50-200 scc/s helium and and/or argon, and 2-25 scc/s methane. For these experiments the glow discharge was typically operated at 1 to 2 A and 1000-1500 VDC, corresponding to 1 to 3 kW power. Except for low substrate temperature depositions described below, it was often necessary to provide additional, electrical substrate heating beyond that provided by the plasma jet. To accomplish this uniformly, the substrate was supported on a heated graphite disk.

We found that substrates to be diamond coated did not require pretreatment, such as scratching with a hard material (e.g. diamond or SiC powder) as reported for a high deposition rate, thermal plasma process [6]. However such treatment was found to affect the deposited diamond film appearance and presumably, its morphology. We have not observed any "induction period" for nucleation, and the deposition rates are comparable for scratched and unscratched substrates.

Deposition rates in mg/hr are determined by substrate weight changes. Film thicknesses have been determined by several methods. Average values can be easily determined based on estimates of the overall film area and the measured substrate weight change. A diamond film density of 3.5 g/cm^3 is assumed. We have also measured thickness profiles with beta back-scattering (β-BS). The instrument (Twin Cities Model TC 2000) was fitted with a Cd^{109} beta source and was calibrated for each substrate material using a series of standard thickness polypropylene films, correcting the thicknesses for the density ratio 0.91/3.5. The thicknesses of fractured films have been directly measured using scanning electron microscopy (SEM) or a surface profilometer (Tencor Instrument, Model 10-00020). Even for films deposited on complex shapes, the estimated average thicknesses were found to consistently scale with the direct thickness measurements within about a factor of two, the <u>average</u> values being less, as expected.

Diamond film quality is routinely determined on the basis of Raman spectroscopy using the 515 nm argon ion laser line (Spectra Physics Model 164). Typical laser power is 200 mW and the beam is only moderately focused, yielding sample-average spectra. The scattered radiation is focused onto the slits of a Spex Model 4 Ramalog laser Raman Spectrometer. The double monochromator has a 0.85 m focal length, is fitted with 1800 groove/mm gratings, and with 150 micron slits provides about 2 cm^{-1} resolution. A cooled photomultiplier (RCA C31034) collects the light from the exit slit, and the signal is stored on a computer for manipulation, storage, and routing to an X-Y plotter. We routinely judge the film quality based on the intensity ratio of a broad Raman feature at 1550 cm^{-1} (attributed to sp^2 carbon) and the narrow diamond peak around 1332 cm^{-1}; and the intensity of continuum emission underlying this region[7]. Since the Raman scattering efficiency of graphite is about 50 times greater than that for diamond, the 1550 cm^{-1} to 1332cm^{-1} intensity ratio can be roughly interpreted as a percentage of non-diamond carbon in the film.

Substrate Materials. Some materials, listed in order of increasing thermal expansion coefficient (TEC), on which we have deposited diamond films are given in Table I. In most cases part of the substrate surface was rubbed with diamond powder or paste to optimize nucleation. All attempts to observe, by Raman scattering, any residual diamond detritus from the pretreatment have failed, probably due to the low amount of diamond remaining and the weak focus of the probe laser beam. The appearance of the diamond peak around 1332 cm^{-1} in the Raman of the deposited films was taken as evidence of diamond deposition. As the table shows, the peak Raman shift of this feature depends on the substrate material. A general correlation between the peak position and the thermal expansion mismatch between diamond and the substrate material is seen. Raman shifts greater than 1332 cm^{-1} are consistent with compressive film stresses induced by TEC mismatch[7]. Interestingly, the film on copper exhibited a shift <u>less</u> than 1332 cm^{-1}, even though its TEC is an order of magnitude higher than that of diamond. Poor adhesion may be responsible.

TABLE I. Some Materials Diamond Coated with NEPJ

Material:	SiO$_2$	Si$_3$N$_4$ Si (Diamond)	SiC	Mo, AlN AlN/Ti/W AlN/Ti/W/Au WC cermet	Al$_2$O$_3$	Cu
Peak Raman Shifts, cm^{-1}	1332±2	1333±1	1336±1	1337±3	1340±4	1329±2

Deposition Rates. Total deposition rates as high as 26mg/hr have been measured using molybdenum plates with exposed 16 cm^2 areas as substrates. A 56 mg diamond film resulted from an 8 hr deposition at a plasma power of 2.4 kW and substrate temperature of 800 C (methane to hydrogen flow ratio of 0.5%). Figure 2 gives the diamond film thickness profile as determined by beta back-scattering and increased by 40% to give an integrated volume of 16 mm^3, corresponding to the 56 mg diamond deposit. Raman spectra as recorded at various positions on the film are presented in Figure 3. The major variation appears to be the intensity of a continuum background which has been attributed to photoluminescence rather than true Raman scattering[7]. From the thickness profile in Figure 2, we estimate a maximum thickness growth rate of 7 micron/hr. This is much lower than the reported rates for thermal plasma jet processes of hundreds of microns an hour. However, these high speed depositions are generally based on smaller area films. If total deposition rates are used we have deposited at 7 to 26 mg/hr (2 to 7 mm^3/hr) using a 2.4 kWh plasma. Values estimated from other reports do not always provide a direct comparison to this, but we find, for example, for a DC thermal plasma[8] that a small area 400 micron/hr rate corresponds to 30 mm^3/hr at 10 kW. Higher rate processes tend to utilize higher powers, 1 kW for MWCVD, 1 to 10 kW equivalent for atmospheric torches, and approaching 50 kW for high pressure thermal plasmas[9]. An encompassing definition of process rate has not yet been developed or adopted.

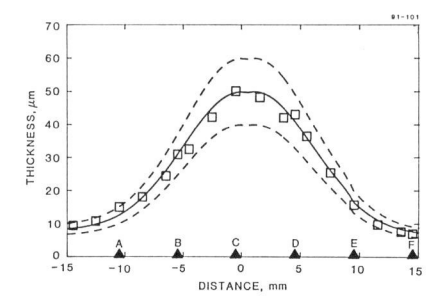

Figure 2. β-BS thickness profile of adherent 56 mg diamond film deposited on molybdenum. Curve through data points is scaled to experimental film mass (see text). Thickness uncertainty (1σ) is indicated based on β-BS counting statistics. Letters A-F refer to Raman spectra in Figure 3.

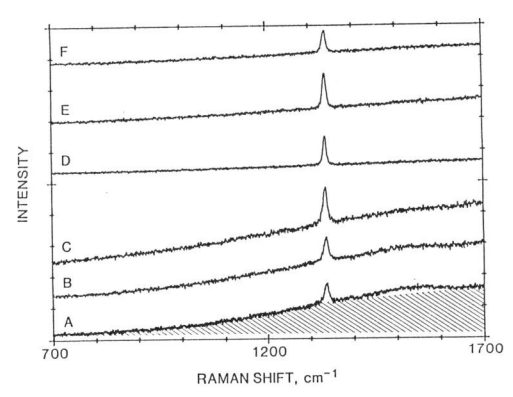

Figure 3. Raman spectra recorded for indicated positions in diamond film in Figure 2. Diamond peak shifts and full width half maxima, in cm^{-1}, are: A-1337,11; B-1338,10; C- 1337,11; D- 1336,8; E- 1336,10; F-1339,9. Spectrum A indicates most scattering by non-diamond carbon: broad peak in range of 1500-1550 cm^{-1} seen above estimated continuum photoluminescence (cross-hatch).

Low Temperature Depositions. The NEPJ has demonstrated the advantage of high power application with relatively little power dissipation by the substrate. For example, we have obtained high deposition rates at low temperature on cooled molybdenum substrates. In one experiment, the molybdenum substrate was mounted on a water-cooled copper support block. Heat flux through the molybdenum substrate was calculated from the temperature rise in the support block cooling water and the water flow rate. This was used to put limits on the temperature gradient across the substrate area upon which diamond was observed to be deposited and indicated a surface temperature of less than 100°C. The deposition rate was 300 μg/hr over an area of 2.3 cm^2 corresponding to an average deposition rate of 0.4 μ/hr. A continous film was indicated by the appearance of color interference rings.

In another experiment, a chromel/alumel thermocouple was welded to the surface of a molybdenum substrate. The wires passed through a hole in the substrate so that only the thermocouple bead extended above the substrate surface (by~0.5 mm). Thus the bead temperature was assumed to be equal to or greater than the substrate surface temperature. During a one hour deposition experiment, the temperature was 292-300°C. Raman spectra confirmed diamond deposition on both sides of the bead location and SEM revealed diamond-like crystals on the thermocouple bead. No attempt was made to measure the weight increase, but SEM revealed near continous film with composite crystals sizes of about 1 micron, as shown in Figure 4. Diamond was confirmed in Raman spectra as shown in Figure 5.

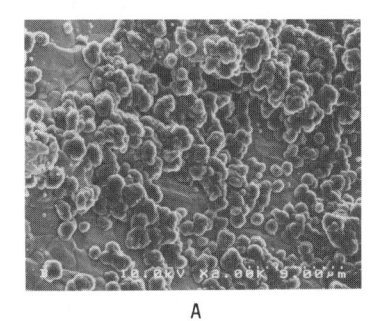

A

B

Figure 4. SEM micrographs of diamond film deposited at about 300C. Growth rate of crystals in B) is about 1 micron/hr.

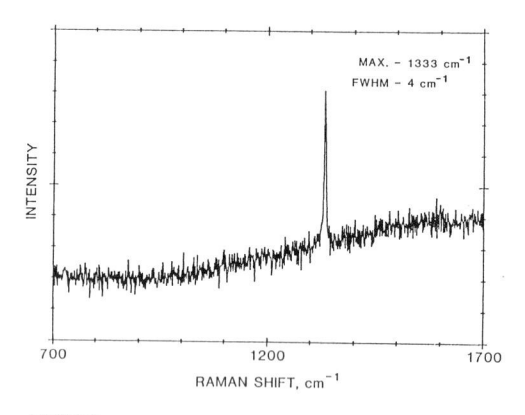

Figure 5. Raman spectrum of diamond film in Figure 4. Peak centered at 1333 cm^{-1}, 4 cm^{-1} wide (FWHM).

SUMMARY

A new high deposition rate process has been developed to produce quality diamond films on a variety of substrate materials. Substrate cooling is not necessary as in other high rate processes. Diamond films have been deposited at high rates at a surface temperature less than 600K, indicating that surface temperature may not be a limiting factor for diamond deposition.

ACKNOWLEDGMENT

This research was funded by Hercules Advanced Materials and Systems Company.

REFERENCES

1. J. C. Angus and C. C. Hayman, Science, **241**, 915 (1988).
2. W. A. Yarbrough and R. Messier, Science, **247**, 688 (1990).
3. P. K. Bachmann, D. Leers, and H. Lydtin, Diamond Rel. Mater., 1, 1 (1991).
4. J. B. Fenn, "Electrical Discharge Jet Streams," U.S. Patent 3,005,762, October 1961.
5. H. F. Calcote, "Process for Forming Diamond Coating Using a Silent Discharge Plasma Jet Process," U.S. patent pending.
6. K. Kuhihara, K. Sasaki, M. Kawarada, and N. Koshino, Appl. Phys. Lett., **52**, 437 (1988).
7. L. H. Robins, E. N. Farabaugh, and A. Feldman, J. Mater Res.,5, 2456 (1990).
8. S. Matsumoto, I. Hosoya, and T. Chounan, Jpn. J. Appl. Phys., **29**, 2082 (1990).
9. C. Tsai, W. Gerberich, Z. P. Lu, J. Heberlein, and E. Pfender, J. Mater. Res., **6**, 2127 (1991).

GROWTH AND CHARACTERIZATION OF PECVD DIAMOND FILMS

J. A. MUCHA AND L. SEIBLES
AT&T Bell Laboratories, Murray Hill, N.J. 07974

ABSTRACT

Polycrystalline diamond films have been deposited on (100) silicon wafers by plasma enhanced chemical vapor deposition (PECVD) using a 1.5 kW microwave source to dissociate dilute gas mixtures of CH_4 and O_2 in H_2. Films as thick as $20\mu m$ covering a 2" diameter area were deposited at 925 °C, 40 Torr total pressure, and 500 sccm total flow. These have been characterized as a function of CH_4 [0.2-4%] and O_2 [0-3%] concentrations by measurements of deposition rate, stress, surface roughness, morphology, and impurity levels (C-H and amorphous-graphitic carbon). Addition of oxygen to the discharge tends to reduce impurity levels in the diamond films; however, this is accompanied by a reduction in deposition rate. When an *effective* CH_4 concentration $[= \%CH_4 - \%O_2]$ is used as a metric for O_2-containing feed compositions, deposition rates as well as film properties are found to agree well with those obtained in the absence of O_2. Thus, 1% CH_4 in hydrogen is nearly equivalent to 4% CH_4, 3% O_2 in hydrogen as feed compositions for depositing diamond.

INTRODUCTION

CVD diamond research which began in earnest in Japan in the 1970's has expanded considerably in the 1980's with large efforts in Japan, the United States, and Europe. Most of the early Japanese work demonstrated that diamond was indeed the dominant phase that formed using hot-filament dissociation of dilute mixtures of methane in hydrogen. During the 1980's, U.S. researchers focused on confirming these results and elucidating the mechanism of diamond formation while their Japanese cohorts focused mainly on developing novel sources for diamond deposition. As a result, DC and microwave plasmas, oxy-acetylene flames, and DC and RF thermal plasmas have all been found to produce diamond and the Japanese have virtually cornered the market on patents relating to sources for the production of polycrystalline, CVD diamond thin films. It was not until the late 80's that attention began to be focused on determining the properties of these materials. Perhaps the most important were the demonstration that CVD diamond can exhibit a thermal conductivity[1,2] near that of bulk diamond at room temperature (~ 10 W/cm K) and resistivities[3] near 10^{16} $\Omega-$ cm suggesting that CVD diamond might be a useful dielectric material in electronic applications. However, none of these studies were comprehensive and it remains unclear whether desirable properties can be achieved in the same film or whether trade-offs might be required in incorporating diamond films in device applications. Furthermore, there has been little information regarding stress, adhesion, surface roughness, cutting, dicing, and etching; all of which are of paramount importance to using CVD diamond in practical applications. Recently, we have established a program to evaluate CVD diamond as an electronic material and examine issues related to manufacturability. In this paper, we present a summary of variations of film properties as a function of methane concentration and describe some of the effects of adding oxygen to the deposition feed gas.

EXPERIMENTAL

An ASTeX (Applied Science and Technology, Inc.) model HPM/M microwave deposition reactor was used to deposit polycrystalline diamond films by exciting a discharge through a dilute mixture of CH_4 and O_2 in H_2. For the results reported here, a 500 sccm flow of hydrogen was used at a pressure of 40 Torr with 1.4 kW of incident microwave power. Methane and oxygen flows were adjusted to give feed compositions of 0.2 - 4% CH_4 and 0 - 3% O_2. The temperature of the rf-heated susceptor was set and controlled at 850 °C, and was invariant to microwave power. Polished (100) silicon wafers were used as substrates for the deposition of diamond and varied in size and shape from 2 - 4" diameter to 1/4 sections of 3" and 4" wafers. The wafers were seeded by polishing with a metallographic paste containing 0.1μm diamond grit using a Dremel tool and cotton polishing pad. The wafer temperature was monitored with a 2μm optical pyrometer using an emissivity estimated by heating a wafer to a temperature of 600 °C in a 50 Torr bath of helium and assuming convection as the dominant mechanism of thermal transport. A value of 0.71 was obtained for the emissivity (ϵ) relative to an ideal black body and is in good agreement with handbook values for silicon at this wavelength. The substrate temperature rose by 50-80 °C (depending on microwave power) when a plasma was struck and the pyrometer temperature 30 minutes into a deposition run was chosen as the best value. For the results reported here the temperature was $925(\pm 10)$ °C. Deposition times varied from 18 to 48 hrs. to allow formation of a film at least 4μm thick.

Raman spectra of the polycrystalline diamond films were obtained in the backscattering configuration using 488.0 nm and 514.5 nm excitation lines from a 5 W argon ion laser and focusing the scattered radiation into a Jobin-Yvon Ramanor U-100 double monochromator equipped with a photomultiplier and photon counting electronics. Raman spectra provided a convenient assessment of the quality of the diamond film by comparing the relative intensities of the characteristic sharp diamond Raman band at 1332 cm^{-1} to those of non-diamond carbon (amorphous and graphitic, a,g-C) which appear as broad bands in the regions of 1330—1380 cm^{-1} and 1520—1580 cm^{-1}. Infrared spectra were recorded using a Mattson Alpha Centauri Fourier transform infrared (FTIR) spectrometer. The interference fringes observed provided a useful estimate of the film thickness, based on the assumed refractive index of 2.42. In addition, the presence of hydrogen in the film was detected by absorption bands in the 2800-3000 cm^{-1} C-H stretching region. Relative C-H concentrations in films was determined by integrating this band in the absorbance spectrum and normalizing to film thickness.

RMS surface roughness (R_A) and stress were measured using the Sloan Dektak model IIA surface profilometer. The radius of bowing (R) of the wafer was computed from the Dektak scan length and deflection, and the film stress was obtained (σ_f, dyn cm^{-2}) using the expression

$$\sigma_f = \frac{E_s}{6(1-\nu_s)} \frac{t_s^2}{t_f} \frac{1}{R} \tag{1}$$

where E_s is Young's modulus of the substrate, ν_s is Poisson's ratio of the substrate, and t_s and t_f are the thicknesses of the substrate and film, respectively. Morphology, grain size, and dominant faceting of the polycrystalline films were examined using an Olympus BH-2 optical microscope. Under maximum magnification (1500 X) it was possible to estimate grain sizes and habit down to about 1 μm.

RESULTS AND DISCUSSION

General Observations

Diamond films up to about 20 μm thick were deposited on the silicon substrates over an area of about 50 mm in diameter. This corresponded closely to our observations of the extent of the active glow region of the plasma ball. The grain sizes observed in these films seemed to be independent of the CH_4 and O_2 content of the feed gas. Usually, grain sizes were 1 - 3 μm; however, grains as large as 5 - 10 μm were observed in many films but were usually not in high concentration. The only exception to this generality was observed with high CH_4 concentrations ($> 2\%$). Here, grains size became small ($<< 1\mu$m) and the films took on a more specular appearance in reflection than the gray "emery cloth" look typical of diamond films.

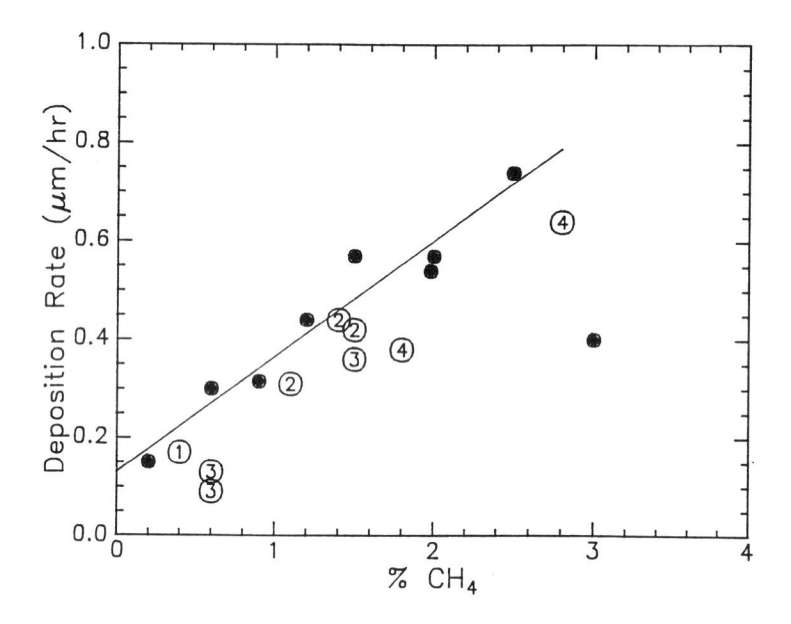

Figure 1. Deposition rate as a function of methane concentration in a 500 sccm flow of hydrogen at 925 °C, 40 Torr, and 1.4 kW incident microwave power. The solid line represent a least-squares fit to the CH$_4$-only data (solid symbols). Open symbols refer to deposition rates obtained by adding oxygen and are plotted against an *effective* methane concentration ($= [\%\mathrm{CH}_4 - \%\mathrm{O}_2]$). The numbers within the symbols refer to the actual %CH$_4$ in the feed gas.

Variations with Methane Concentration

The solid symbols in Figure 1 show the variation of deposition rate with methane concentration. The solid line represents a least-squares fit to the data with the exception that the highest CH$_4$ point (3%) was excluded from the fit since the Raman spectrum of this film suggested it was non-diamond. The overall growth rate, based on the slope of the least-squares fit, is 0.235μm/hr per %CH$_4$ and the correlation coefficient is 0.969. The observed rates are of similar magnitude to those reported by others using hot-filament and low-pressure microwave sources.[4] The apparent positive intercept (0.13 μm/hr) of the fit to the data looks statistically significant, and it is likely that etching of the surrounding graphite susceptor is supplying gas-phase carbon for diamond deposition.

Below 2% CH_4, polycrystalline diamond was the main form of carbon deposited based on the characteristic $1332\,cm^{-1}$ diamond band observed by Raman spectroscopy and the cubic faceting of grains seen using optical microscopy. While the diamond lines generally dominated the low frequency range of the spectra, they were superimposed on a broad, intense photoluminescence (PL) background and always exhibited some contribution from a,g-C near $1560\,cm^{-1}$.

Between 2% and 3% CH_4 the morphology of the films became heterogeneous with diamond concentrated in the center (1" diameter) surrounded by amorphous and graphitic carbon (a,g-C), and some soot forming on the outermost regions the same wafer. At 3% CH_4, diamond crystallites were no longer observed and the regions of soot formation on a smooth a,g-C film became more pronounced until at 4%, soot was the only form of carbon deposited. The Raman data for the films grown with 3% CH_4 displayed two broad bands near $1340\,cm^{-1}$ and $1580\,cm^{-1}$, indicative of disordered microcrystalline graphite and some amorphous carbon, and no characteristic diamond feature at $1332\,cm^{-1}$.

Along with the variations in carbon allotrope, we also noted changes in crystal habit as CH_4 concentration was increased. Below 1.5% CH_4, triangular, (111) faceting of the diamond crystallites was dominant. At higher concentrations, the films began to display increasing numbers of square, (100) facets until at 2.5% CH_4 a nearly complete conversion to (100) platelets with their normals oriented strongly along an axis perpendicular to the wafer surface was observed. Figure 2a shows that the surface roughness changes in accord with these observations. At low methane the RMS roughness was nearly 3000 Å and continuously decreased with increasing methane concentration. This is consistent with the formation of more (100) platelets and smaller grained a,g-C which begin to dominate the morphology of the film at higher CH_4 concentrations. The low-methane roughness corresponds to about 10% of the film thickness and is comparable to the grain size of the films. This seems to be typical of other CVD diamond films based on reported SEM data. For most thermal applications the roughness reported here are acceptable; however, polishing may become necessary with thicker films.

Figure 2b shows the stress in these films is strongly tensile, ranging from a low of $3 \times 10^9\,dyn\,cm^{-2}$ to a high near $1.6 \times 10^{10}\,dyn\,cm^{-2}$ as methane concentration is varied. The origins of stress are unclear but may be indicative of the mechanism of grain growth. For example, nuclei that grow hemispherically until they touch, and then undergo columnar growth are likely to induce tensile stress through interatomic forces at the grain boundaries. Thermal mismatch between the substrate and film cannot account for our observations since differential thermal expansion would result in compressive stress on cooling. Interestingly, stress appears to maximize at nearly the same CH_4 concentration that we observe the onset of formation of other

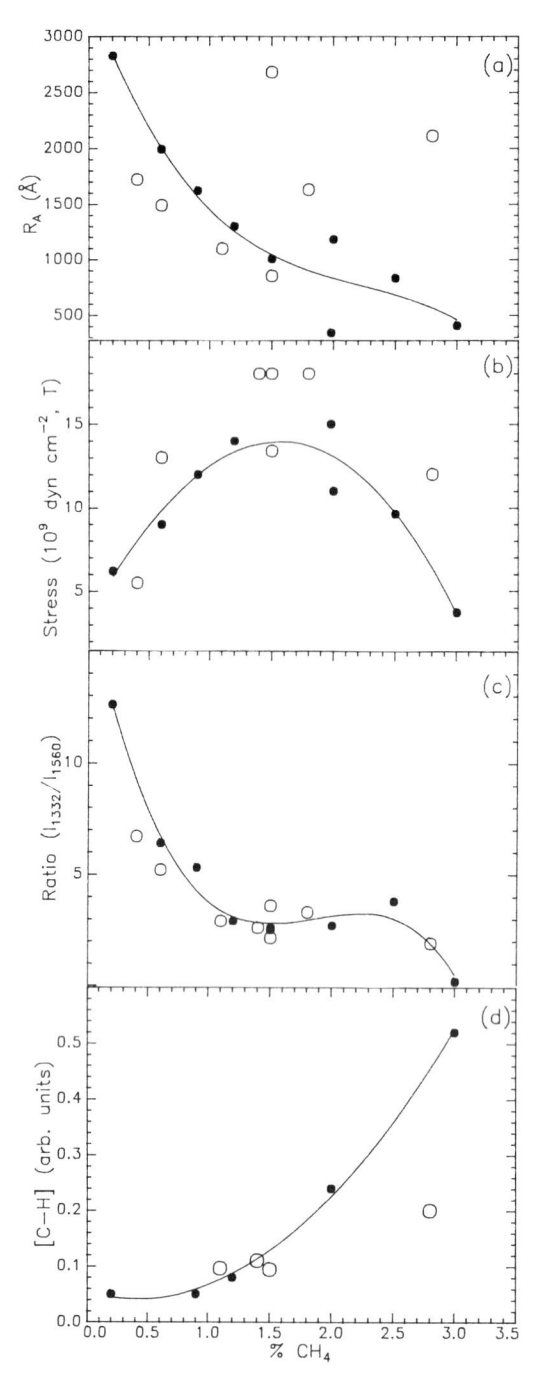

Figure 2. CH$_4$ concentration dependence of (a) RMS surface roughness (R$_A$), (b) tensile stress, (c) ratio of diamond/non-diamond Raman intensities, and (d) relative C-H content of films deposited using the conditions noted in Figure 1. Solid symbols refer to CH$_4$-only feeds while open symbols refer to mixtures containing O$_2$ that are plotted against *effective* methane ($= [\%CH_4 - \%O_2]$). Solid lines are polynomial fits to the data and meant to guide the eye.

carbon allotropes and decreases as the films become more amorphous in character. The high stress is certainly a limitation to further processing such as metallization and patterning since lithography requires a flat working surface. Furthermore, the high stress may also limit the adhesion of thicker films to the substrate. Thus, stress control will be an extremely important issue in the technological applicability of CVD diamond.

Purity is another factor in determining whether CVD diamond is a useful dielectric material. It is well established with oxide and nitride dielectrics in silicon technology that impurities, hydrogen in particular, are deleterious to the reliability and performance of microelectronic devices operated under electrical stress. CVD diamond is a more complex material since graphitic carbon contamination as well as hydrogen can accumulate at grain boundaries or form defect sites within crystallites. Raman spectra were invaluable for identifying non-diamond carbon as the origin of changes in morphology with CH_4 concentration. The non-diamond carbon structures represent a major source of contamination in CVD diamond films, and since the intrinsic Raman intensity of graphitic carbon is significantly greater than diamond, Raman characterization is an effective means for detecting low-level a,g-C domains. We have adopted the intensity of the diamond line at 1332 cm^{-1} relative to the intensity of the major a,g-C band near 1560 cm^{-1} as a convenient assessment of quality of our films. Figure 2c shows that this ratio decreases from 12.6 to about 3 as CH_4 concentration in the feed gas mixture is increased from 0.2% to 1.5%, falling again when CH_4 concentration is increased above about 2.5% (*i.e.*, when no diamond is deposited). It is clear that film quality, based on a,g-C contamination, is best at the lowest CH_4 concentrations where the deposition rate is low. This is not surprising since in dilute CH_4/H_2 mixtures, the amount of H atoms relative to film-forming carbon species is likely to be higher, thus allowing the more complete removal of a,g-C during diamond film growth.

As important as a,g-C is as a source of contamination, hydrogen may be more important. Infrared spectroscopy is useful for detecting bound hydrogen in thin films, and we have detected C-H in our films by broad absorption bands centered near 2830 cm^{-1} and 2920 cm^{-1}. These have been assigned to the symmetric and antisymmetric stretching modes of hydrogen bound to sp^3 carbon. The variation of the integrated absorption (normalized to constant film thickness) of these bands with CH_4 concentration is shown in Figure 2d. As can be seen there is a smooth increase in hydrogen incorporation with increasing CH_4, spanning films that are nearly pure diamond (low CH_4) to a film that is non-diamond (3% CH_4, see above). Using the C-H absorptivity reported for hydrogenated silicon carbide films $(1.7 \times 10^{21}$ cm$^{-2})^5$ and the atom density of diamond $(1.76 \times 10^{23}$ cm$^{-3})$, we estimate the C-H concentrations of Figure 2d range from a low of 0.4 at-% for feeds containing <1% CH_4 to 4 at-% at 3% CH_4. Based on our present understanding of the mechanism of diamond film growth, this trend is not unusual. Since one key role of atomic hydrogen is to abstract hydrogen and form carbon-centered radical sites both in the gas phase and on the surface of the film, increased CH_4 concentration will tend to deplete the system of atomic hydrogen. This results in incomplete removal of C-H on the growing surface and inclusion of these "defects" in the crystallites.

Generally, for most dielectrics used in devices, hydrogen concentrations less than 1% tend to be benign for low voltage applications. Thus, diamond deposited under the present conditions using CH_4 feeds less than about 1.4% may be suitable for laser heat sink applications. Applications involving high potentials, high frequencies, or exposure to ionizing radiation may require films with lower hydrogen concentrations. However, our analysis here is deliberately conservative, erring toward higher concentrations of hydrogen in the diamond. The smooth transition in C-H content as carbon allotrope changes suggests that the hydrogen may be concentrated at grain boundaries in a-C rather than in the crystallites themselves. Thus, by developing chemistries that minimize a,g-C formation we might expect considerable improvement. One possibility is the use of additives such as oxygen and some preliminary results are presented in the next section.

Effects of Oxygen

Generally, oxygen additions appeared to improve film quality, particularly in terms a spectroscopic purity based on FTIR and Raman data. For example, using a 2% CH_4 feed composition adding 0.6% oxygen reduced the hydrogen concentration of the film from about 1.9 at-% to 0.8 at-%. Further increases in O_2 concentration resulted in only a minor improvement, leveling off near 0.7 at-% hydrogen for 1% added oxygen. Also, with methane feeds greater than 2%, we consistently were able to deposit diamond rather than a,g-C when oxygen was introduced along with methane. However, we also noted that oxygen tended to suppress the deposition rate as well. By making the assumption that addition of oxygen did little more then consume carbon through the net reaction

$$C + O_2 \rightarrow CO_2 \qquad (2)$$

we find strong correlations between the growth rates and film properties, with and without oxygen. Since CH_4 is the primary source of carbon in our system the difference [%CH_4 − %O_2] has been found to a useful metric for making these correlations. The dependence of the deposition rate on this *effective* methane concentration (=[%CH_4 - %O_2]) is shown in Figure 1 by the open symbols. Thus, for example, a feed composed of 4% CH_4 and 2.2% O_2 is plotted as an *effective* methane concentration of 1.8%. For completeness, the number within the symbol gives the true CH_4 concentration in the feed. As can be seen, the agreement with the CH_4-only data is good over the range of CH_4 concentrations actually used — 0.9 - 3.9 %. In fact, a least-squares fit of this data yields a correlation coefficient as good as the CH_4-only data with a overall rate (0.222 μm/hr per [%CH_4 − %O_2]) within 10% of the CH_4-only data.

Moreover, when stress, Raman, and FTIR data for films deposited with oxygen are plotted using this metric, they tend to cluster around the CH_4-only data and exhibit similar trends. This is shown by the open symbols in Figures 2b - 2d. There is too much scatter in the roughness data for films grown with added O_2 (Figure 2b, open symbols) to draw the same conclusion; however, films were always rougher than predicted solely from the CH_4 content of the plasma. It is clear from the present results that, with the exception of surface roughness, the properties of the CVD diamond films are improved by *lowering* the growth rate whether by reducing CH_4 concentrations or by adding O_2 to methane-rich feeds.

SUMMARY

Polycrystalline diamond films have been deposited on (100) silicon wafers by microwave enhanced chemical vapor deposition gas mixtures of CH_4 and O_2 in H_2 at 40 Torr and 925 °C. These have been characterized as a function of CH_4 [0.2-4%] and O_2 [0-3%] concentrations by measurements of deposition rate, stress, surface roughness, morphology, and impurity levels (C-H and a,g-C). Deposition rates varied from 0.15 to 0.74μm/hr. RMS surface roughness [< 3000 Å] appears acceptable for heat sink applications calling for film thicknesses under 20 - 30 μm; however, tensile stress levels [> 6x10^9 dyn cm^{-2}] must be reduced to improve adhesion and permit further processing.

Addition of oxygen tends to reduce impurity levels in the diamond films; however, this is accompanied by a reduction in deposition rate. When an *effective* CH_4 concentration [= %CH_4 − %O_2] is used as a metric for O2-containing feed compositions, deposition rates as well as film properties are found to agree well with those obtained in the absence of O_2. Thus, 1% CH_4 in hydrogen is nearly equivalent to 4% CH_4, 3% O_2 in hydrogen as feed compositions for depositing diamond.

References

1. D. T. Morelli, C. P. Beetz and T. A. Perry, *J. Appl. Phys.*, **64**, 3063 (1988).

2. A. Ono, T. Baba, H. Funamoto and A. Nishikawa, *Japan. J. Appl. Phys.*, **25**, L808 (1986).

3. M. I. Landstrass and K. V. Ravi, *Appl. Phys. Lett.*, **55**, 1391 (1989).

4. For a review, see K. E. Spear, *J. Am. Cer. Soc.*, **72**, 171 (1989).

5. K. Nakazawa, S. Ueda, M. Kumeda, A. Morimoto and T. Shimizu, *Japan. J. Appl. Phys.*, **21**, L176 (1982).

Correlation of Raman Spectra and Bonding in DLC Films Deposited by Laser Ablation and Laser-Plasma Ablation Techniques

A. Rengan, J. Narayan., Dept of Materials Science and Engineering, North Carolina State University, Raleigh, N. C. - 27695.
J. L. Park [1]., Solid State Div., Oak Ridge National Laboratory, Oak Ridge, TN-37831, and,
M. Li., Chemistry Dept., University of North Carolina, Chapel Hill, NC-27514

ABSTRACT

We have deposited diamondlike carbon (DLC) films on a variety of substrates from 25° C and higher. The effects of deposition temperature on the properties of DLC films deposited by a conventional laser ablation technique are compared with that of a unique laser-plasma deposition scheme. The calculated values of n_{eff}, the effective number of valence electrons, suggest that, with the increase in the deposition temperature, the diamondlike component (sp^3 bonds) remains invariant for the laser deposited samples, and increases for the laser-plasma deposited films. Raman measurements show that the Raman allowed 'G' band upshifts to ~1600 cm^{-1} for both deposition schemes. However, the disorder induced 'D' band remains invariant at ~1370 cm^{-1} for the laser ablated samples, and downshifts to ~1350 cm^{-1} for the laser-plasma deposited samples. These results suggest a correlation between the diamondlike content (sp^3 bonds) and the Raman shift of the 'D' band.

I. INTRODUCTION

The laser deposition process manifests nonequilibrium features during deposition such as atomically and electronically excited (as well as ionized) species and fast quenching, which lead to the formation of films in metastable states with unique properties. Malshe et al [1] and others [2, 3] have shown that the physical evaporation of graphite atoms has resulted in the deposition of hard diamondlike films. Herein, it is shown that further nonequilibrium features can be incorporated into laser ablated plasma by coupling capacitively stored energy to the laser ablated spot in synchronism with the laser pulse. Uniform depositions over a significantly larger area are characteristic of this deposition technique, and the amorphous hard carbon films deposited by this method show improved optical properties as well as increased hardness as compared to films deposited by the conventional laser ablation method [4]. These improvements result from an increase in the sp^3 to sp^2 ratio in the DLC films, as estimated from n_{eff} values originally derived by Savvides [5].

Briefly, since most $\pi \rightarrow \pi*$ transitions occur at energies below ~7 eV, and $\sigma \rightarrow \sigma*$ transitions make no contributions below ~ 9eV, it is possible to estimate the fraction of sp^2 bonds (and hence sp^3 bonds = 1- sp^2) using 'n_{eff}' data at energies

---
[1]Current address: 706 Southwinde Dr., Bryn Maur, PA - 19010.

below 5 eV. The contributions of the $\pi \rightarrow \pi*$ transitions as calculated for graphite and $\sigma \rightarrow \sigma*$ transitions as determined for diamond are shown in Fig 1 [5].

The DLC films are characterized by spectroscopic ellipsometry and Raman scattering spectroscopies.

II. FILM PREPARATION BY LASER ABLATION

The simplest scheme for the laser deposition of DLC films is the direct ablation of a graphite target. The apparatus used for laser evaporation and deposition consists of a chamber maintained at a base pressure of 10^{-6} torr. The substrates were single crystal silicon mounted on a heater. The target was a mostly single crystal mineral graphite. The laser used for these experiments was a XeCl excimer laser (λ = 308 nm, Pulse width = 40 nsec) at a constant fluence of ~5 J/cm^2 (estimated power: ~ 1.25 x 10^8 W/cm^2). The substrate to target distance was ~ 3.0 cm.

A. Optical Properties

The optical properties of the DLC films on <100> Si substrates were analyzed by spectroscopic ellipsometry. Briefly, an automated rotating analyser ellipsometer with a Xe lamp-monochromator combination for producing a spectrally separated probing beam for sample analysis was used. A two phase ambient-substrate model was adequate to compute the refractve index, 'n', and the extinction coefficient, 'k'.

The refractive index 'n' of films deposited at 25°, 50° and 100° C are shown in Fig. 2 (a). The 'n' values are smoothly varying as a function of hv, which is typical for amorphous materials possessing short range order (SRO). It is interesting to note that the films show little difference in 'n' as a function of increasing substrate temperature. The imaginary part of the refractive index also known as the extinction coefficient is shown in Fig. 2 (b). The sequence of curves are smoothly varying with hv. The results for n_{eff} are shown in Fig. 2 (c), and indicate a diamondlike content (sp^3 bonds) of 65%, 65% and 63% for the films deposited at 25°, 50° and 100° C respectively.

B. Structural Properties:

Raman characterization was carried out with an Ar-ion laser (2.41 eV) in the usual Raman scattering geometry. The incident power at the substrate was below 70 mW. The Raman spectra with the 'D' and 'G' bands are shown in Fig. 3 (a-c) for DLC films deposited at 25°, 100° and 500° C. The Raman bands were deconvoluted into the 'D' for disorder mode and 'G' for the Raman allowed bands using Gaussian fits for the spectra. The 'D' band has been attributed to scattering by disorder activated optical zone edge phonons and the 'G' band to graphitic optic zone center phonons [6].

The calculated curve for the 25° C deposition, shows an excellent fit to the data (Fig. 3 (a)). A broad spectra is observed showing the Raman allowed 'G' band at ~1537 cm^{-1}. The disorder induced 'D' band with a smaller intensity is observed near 1400 cm^{-1}. The calculated curve for the 100° C deposition also shows an good fit to the data. The 'G' band has upshifted to ~ 1555 cm^{-1} and the 'D' band has downshifted to ~ 1370 cm^{-1}. The spectra at 500° C exhibits a 'G' band with a narrower half width upshifted to ~ 1596 cm^{-1}, moreover the 'D' band with higher intensity, is now evident near ~ 1373 cm^{-1}. In other words, no change was

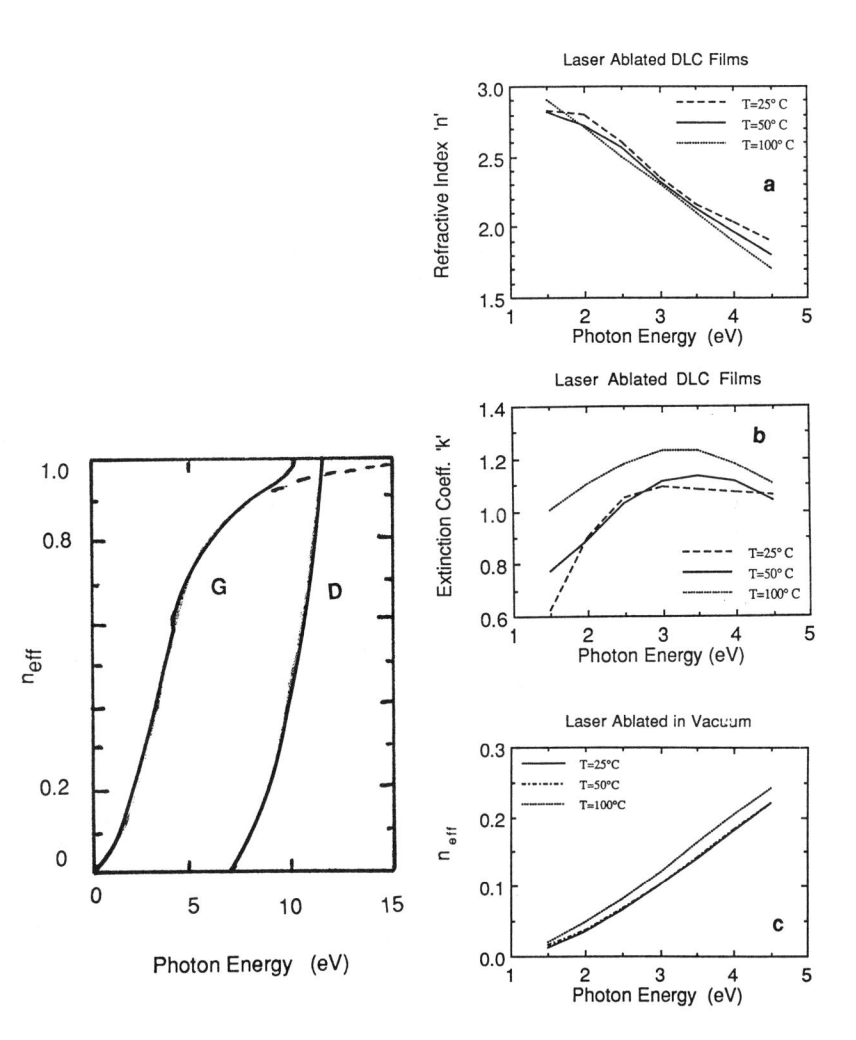

Fig 1. The effective number of valence electrons, n_{eff}, per carbon atom taking part in optical transitions vs photon energy. Curve G represents graphite, curve D represents diamond. The dashed curve shows the π component for graphite.

Fig 2. Optical properties of DLC films deposited by laser ablation, showing the effects of increasing substrate temperature. (a) Refractive index values 'n', show little change. (b) Extinction coefficient shows slight increase, and (c) n_{eff} values for the 25°, 50° and 100°C films. The estimated sp^3 bonding are 65%, 65% and 63% respectively.

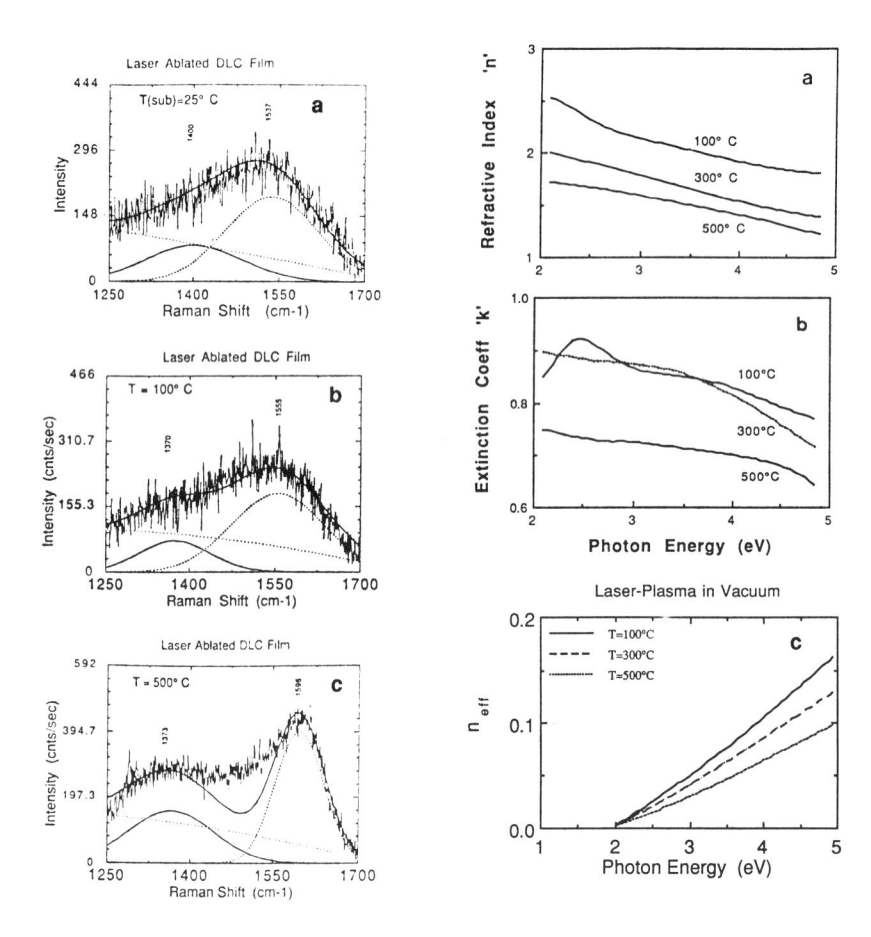

Fig 3. Raman spectra of DLC films deposited by laser ablation of graphite, at substrate temperatures of (a) 25°, (b) 100° and (c) 500° C. As substrate temperature increases, the 'G' band upshifts to ~1600cm^{-1}, however, the 'D' band is relatively constant at ~1370 cm^{-1}.

Fig 4. Optical properties for DLC films deposited by a laser-plasma hybrid scheme, showing the effects of increasing substrate temperature. (a) Refractive index 'n', decreases in value. (b) Extinction coefficient values are lower than the laser ablated DLC films. (c) n_{eff} values for the 25°, 100°, 300° and 500° C films. The estimated sp^3 bonding are 77%, 81% and 86% respectively.

observed in the 'D' band position over a 400 degree increase in substrate temperature whereas, the 'G' band upshifted. The discrepancy in the calculated fit and the data is due to a broadening of the tails of the 'D' band. A similar tail broadening has been observed in the Raman spectra of highly defective CVD diamond films and has been ascribed to regions of sp^3 adjacent to sp^2 bonded carbon, whereas the central peak arises from regions of sp^3 further away from sp^2 structures [7].

III. FILM PREPARATION BY A LASER-PLASMA HYBRID TECHNIQUE

In the new laser ablation/plasma hybrid technique a 0.5 µF capacitor, placed outside the vacuum chamber is connected by feedthroughs between a 20 mm-diam graphite ring electrode (0.65 cm from the target) and the graphite target. The capacitor is charged to 1.5 KeV by a stabilized DC source. Synchronously with the impingement of the laser pulse on the target, the capacitor is automatically discharged and a large discharge plume which extends from the ablated spot to the ring and beyond upto the substrate is formed. The inductance of the capacitor-ring-target circuit was not very critical to discharge the capacitor. Details of the experimental setup have been described elsewhere [6]. A KrF laser (λ = 248 nm, Pulse Width = 40 ns) was used at 10 Hz repetition rate to give a power density of ~1 x 10^8 W/cm^2 at the target surface.

A. Optical Properties

The refractive index 'n' as a function of photon energy, for substrate temperatures 25°, 100°, 300° and 500° C are shown in Fig. 4 (a). The results show a family of curves that display a clear dependence of 'n' on substrate temperature. The DLC films deposited at 25° C shows the highest 'n' values, and that deposited at 500° C the least. The spectral dependence of 'k', the imaginary part of the complex refractive index, (also known as the extinction coefficient), is shown in Fig. 4 (b). The n_{eff} values are shown in Fig 4 (c). The sp^3 ratios are estimated to be 77%, 81% and 86 %, for the films deposited at 100°, 300° and 500° C, respectively.

B. Structural Properties

The Raman spectra of the laser-plasma deposited samples in a vacuum ambient of 10^{-3} torr are shown in Fig. 5 (a-c). The depositions were done for substrate temperatures of 25°, 300° and 500° C and the effects are clearly revealed through the variations in the Raman spectral profiles.

The Gaussian deconvolution exhibits excellent fits for the films deposited at 25° and 300° C. In the 500° C film, the 'G' band shows an excellent fit on the high energy side, however, a difference is observed between the deconvoluted spectra and the experimental curve for the 'D' band.

On increasing the substrate temperature the 'D' band downshifts to 1350 cm^{-1} and the 'G' band upshifts to 1600 cm^{-1}. Moreover the halfwidths of both bands decrease, suggesting a local ordering of the sp^3 and sp^2 bonding components of the DLC films to perhaps intermediate range ordering. The 'D' band increases in intensity, relative to the Raman allowed 'G' band. The biggest indicator of change however is the large 'D' band shift. In the DLC films deposited by laser ablation the 'D' band shift was minimal.

IV. CONCLUSIONS

Two deposition schemes were utilized for the deposition of diamondlike carbon films. The first and simplest was the laser ablation of graphite to form diamondlike carbon films, using a medium intensity XeCl excimer laser. The second technique used a novel hybrid laser-plasma scheme wherein the the additional charge from a capacitor was added to the ablated vapor plume.

The most interesting factor in the optical measurements concerns the change in the effective number of valence electrons, n_{eff}. With increasing deposition temperature, the diamondlike component (sp^3 bonds) was invariant for the laser ablated films and increased for the films deposited by the laser-plasma technique.

In both the deposition schemes the Raman allowed 'G' band upshifted with increasing deposition temperature. However, in the films deposited by the laser ablation technique, the 'D' band remained invariant and in the films deposited by laser-plasma ablation, the 'D' band downshifted. Based on these observations, one could suggest that the 'D' band shift is indicative of the sp^3 concentration.

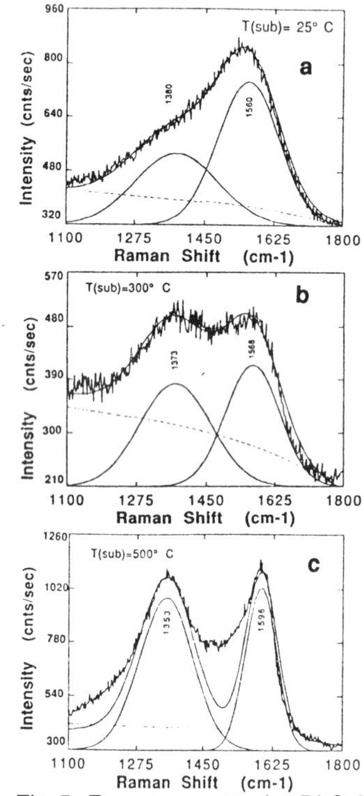

Fig 5. Raman spectra for DLC films deposited by laser-plasma hybrid scheme, at substrate temperatures of (a) 25, (b) 300 and (c) 500° C. The 'G' band shows an upshift in values to ~1600 cm^{-1}, however, the 'D' band downshifts in value to ~1350 cm^{-1}.

Acknowledgements:

Partial support of this research is credited to the Army Research Office under contract No. DAALO3-89-K-0118. The assistance of J. Rengan in the preparation of this manuscript is also appreciated.

References:
1. A. P. Malshe, S. M. Chaudhari, S. M. Kanetkar, S. B. Ogale, S. V. Rajarshi and S. T. Kshirsagar., J. Mater. Res., 4, 1238 (1989).
2. T. Sato, S. Furuno, S. Iguchi and M. Hanabusa., Jpn. J. Appl. Phys., 26, L1487 (1987).
3. A. Rengan, A. R. Srivatsa, J. Krishnaswamy, J. Narayan, K. Vedam, K. V. Ravi and M. A. Caolo., Proc. of the First Int. Symp. on Diamond and Diamondlike Films. 175 th Meeting of the ECS, Los Angeles, CA. May 7 - 12, p 456 (1989).
4. J. Krishnaswamy, A. Rengan, J. Narayan, K. Vedam and C. J. McHargue., Appl. Phys. Lett., 54, 2455 (1989).
5. N. Savvides, J. Appl. Phys., 59, 4133 (1986).
6. M. Ramsteiner and J. Wagner., Appl. Phys. Lett., 51, 1355 (1987).
7. L. H. Robbins, E. N. Farabaugh and A. Feldman., J. Mater. Res., 5, 11, 2556 (1990).

Author Index

Akerman, M. Alfred, 173
Allard, Lawrence F., 275
Allendorf, Mark D., 29, 59
Amato, C., 283

Badzian, Andrzej, 339
Badzian, Teresa, 339
Bae, Y.W., 107
Barbero, R.S., 245
Bernard, C., 3
Besmann, Theodore M., 93, 233
Bhatt, R.T., 187
Breunig, T.M., 215
Butts, M.D., 215

Caballero, Celia R., 297
Calcote, H.F., 351
Castillo, J., 53
Chang, Yuneng, 291
Chen, Arthur, 331
Chen, Jian, 303
Chiu, Chien C., 179
Chiu, Hsin-Tien, 317
Choy, K.L., 257
Chuang, Shiow-Huey, 317
Collins, J., 53
Currier, R.P., 245

Datye, Abhaya K., 275
Derby, B., 257
Desai, Hemant D., 145
Desu, Seshu B., 65, 179, 227, 323
Devlin, D.J., 245
Dobbins, Richard A., 101
Dowell, M.B., 161
Dyer, P.N., 193

Elliott, N., 245
Espinoza, B.F., 245

Felder, W., 351
Fiordalice, Robert W., 199
Freeman, Dean W., 331

Gallois, Bernard, 85, 107, 167
Garg, D., 193
Goela, Jitendra S., 145
Gokoglu, Suleyman A., 17, 251
Golda, E. Michael, 167
Goodman, D.W., 131
Guth, Jason R., 263
Guvenilir, A., 215

Hallock, Robert B., 303
Halverson, W., 193
Hay, Stephen O., 113
Hegde, Rama I., 199
Hurt, Robert H., 59

Interrante, L.V., 383
Ismail, Ismail M.K., 71

Joklik, R.G., 119

Keil, D.G., 351
Kher, Shreyas, 311
Kinney, J.H., 215
Kirss, Rein U., 303
Klein, Max, 85
Kwok, Chi Kong, 179

Lane, Andrew P., 331
Larkin, D.J., 283
Lee, W.Y., 47
Lee, Wei William, 137
Lennartz, J.W., 161
Li, M., 367
Lowden, Richard A., 173, 233
Lundgren, C.A., 215

Madar, R., 3
Melius, Carl F., 29
Moore, Arthur W., 269
Mucha, J.A., 357

Narayan, J., 125, 367
Neuschütz, Dieter, 41
Nichols, M.C., 215
Ning, X.J., 187
Norton, Kirk P., 239

Outka, Duane A., 79

Page, Barry S., 331
Paine, Robert T., 275
Papasouliotis, George D., 35
Park, J.L., 367
Peng, Chien H., 323
Pickering, Michael A., 145
Pirouz, P., 187
Pramanick, S., 125

Qiu, Xiaomei, 275

Rees, Jr., William S., 297
Reeves, Robert R., 137
Reichle, Philip A., 93
Rengan, A., 367
Riester, Laura, 173
Rodriguez, José A., 131
Roman, Ward C., 113
Rosner, D.E., 53

Salehomoum, Farzin, 41
Sandler, Neal P., 331
Schrader, Glenn L., 291
Seibles, L., 357
Sheldon, Brian W., 93

Si, Jie, 323
Sotirchos, Stratis V., 35, 221
Spencer, James T., 311
Starr, Thomas L., 207, 215
Stinton, D.P., 233
Stock, S.R., 215
Streckert, Holger H., 239
Strife, J.R., 47
Sun, H.C., 107

Taylor, Raymond L., 145
Terepka, Francis M., 251

Tobin, Philip J., 199
Tomadakis, Manolis M., 221
Travis, Edward O., 199
Truong, Charles M., 131
Tsai, Ching-Yi, 65, 227

Vakerlis, G.D., 193
Veitch, Lisa C., 251
Veltri, R.D., 47

Whittaker, E.A., 107
Wu, Ming-Cheng, 131

Subject Index

AACVD, 137
active surface area, 71
aggregate structures, 101
AlN, 283
annealing, 187

baffle, 173
barriers, 199
bisimido complex, 317
   metal, 311
   nickel, 311
   tungsten, 317
boron(-)
   carbide, 173
   carbon-silicon-ceramics, 167
   nitride, 131, 239, 269, 275
      cubic, 339
(^{t}BuN)$_2$W(NH^{t}Bu)$_2$, 317

carbon, 35, 263
   films
      carbon-carbon composites, 269
      fibers, 71
      films, 187
      yarns, 269
CARS, 113
ceramic, 257
   fibers, 269
   films, 59
   matrix composites, 227, 239
      fiber re-inforced, 245
   protective coatings, 257
chemical vapor
   deposition, 3, 17, 47, 59, 65, 71, 113, 119, 145, 161, 179
      atom assisted, 137, 291
      coatings, 269
      diamond, 339, 357
      low-pressure, 317, 331
      metallo-organic (MOCVD), 311, 317
      plasma assisted, 193
   infiltration, 41, 59, 207, 215, 221, 233, 245, 263
      BN, 239, 269
      carbon, 263
      composites, 17, 227, 233, 263
      CVD on fibers, 251, 257, 263, 269
      3-D finite element, 207
      diffusion, 221
      forced flow, 227
      interface, 239
      mass transport, 221
      metals, 263
      microwave, 245
      modelling, 207
      Nicalon, 215, 233, 245, 269
      nondestructive testing, 215
      SiC, 215, 233, 245, 251
         monofilaments, 257

      SiH$_4$, 251
      Si$_3$N$_4$, 239
      TiB$_2$, 257
      TiC, 257
      TiN, 251
      tubes, 233
      x-ray tomography, 215
coherent anti-Stokes Raman, 113
cold wall CVD reactor, 85
composites, 17, 227, 233, 263
copper, 291, 297
   oxide, 291, 297
Cu(acac)$_2$, 297
Cu$_2$O copper acetylacetonate, 291
Cu(tmted)$_2$, 297

3-D finite element code, 207
deposition rates, 41, 137, 199
diamond(-), 357
   heteroepitaxy, 339
   like carbon, 367
   polycrystalline, 357
diborane, 131
dielectric(s), 331
   constant, 323
diffusion, 221
disilacyclobutane, 283

enthalpies, 137
equilibrium calculations, 145

fibers, 17, 251, 263
   ceramic, 269
film, 199
   resistivities, 199
   thin, 311
   ZrO$_2$, 311
finite element model, 65
forced-flow CVI, 227

gas-phase thermodynamics, 29
Gibbs free energy, 137

halide, 3
HF, 79
homogeneous reaction, 53
hydrazine, 131

infrared, 173
interface coatings, 239

kinetic(s), 41, 47
   model, 35
Knudsen diffusivities, 221

laser(-), 119
   ablation, 367
   light scattering, 85
   plasma ablation, 367

LPCVD, 199, 317, 331
LSE, 101

mass
  spectroscopy, 47
  transport, 221
metalorganic chemical vapor deposition, 107
metals, 263
methylsilazane, 107
methyltrichlorosilane, 35, 41
microwave, 245
mirrors, 145
MOCVD, 311, 317, 323
modeling, 17, 207
molybdenum carbide, 303
multiphase coatings, 167
multiple correlation analyses, 161

NH_3 decomposition, 113
Nicalon, 233, 245
  /SiC composites, 215, 269
nickel hydride, 339
nido-pentaborane, 311
nondestructive testing, 215
nonequilibrium plasma jet, 351
nucleation, 93

OH, 119
optical, 125, 145, 173, 323
  baffle, 173
  elements, 145
  properties, 323
  Ti, 125
  TiN, 125
  $TiSi_2$ plasma, 125
optics, 173
organometallic, 283, 291
oxidation, 275

PACVD, 193
particle, 53
  aerosol, 59
  nucleation, 53
phase transformed, 323
plasmons, 179
polydisperse aggregates, 101
pulse-CVI, 221
pyrolytic
  BN, 275
  carbon, 71

Raman spectroscopy, 367
reaction mechanism, 41
refractory metal, 125

roughness, 85

scattered light, 85
scattering/extinction tests, 101
SCS-6 SiC fiber, 187
SiC, 29, 41, 193, 233, 245, 251, 283
  coatings, 251
  monofilaments, 257
SiF_4, 79
silicide, 3
silicon(-), 35
  based ceramic thin films, 107
  boride, 167
  carbide, 35, 41, 93, 145, 167, 193, 251
  nitride, 79, 113, 161
  tetrafluoride, 79
Si_3N_4, 47
  CVD, 47
  matrix, 239
single-source precursors, 317
spectroscopy, 113
step coverage, 65
surface analysis, 199

Ta_2O_5, 331
TEM, 101
temperature, 119
thermochemistry, 3
thermodynamics, 47
thermophoretic sampling, 101
thin films, 275, 311
  ZrO_2, 275, 323
TiO_2, 53
$TiCl_4/O_2$, 53
TiN, 85
tunable diode laser absorption spectroscopy, 107
tungsten, 137, 303
  oxide, 137
  tetra, 303

W, 137
WO_3, 137
windows, 145
$WN_x (x=0.8-1.8)$, 317

XeF_2, 79
x-ray
  diffraction, 339
  tomographic, 215
XRD patterns, 137

$Zr(thd)_4$, 323